W0112585

Smart Polymers for Bioseparation
and Bioprocessing

Smart Polymers for Bioseparation and Bioprocessing

Edited by
Igor Galaev and Bo Mattiasson

London and New York

First published 2002
by Taylor & Francis
11 New Fetter Lane, London EC4P 4EE

Simultaneously published in the USA and Canada
by Taylor & Francis Inc,
29 West 35th Street, New York, NY 10001

Taylor & Francis is an imprint of the Taylor & Francis Group

© 2002 Igor Galaev and Bo Mattiasson

Typeset in 10/12 pt Sabon
by Newgen Imaging Systems (P) Ltd. Chennai, India
Printed and bound in Great Britain by TJ International Ltd, Padstow, Cornwall

All rights reserved. No part of this book may be reprinted or
reproduced or utilised in any form or by any electronic,
mechanical, or other means, now known or hereafter
invented, including photocopying and recording, or in any
information storage or retrieval system, without permission in
writing from the publishers.

Every effort has been made to ensure that the advice and information
in this book are true and accurate at the time of going to press. However,
neither the publisher nor the authors can accept any legal responsibility
or liability for any errors or omissions that may be made. In the case of
drug administration, any medical procedure or the use of technical
equipment mentioned within this book, you are strongly advised to
consult the manufacturer's guidelines.

British Library Cataloguing in Publication Data
A catalogue record for this book is available
from the British Library

Library of Congress Cataloging in Publication Data
A catalogue record for this book has been requested

ISBN 0-415-26798-6

Contents

vi   *Contents*

Contributors

Mitsuru Akashi, Department of Applied Chemistry and Chemical Engineering, Faculty of Engineering, Kagoshima University.

Guohua Chen, Alza Corporation.

Jyh-Ping Chen, Department of Chemical and Materials Engineering, Chang Gung University.

Boris B. Dzantiev, Immunobiochemistry Lab., Institute of Biochemistry, Russian Academy of Sciences.

Keiji Fujimoto, Department of Applied Chemistry, Keio University.

Igor Yu. Galaev, Department of Biotechnology, Lund University.

Allan S. Hoffman, Bioengineering Department, University of Washington.

Kazuhiro Hoshino, Department of Material Systems Engineering and Life Science, Toyama University.

Alexander E. Ivanov, Shemyakin & Ovchinnikov Institute of Bioorganic Chemistry, Russian Academy of Sciences.

Vladimir A. Izumrudov, Polymer Chemistry Department, Moscow State University.

Hans-Olof Johansson, Department of Biochemistry, Lund University.

Baghwati Kabra, Alcon Laboratories.

Haruma Kawaguchi, Department of Applied Chemistry, Keio University.

Jung Ju Kim, School of Pharmacy, Purdue University.

Akio Kishida, Department of Applied Chemistry and Chemical Engineering, Faculty of Engineering, Kagoshima University.

Etsuo Kokufuta, Institute of Applied Biochemistry, University of Tsukuba.

Bo Mattiasson, Department of Biotechnology, Lund University.

Kinam Park, School of Pharmacy, Purdue University.

Josefine Persson, Department of Biochemistry, Lund University.

Mårten Svensson, Department of Biochemistry, Lund University.

Masayuki Taniguchi, Department of Material Science and Technology, Niigata University.

Folke Tjerneld, Department of Biochemistry, Lund University.

Elena V. Yazynina, Immunobiochemistry Lab., Institute of Biochemistry, Russian Academy of Sciences.

Vitali P. Zubov, Shemyakin & Ovchinnikov Institute of Bioorganic Chemistry, Russian Academy of Sciences.

Anatoliy V. Zherdev, Immunobiochemistry Lab., Institute of Biochemistry, Russian Academy of Sciences.

Preface

Life is polymeric in its essence. The most important components of living cells, pro-teins, carbohydrates and nucleic acids are polymers. Even lipids, which are of lower molecular weight, could be regarded as methylene oligomers with polymerization degree around 20. Furthermore, the spontaneous aggregation of lipids contributes to even larger supramolecular structures. Nature uses polymers both as constructive elements and as parts of a complicated cell machinery. The salient feature of func-tional biopolymers is their all-or-nothing, or at least highly nonlinear, response to external stimuli. Small changes happen in response to varying parameters until the critical point is reached. The subsequent transition occurs within a narrow range of the parameter varied. After the transition is completed, there is no significant further response from the system. Such nonlinear responses of biopolymers is warranted by highly co-operative interactions. Despite the weakness of each particular interaction taking place in a separate monomer unit, these interactions, when summed through hundreds and thousands of monomer units, can provide significant driving forces for the processes occurring in such systems.

Not surprisingly, the understanding of the mechanisms of co-operative interactions in biopolymers has generated a large amount of research. Many attempts have been made to mimic the co operative behaviour of biopolymers in synthetic systems. Recent decades have witnessed the appearance of synthetic functional polymers which respond in some desired way to a change in temperature, pH, electric or magnetic fields or other stimuli. These polymers were nick-named *stimuli-responsive*.The name *smart polymers* was coined due to the similarity between stimuli-responsive polymers and biopolymers.

The polymer systems considered in this book under the name smart polymers are defined as *macromolecules that undergo fast and reversible changes from hydrophilic to hydrophobic microstructures triggered by small changes in their environment. These microscopic changes are apparent at the macroscopic level as the formation of a precipitate of the smart polymer in solution, changes in wettability of the surface to which the smart polymer is grafted or dramatic shrinking/swelling of the hydrogel. The changes are relatively fast and reversible i.e. the system returns to its initial state when the trigger is removed.*

The driving force behind these transitions could be of a different nature e.g. neu-tralization of charged groups, either by pH-shift or addition of oppositely charged polymer, changes in the efficiency of hydrogen bonding with the increase in tem-perature or ionic strength, critical phenomena in hydrogels and interpenetrating polymer networks. An appropriate balance of hydrophobicity and hydrophilicity in the molecular structure of the polymer is believed to be of key importance in demonstrating the phase transition. Unique properties of water as a solvent are the indispensable condition of stimuli responsive polymer systems.

The **applications** covered in the book include protein purification using affinity precipitation with macroligands formed by thermo- or pH-sensitive polymers, aqueous two phase polymer systems formed by thermosensitive polymers or chromatographic matrices with physically or covalently attached smart polymers; reversibly soluble biocatalysts immobilized in smart hydrogels with catalytic activity responding to the environmental changes; the systems with regulated porosities, so called *chemical valves*; new immunoanalytical systems. We strongly believe that nature has always striven for smart solutions in creating Life. The goal for the scientists is not only to mimic biological processes, and therefore understand them better, but also to create novel species and invent new processes.

The field of smart polymers and their applications is developing very rapidly. Reports on new smart polymer systems and original applications appear literally every week. In 1996 the first product using a thermoresponsive polymer was commercialized by Gel Sciences/GelMed (Bedford, Massachusetts). The product, Smart-Gel, is a viscoelastic gel that is soft and pliable at room temperature but becomes much firmer when exposed to body heat and is being used as shoe insert to make shoes conform to the wearer's feet.

To the best of our knowledge, there is no book covering smart polymers and their applications in bioseparation and bioprocessing. The main objective of the book is to fill this gap.

The **polymer systems** considered are those where the highly non-linear response of a smart polymer to small changes in the external medium is of critical importance for the successful functioning of the system. The main applications of smart polymers in biotechnology and medicine include biorecognition and/or biocatalysis, which take place principally in aqueous solutions. Therefore only water-compatible smart polymers will be considered and the smart polymers in organic solvents or water/organic solvent mixtures are beyond the scope of this volume. The systems discussed in the book are based on soluble/insoluble transitions of smart polymer in aqueous solution, conformational transitions of the macromolecules physically attached or chemically grafted to the surface and on the shrinking/swelling of covalently cross-linked networks of macromolecules, i.e. *smart hydrogels*.

One of the most frequently used smart polymers is a thermosensitive polymer, poly(*N*-isopropylacrylamide), which is very often referred in this book. At present in the original literature, there is no consistency in the abbreviation of the name of this polymer. One can come across different notations like PNIPAAm, polyNIPAAm, polyNIPAM, PNIAAm, pNIPAm. As the editors, we have left these notations the way they were used by the authors, despite the temptation to unify the abbreviation throughout the book. The use of different notations in different chapters will help the reader later on to recognize this polymer when he or she comes in the original literature across one of the abbreviations used. The reader is kindly directed to use Abbreviation List provided at the end of each chapter.

The book, with contributions from leading researchers in the field of smart polymers and their applications, aims to summarize the current state of affairs and promote further developments in the field, by attracting fresh recruits, including both polymer chemists capable developing new advanced polymers and biotechnologists that are ready to use these polymers for new applications.

Igor Galaev
Bo Mattiasson

1 Temperature- and pH-sensitive graft copolymers

Guohua Chen, Allan S. Hoffman and Baghwati Kabra

1.1. Introduction

Graft and block copolymers are of interest as they usually retain the unique properties of their individual components, in contrast to random copolymers that exhibit 'average' properties of two components. Graft and block copolymers also represent a way to combine two different and unusual properties in one polymer structure. Two different moieties in graft copolymers could provide them with a combination of properties desirable for particular applications.

Significant interest in mucoadhesive polymers has developed in the last decade (Gu, 1988). The mucosae represent a very large and readily accessible surface area of the body for drug delivery, especially topical surfaces such as ophthalmic, nasal, buccal and vaginal surfaces, or the gastro-intestinal tract. Such polymers are designed to prolong the residence time of a delivery system by adhering to mucosal surfaces at the site where the therapeutic drug is to be delivered. Polyacrylic acid (PAAc) is well-known as a bioadhesive polymer and widely used for the development of a controlled release bioadhesive drug delivery system (Park and Robinson, 1987). It is often applied in lightly crosslinked forms available commercially, known as Polycarbophil® or Carbopol®. However, PAAc is also a pH-sensitive polymer in which the carboxyl groups are ionized at physiological condition (pH 7.4). This can lead to high swelling and rapid disintegration of PAAc formulations at pH 7.4, causing drug to be released as a burst from the polymer matrix upon contact with the mucosal surface (other than in the stomach). Such an initial burst of drug usually results in a low bioavailability, as well as a lower overall efficacy of the formulation. It can also sometimes lead to undesirable side effects. Therefore, modifications of the mucoadhesive polymer, of the formulation, or of delivery conditions are necessary if one wants to avoid these problems.

Recently many publications have appeared concerning 'intelligent' polymers that are responsive to external stimuli such as temperature (Tanaka, 1979; Hoffman, 1987; Bae, 1987; Chen and Hoffman, 1993), electric fields (Kwon *et al.*, 1991; Kishi *et al.*, 1991) and pH (Siegel and Firestone, 1988; Peppas and Buri, 1985; Park and Robinson, 1985) for applications ranging from drug delivery to diagnostics, separations and robotics. In the most interesting cases, the response to small changes in conditions can stimulate a sharp, or discontinuous, phase transition. For example, when used in a drug delivery system, stimuli-responsive polymer systems can provide time-dependent pulses of drug not achievable by more common polymer-based delivery systems (Hoffman, 1987).

Our goals have been to design and synthesize polymeric carriers for mucosal drug delivery that combine a mucoadhesive component, to increase the residence time on mucosal surfaces, with a stimuli-responsive component, that 'thickens' the overall formulation and thereby retards its swelling and dissociation, resulting in a slower rate of drug release and an increased drug bioavailability at that site.

Initially, a series of random copolymers containing poly(acrylic acid) (PAAc) and poly(N-isopropylacrylamide) (PNIPAAm) was synthesized. PAAc was chosen because of its well-known bioadhesive properties. PNIPAAm is a well-known thermally-reversible water-soluble polymer exhibiting a sharp phase transition at 32 °C (called the lower critical solution temperature or LCST). PNIPAAm was incorporated in order to retard the swelling and erosion rate of the copolymer at 37 °C, which should prolong drug release from the copolymer matrix. However, it was soon apparent that the random copolymer of AAc with NIPAAm cannot provide both of the desired properties of bioadhesion and temperature-sensitivity in the same random copolymer (Dong and Hoffman, 1991; Chen and Hoffman, 1995a). That is, not only does the bioadhesive property of PAAc polymers depend on high frequency of carboxyl groups along the polymer chain, but the thermal phase transition property of PNIPAAm likewise requires long sequences of the homopolymer of PNIPAAm. Therefore, it was reasoned that polymer compositions which contained long sequences of both PAAc and PNIPAAm were required in order to retain both bioadhesion of the formulation, as well as resistance to swelling and drug release derived from the thermal phase transition properties.

Graft and block copolymers allow combination of two different and unusual properties into one polymer structure. For instance, it is often desirable for polymeric drug carriers to provide more than one important property. In the case of delivery of drugs to mucosal surfaces, both prolonged residence time on the mucosal surface, as well as prolonged release kinetics of the active agent are necessary properties for a successful system, but they can often conflict, as described above.

Therefore, for this purpose a series of graft copolymers containing PNIPAAm (Chen and Hoffman, 1995b and c), as the temperature-sensitive component, grafted to a PAAc backbone, as the bioadhesive component was synthesized and investigated as a vehicle for erosion-controlled, mucosal drug delivery. To further enhance the desired action of the thermally-sensitive polymer component to retard the swelling and erosion rate of the graft copolymer matrix, a copolymer of NIPAAm with butyl methacrylate (BMA), a more hydrophobic component, was also grafted onto the PAAc backbone (Chen and Hoffman, 1995d). The thermal and pH behavior of those two types of graft copolymers was investigated. The rates of drug delivery were also evaluated in *in vitro* release studies. These results are presented and discussed later in this chapter.

In related publications, it has been shown that, at low pHs, polymers, copolymers (Klier *et al.*, 1990) and inter-penetrating networks of AAc (Katono *et al.*, 1991) or methacrylic acid (MAAc) form complexes with poly(ethylene glycol) (PEG) (Klier *et al.*, 1990) or polyacrylamide (PAAm) (Katono *et al.*, 1991). This is due to formation of H-bonds between the −O− groups of the PEG chains or the −CONH− groups of the PAAm chains and the −COOH groups of the PAAc or PMAAc chains. Such complexes exhibit gradual deswelling with increasing temperature as the H-bonds are dissociated, but this gradual temperature-sensitivity disappears when the pH is raised above where a significant fraction of the carboxyl groups become ionized. For example, Klier *et al.* (1990) prepared hydrogels based on copolymers of PEG methacrylate

(PEGMA) and MAAc, which form H-bonded complexes at low pHs between the $-O-$ groups of the PEG side chains and the $-COOH$ groups of the poly(methacrylic acid) (PMAAc) backbone. Although at pH 4.0 these gels shrank as temperature was raised, this behavior would not exist at pH 7.4, because at that pH the carboxyl groups are ionized, the complex is disrupted, and the gels become highly swollen. Such a gel might also release drug too rapidly at pH 7.4 because of its rapid swelling.

Since NIPAAm polymers are not 'generally regarded as safe' (GRAS), our design concept of the graft copolymers was further extended to the synthesis of copolymer structures using thermally-sensitive 'GRAS' polymers as the grafted side chains. For this purpose, we chose to graft a temperature-sensitive triblock copolymer of poly(ethylene oxide (PEO) and poly(propylene oxide) (PPO) (or PEO-PPO-PEO, known as Pluronic® polyol surfactants) to the PAA backbone. Several of the Pluronic® polyols have been approved by the FDA for *in vivo* use, and are therefore GRAS polymers.

In this chapter we describe and discuss the design, synthesis and properties of graft copolymers having PAAc as the backbone polymer, and PNIPAAm, P(NIPAAm-co-BMA), or Pluronic® as grafted side chains.

Graft copolymers can be synthesized by several methods. We have used two different methods to prepare graft copolymers of thermally-sensitive polymers grafted to a bioadhesive, pH-sensitive backbone, polyacrylic acid (PAAc). The first method was through copolymerization of the backbone monomer, AAc, and the macromonomer of oligomers of NIPAAm or its co-oligomer with BMA (Figure 1.1). One advantage of this method is that the frequency of grafting can be readily and simply varied by varying the ratio of the two monomers in the reaction mixture. A potential disadvantage is that in some cases the macromer copolymerization reactivity can be significantly reduced due to steric interference of its oligomer chain. This can lead to limited incorporation of the macromer in the copolymer. Therefore, because of this disadvantage, the coupling method was preferred for synthesis of the graft copolymers.

The second method we used to prepare the graft copolymers was through covalent coupling of the thermally-sensitive prepolymers that had a terminal reactive amino group, onto the pendant carboxyl groups of the PAAc backbone (Figures 1.2–1.4). One advantage of this method is the relatively mild reaction conditions compared to the copolymerization method, which has to be run in an inert atmosphere. A potential disadvantage is that the reaction between an end group of one polymer and a side group of another polymer can be relatively inefficient due to the large sizes of the reactants.

Because of different reactivity between the macromonomer and AAc, the polymerization cannot proceed in very high conversion in order to get the unique distribution

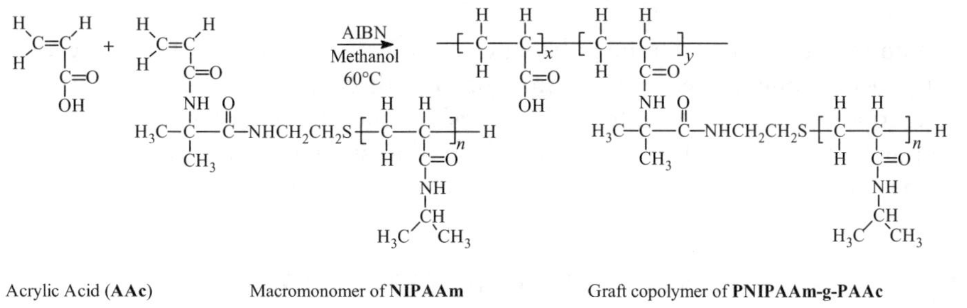

Acrylic Acid (**AAc**)   Macromonomer of **NIPAAm**   Graft copolymer of **PNIPAAm-g-PAAc**

Figure 1.1 Schematic synthesis of the graft copolymers by copolymerization of the macromonomer of NIPAAm with Aac.

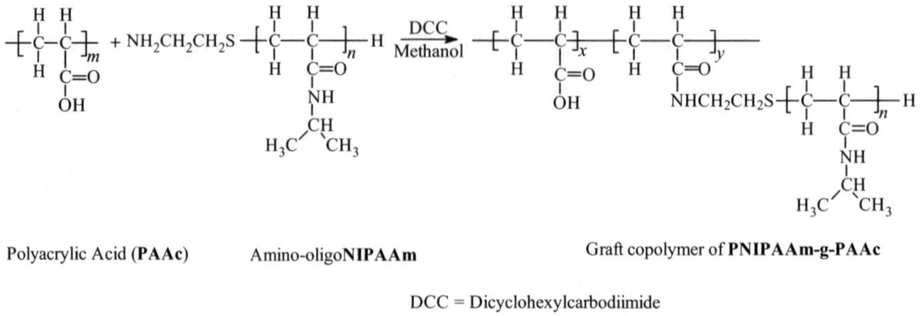

Polyacrylic Acid (**PAAc**)   Amino-oligo**NIPAAm**   Graft copolymer of **PNIPAAm-g-PAAc**

DCC = Dicyclohexylcarbodiimide

Figure 1.2 Schematic synthesis of the graft copolymers by coupling oligoNIPAAm onto PAAc.

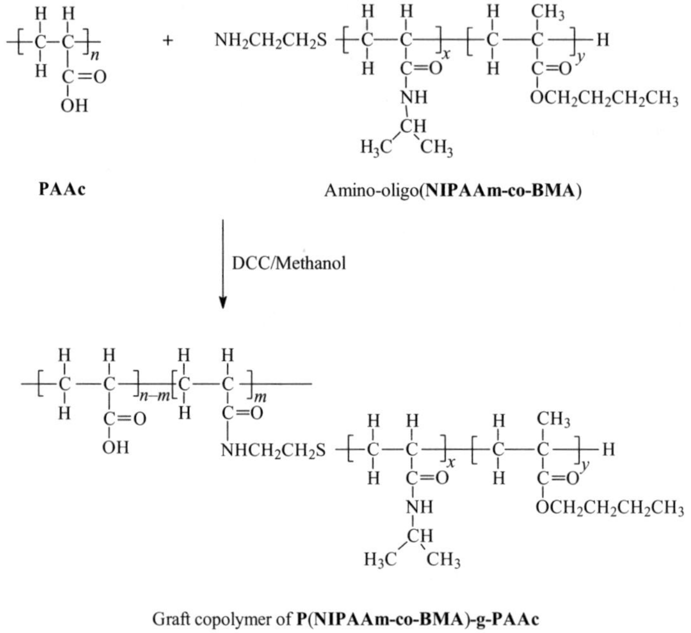

PAAc   Amino-oligo(**NIPAAm-co-BMA**)

DCC/Methanol

Graft copolymer of **P(NIPAAm-co-BMA)-g-PAAc**

Figure 1.3 Schematic synthesis of the graft copolymers of P(NIPAAm-co-BMA)-g-PAAc by coupling oligo(NIPAAm-co-BMA) onto PAAc.

of monomer units through the polymeric chain. Besides, because of the very low reactivity of macromonomer, limited amount of graft chains could be introduced to the graft copolymer. Therefore, for the synthesis of the graft copolymers reported in this chapter, the coupling method was mostly used to synthesize the graft copolymers.

The details of synthetic procedures are presented in Appendix.

1.2. Graft Copolymers of P(NIPAAm-BMA)-g-PAAc

Although the random and graft copolymers have the same basic monomer units, their molecular structures are quite different (Figure 1.5 vs. Figure 1.1 or 1.2). Furthermore,

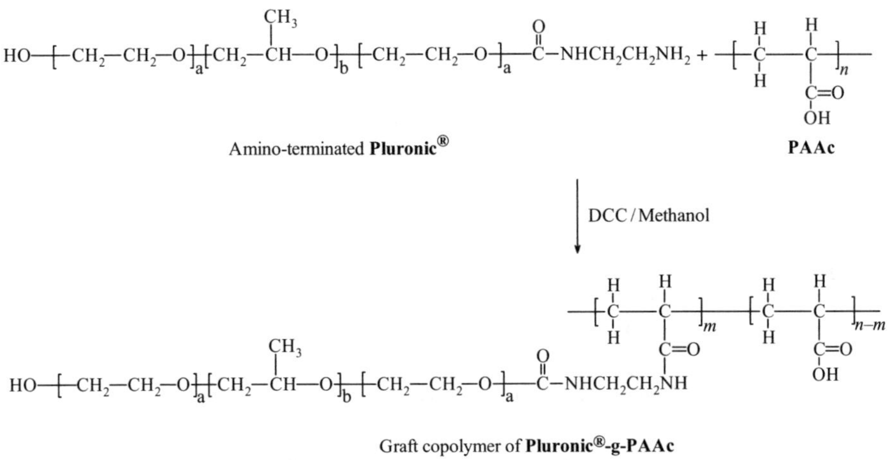

Figure 1.4 Schematic synthesis of graft copolymers of Pluronic®-g-PAAc.

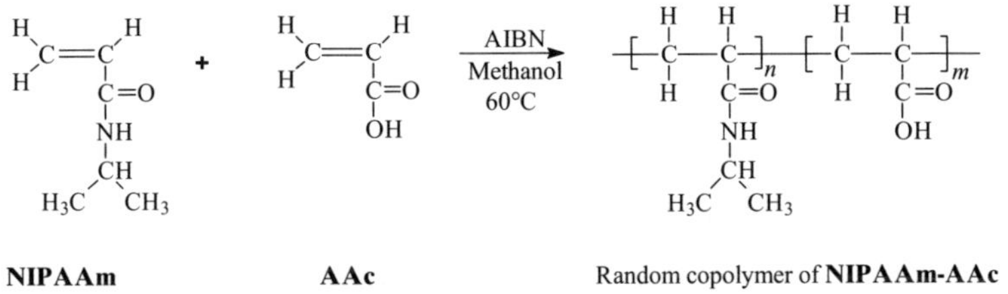

Figure 1.5 Schematic synthesis of the random copolymers of NIPAAm and AAc.

the same average composition of random vs. graft copolymer is expected to exhibit different temperature- or pH-induced phase transition behavior. Figure 1.6 shows the effect of temperature on light transmittance of aqueous solutions of random copolymers at different pHs. At pH 4.0 and 7.4, the random copolymers all exhibit a higher phase transition temperature than PNIPAAm. It is well-known that copolymers of NIPAAm with more hydrophilic monomers usually show a higher LCSTs (Priest *et al.*, 1987). At pH 4.0, even though the –COOH groups of AAc are not ionized, the LCSTs of the random copolymers increase with increase of the AAc content in the copolymer because of the exothermic water interaction of the –COOH group which competes with the entropic driving force for PNIPAAm phase separation due to the hydrophilically bound water. At this low pH, when the AAc content in the copolymer is higher than 40 wt% no LCSTs can be found. However, at pH 7.4, with only 7 wt% of AAc in the random copolymer (i.e. 93 wt% NIPAAm), the LCST dramatically increases from ca. 32 °C to above 60 °C (Figure 1.6). At pH 7.4, the copolymers with more than 7 wt% of AAc no longer show a phase transition.

Unlike the random copolymer, however, the graft copolymers with PAAc as the backbone and oligoNIPAAm as the graft chain show very different phase transition behavior from the random copolymers, as shown in Figure 1.7. Although at pH 7.4 the carboxyl groups in PAAc are ionized, the graft copolymers all start to show an LCST phase transition at almost the same temperature as PNIPAAm, independent of

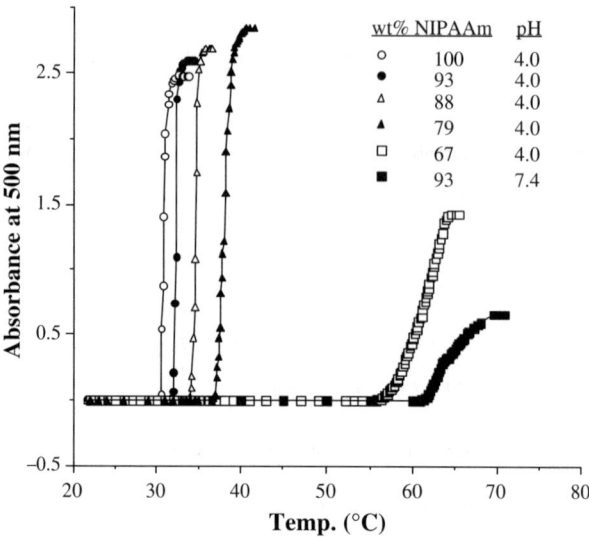

Figure 1.6 Light transmittance (absorbance) of 0.2% solutions of random copolymers of NIPAAm and AAc at pH 4.0 (or pH 7.4) vs. temperature.

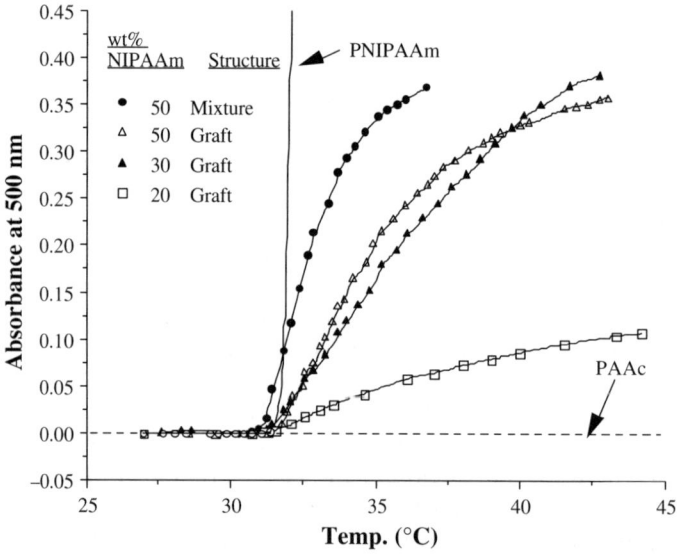

Figure 1.7 Light transmittance (absorbance) of 0.2% polymer solutions in PBS buffer, pH 7.4 vs. temperature. (The graft copolymers of PNIPAAm-g-PAAc were prepared by coupling oligoNIPAAm onto PAAc).

the overall composition. With as high as 80 wt% of AAc, the graft copolymer still shows the onset of turbidity at pH 7.4. The presence of the ionized PAAc backbone seems only to reduce the magnitude of the turbidity and to widen the temperature range of the response, indicating that even when it is ionized, the PAAc backbone does not change the phase transition property of the oligoNIPAAm graft chain.

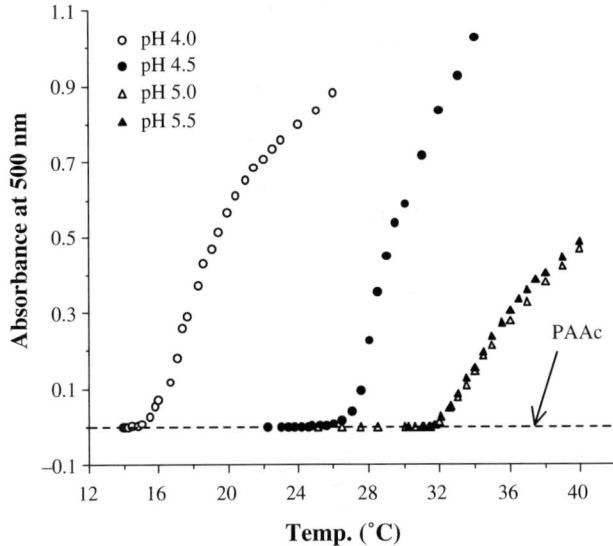

Figure 1.8 Light transmittance (absorbance) vs. temperature of 0.2% solution of the graft copolymer (containing 50 wt% of NIPAAm) in citric + phosphate buffers at different pH values.

In contrast with the effect of ionized PAAc at pH 7.4, when the pH of the solution is sufficiently low, e.g. lower than 5.0, the phase transition of the graft copolymer shifts to lower temperatures and decreases to ca. 16 °C with decrease in pH to 4.0 (Figure 1.8). The same phenomenon is also found for physical mixtures of PAAc with oligoNIPAAm (data not shown). This pH induced decrease of the phase transition temperature for the graft copolymer might be due to formation of inter- or intra-graft-backbone chain complexes caused by hydrogen bonding between the backbone –COOH and graft chain –CONH– groups (as shown in Figure 1.9). The formation of hydrogen bonds between the –COOH and –CONH– groups limits the accessibility of water to the NIPAAm units, leading to an increase in hydrophobicity of the copolymer microenvironment. The lower the pH, the greater the –COOH/–COO⁻ ratio, and thus, the greater the possibility to form these hydrogen-bonded sequences. The effect has been noted by others in different polymer systems (Klier *et al.*, 1990; Katono *et al.*, 1991). In fact, it was found that at pH 3.0, the graft copolymers were no longer soluble in water, even at temperatures as low as 4 °C. At higher pH values, when sufficient number of carboxyl groups are ionized, the hydrogen bonding is disrupted and the graft copolymer becomes soluble once more, as long as it is below 30 °C. These complexes formed by hydrogen bonding between the –COOH and –CONH– groups in the longer homopolymer sequences probably do not occur to any great extent with random copolymers of NIPAAm and AAc, indicating that the unique structure of the graft copolymer chain is required for the formation of the complexes through hydrogen bonding.

An ophthalmic drug, timilol hydrogen maleate ('timolol'), was used as a model drug to evaluate the copolymers as matrices for controlled drug release. As can be seen in Figure 1.10, the rate of release of timolol from homopolymer of AAc (MW 250,000) is compared to different polymer-drug mixtures each containing 30 wt% of NIPAAm (i.e., random copolymer, graft copolymer, and physical mixture of

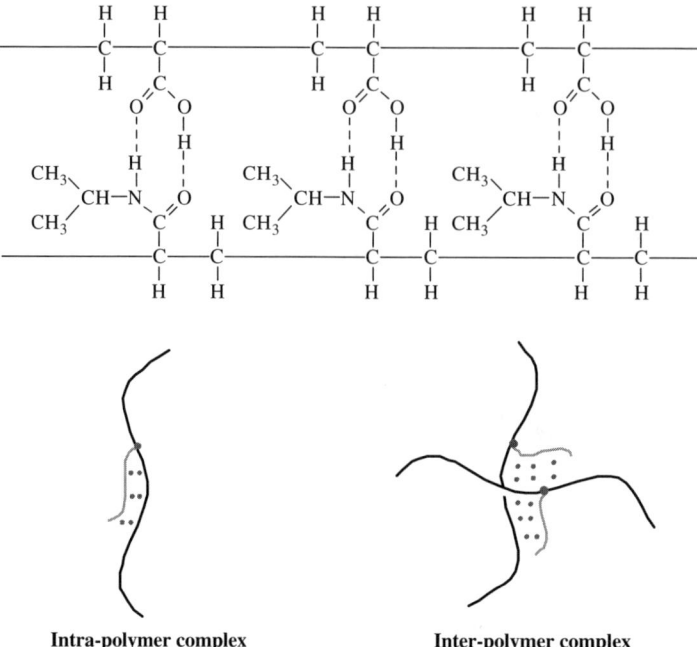

Intra-polymer complex   **Inter-polymer complex**

Figure 1.9 Proposed inter- or intra-polymeric H-bond between the −COOH groups of the PAAc backbone and the −CONH− groups of the oligo NIPAAm graft chain.

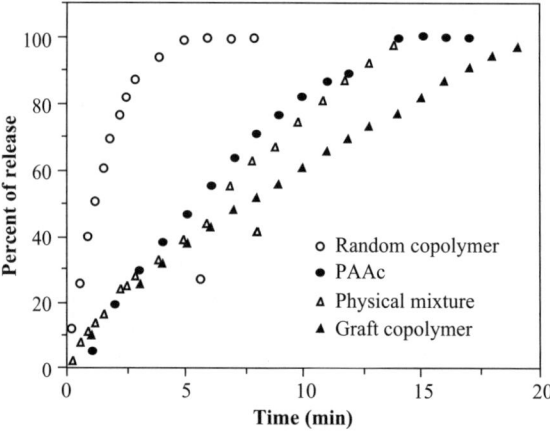

Figure 1.10 Percent of timolol released from PNIPAAm-g-PAAc graft copolymer matrix as a function of time.

homoPAAc and homoPNIPAAm). These results demonstrate that release from the random copolymer is essentially complete within about 5 min, while release from the graft copolymer is the most prolonged, with nearly complete release taking about 20 min. Comparison of the rate of drug release from a physical mixture of homoPAAc and homoPNIPAAm with that from homoPAAc indicates that the physical mixture of temperature-sensitive and pH-sensitive homopolymers is no more effective than the use of the pH-sensitive homopolymer alone, and significantly less effective than the corresponding graft copolymer.

It can be concluded that graft copolymers of PNIPAAm-g-PAAc show unique thermal and pH responsive behaviors in comparison to the random copolymer, homoPAAc, or physical mixtures of homopolymers of the same overall composition. This graft copolymer combines both temperature-sensitive and pH-sensitive properties into one polymer structure. When used as a matrix for mucosal drug delivery, this graft copolymer should be mucoadhesive, and should also provide a prolonged duration of drug release.

1.3. Graft Copolymer of P(NIPAAm-BMA)-g-PAAc

Although the graft copolymers of NIPAAm-g-AAc prolonged the drug delivery compared with either NIPAAm-co-AAc random copolymers or homoPAAc, the prolongation time was only ca. 10–15 min for a 30% graft copolymer. In order to prolong drug delivery for up to ca. 60 min at 34 °C (the temperature of the surface of the eye) it seemed desirable to lower further the transition temperature of the grafted chain. We expected that this would lead to the formation of more swelling-resistant hydrophobic domains, which should more significantly slow the release of drug at eye temperature. For the temperature-sensitive graft copolymer, we synthesized a series of random copolymers of NIPAAm with butyl methacrylate (BMA), a more hydrophobic comonomer, and then grafted this copolymer to a PAAc backbone (Figures 1.3 and 1.11).

With 4 mole% of BMA in P(NIPAAm-co-BMA), the LCST was lowered to 30 °C as opposed to 34 °C for homoPNIPAAm (Figure 1.12). The lower LCST of P(NIPAAm-co-BMA) is due to the introduction of the more hydrophobic comonomer BMA to the

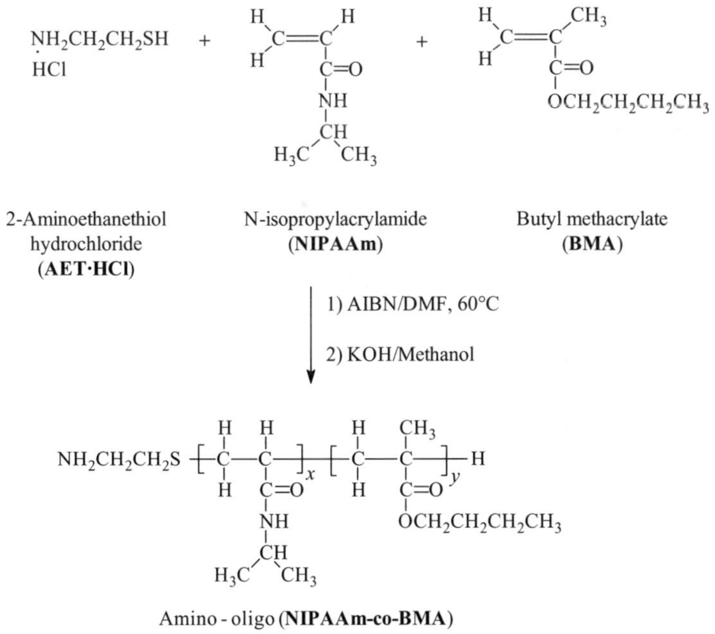

Figure 1.11 Schematic syntheses of the amino-terminated oligo(NIPAAm-co-BMA).

copolymer, as expected (Priest *et al.*, 1987). At pH 7.4, where the PAAc backbone will be almost completely ionized, the graft copolymers all exhibited phase transitions at temperatures close to that for P(NIPAAm-co-BMA), independent of the graft content in the P[NIPAAm-co-BMA]-g-PAAc graft copolymer (Figure 1.13). Furthermore, at 34 °C the phase transitions for this graft copolymer were nearly complete, whereas the less hydrophobic NIPAAm-g-AAc graft copolymers showed partial phase transitions at that temperature (Figure 1.14). The same trends were noted for the effect of pH on

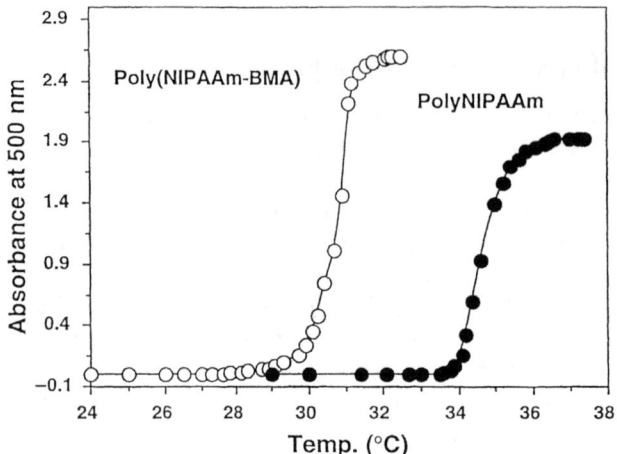

Figure 1.12 Light transmittance (absorbance) of 0.2% polymer solutions in Water vs. temperature. (Comparison of oligo(NIPAAm-co-BMA) having 4 more% BMA with oligoNIPAAm.)

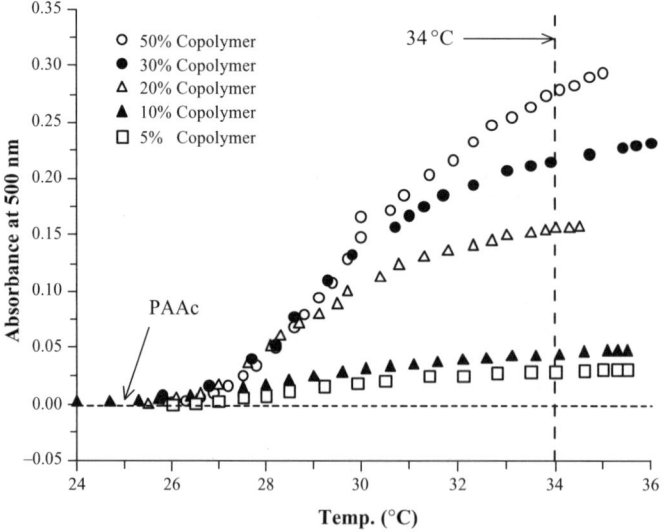

Figure 1.13 Light transmittance (absorbance) of 0.2% polymer solutions in PBS buffer, pH 7.4 vs. temperature. (The graft copolymers of P(NIPAAm-co-BMA)-g-PAAc were prepared by coupling oligo(NIPAAm-co-BMA) onto PAAc.)

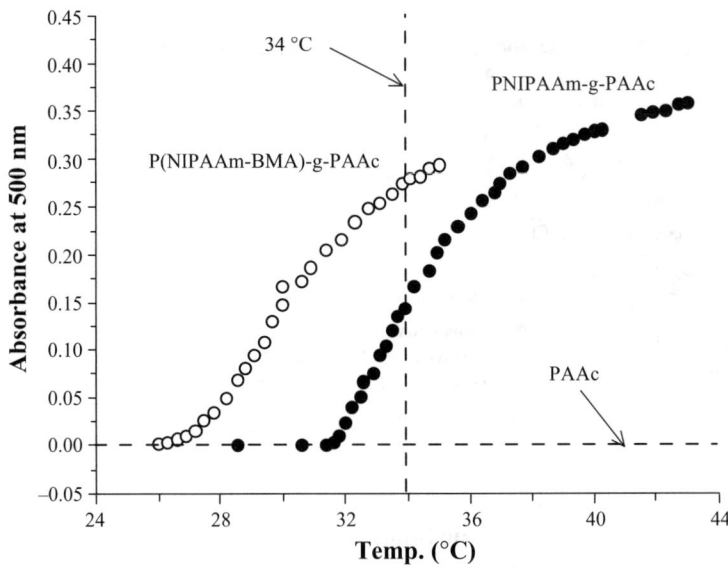

Figure 1.14 Light transmittance (absorbance) of 0.2% polymer solutions in PBS buffer, pH 7.4 vs. temperature. (Comparison of P(NIPAAm-co-BMA)-g-PAAc with PNIPAAm-g-PAAc.)

the LCST behaviors of the P[NIPAAm-co-BMA]-g-PAAc copolymers as for the graft copolymer of PNIPAAm-g-PAAc.

It was found that the P[NIPAAm-co-BMA]-g-PAAc graft copolymer showed significantly more prolonged duration of drug release than the graft copolymer of PNIPAAm-g-PAAc, due to the more hydrophobic character of the former graft chain. In the physical mixtures of PAAc homopolymer with either the [NIPAAm-co-BMA] copolymer or homoPNIPAAm, however, the temperature-sensitive component physically separated from the formulation, and did not prolong the drug release at all (Figure 1.15). The graft copolymers containing 10–20 wt% of P[NIPAAm-co-BMA] seem to prolong the drug release more significantly than equivalent compositions of PNIPAAm-g-PAAc (Figure 1.16). However, the graft copolymers containing more than 20 wt% of P[NIPAAm-co-BMA] are so hydrophobic that they did not erode under the conditions of drug release. In conclusion, by introducing a moderate amount of hydrophobic comonomer units into the PNIPAAm grafted side chains, the thermally induced phase transition temperature is lowered enough to significantly reduce the drug release rate, compared with homoPNIPAAm as the grafted chain.

1.4. Graft Copolymers Pluronic®-g-PAAc

Pluronic® polyols are also temperature-sensitive, but in contrast to NIPAAm and its copolymers, some Pluronics® are approved for use by the FDA as food additives. The Pluronic® polyols used in this study have cloud points (CP) (which indicate the temperature-induced phase transition) between 19 and 26 °C for 1 wt% solutions in water (see Table 1.1) (BASF Corporation, 1989; Yoshioka *et al.*, 1994).

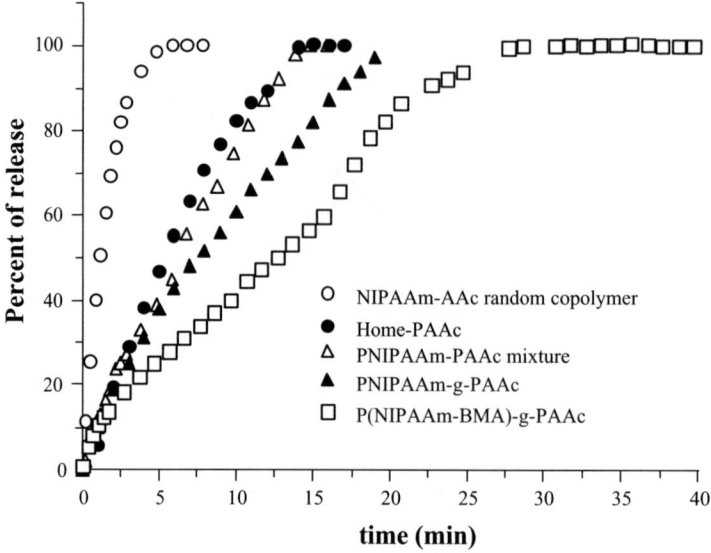

Figure 1.15 Percent of timolol released from P(NIPAAm-co-BMA)-g-PAAc graft copolymer matrix as a function of time.

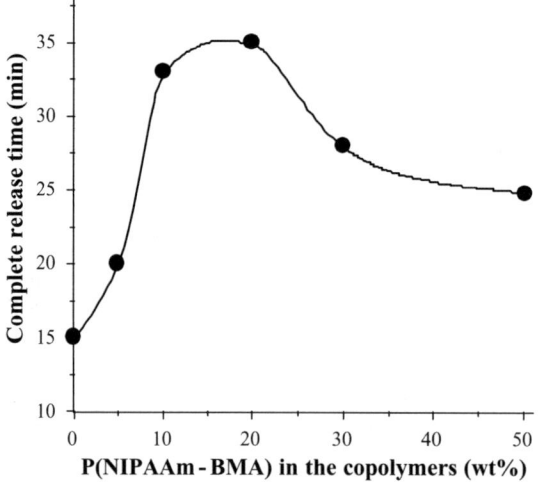

Figure 1.16 Time for 100% timolol release from P(NIPAAm-co-BMA)-g-PAAc matrix as
a function of P(NIPAAm-co-BMA) composition in the graft copolymers of
(P(NIPAAm-co-BMA)-g-PAAc.

In order to graft the Pluronic® onto the backbone of PAAc, a reactive amino group
is required at one end of the Pluronic® PEG blocks. Thus, amino-terminated Pluronic®
derivatives were prepared in the following two steps: (1) reacting one of the hydroxyl
end groups with *p*-nitrophenyl chloroformate by adjusting the stoichiometry (molar
ratio of 2 : 1) to result in a *p*-nitrophenyl carbonate-derivatized Pluronic®; (2) reacting
the latter with an excess amount of diaminoethylene to result in an amino-terminated

Table 1.1 Structure and selected properties of Pluronic® used in this study*
Polyoxyethylene-polyoxypropylene-polyoxyethylene triblock copolymer

(Pluronic®)

$$\text{HO}\underbrace{-\text{CH}_2-\text{CH}_2-\text{O}}_{a}\underbrace{-\text{CH}_2-\overset{\overset{\displaystyle \text{CH}_3}{|}}{\text{CH}}-\text{O}}_{b}\underbrace{-\text{CH}_2-\text{CH}_2-\text{O}}_{a}-\text{H}$$

EO	PO	EO

Pluronic®	*EO-PO-EO*	*Total MW (Avg.)*	*Cloud point in water (°C)*	
			1 wt%	*10 wt%*
L61	4/30/4	2,000	24	17
L81	7/38/7	2,750	20	16
L92	10/50/10	3,650	26	16
L122	13/69/13	5,000	19	13

*Data from BASF brochure.

Pluronic® derivative (Figure 1.17). In the first step, it is important to control the stoichiometry of the reaction so that on the average only one terminal hydroxyl group of the Pluronic® PEG block is derivatized. In the second step, an access of diaminoethylene is needed to ensure that all the activated p-nitrophenyl carbonate groups are converted to amine groups. Using this method, we achieved an average functionality (f) of $f = 0.91 \pm 0.1$, as determined by back-titration, and those amino-terminated Pluronic® polymers were used for synthesis of the Pluronic®-g-PAAc graft copolymers. Statistically, it is possible that some unreacted $f = 0$ or doubly derivatized, $f = 2$ derivatives were formed, but we made no special effort to separate them from the product. However, it is probable that the $f = 0$ Pluronic® was removed during the separation of the graft copolymer (see below). Furthermore, reaction of the $f = 2$ derivatives with PAAc should result in crosslinked PAAc chains, which might have reduced solubility; however, the graft copolymers we synthesized were completely soluble, indicating that the amount of double derivatives in the product was probably small (see below).

   The graft copolymers of Pluronic®-g-PAAc were prepared using the coupling method described in the previous sections. The Pluronic® L-61, or L-92, or L-122 chains were coupled onto the PAAc backbone via the reaction of the amino terminal group of the Pluronic® with the carboxyl groups of PAAc in the presence of dicyclohexyl carbodiimide (DCC). Reaction was carried out in methanol at room temperature for 24 h. The graft copolymers were recovered from the reaction solutions by precipitation into THF or diethylether, both of which are solvents for Pluronic®, therefore, $f = 0$ and unreacted Pluronic® derivatives can be removed in the supernatant from the precipitated graft copolymer product. In this process, we cannot separate ungrafted or free homoPAAc from the graft copolymer product, but statistically the chance for a PAAc chain to have absolutely no Pluronic® grafted is very small, especially for high degrees of grafting. Graft copolymers were synthesized at varying ratios of components, using the different Pluronics®, as summarized in Table 1.2.

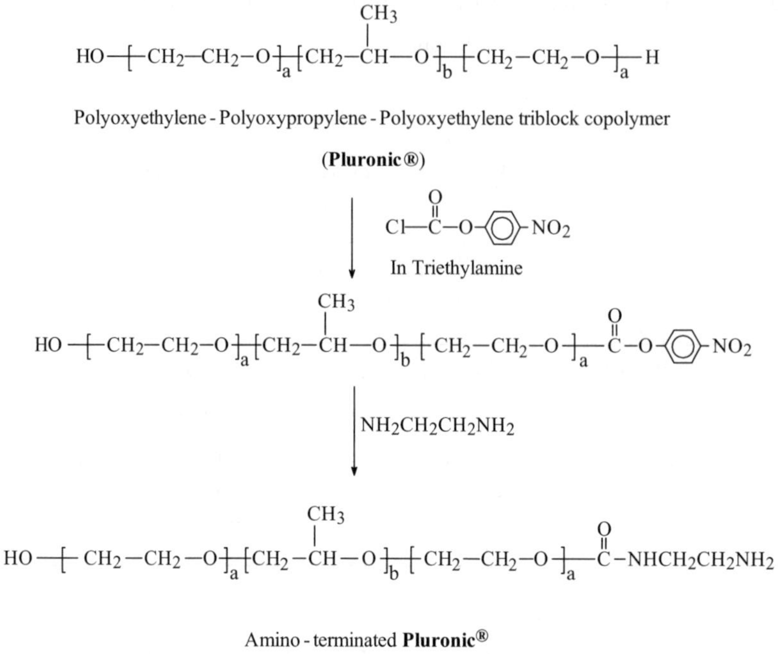

$$HO-\left[CH_2-CH_2-O\right]_a\left[CH_2-\overset{\overset{\displaystyle CH_3}{|}}{CH}-O\right]_b\left[CH_2-CH_2-O\right]_a H$$

Polyoxyethylene - Polyoxypropylene - Polyoxyethylene triblock copolymer

(Pluronic®)

$$Cl-\overset{\overset{\displaystyle O}{\|}}{C}-O-\!\!\bigcirc\!\!-NO_2$$

In Triethylamine

$$HO-\left[CH_2-CH_2-O\right]_a\left[CH_2-\overset{\overset{\displaystyle CH_3}{|}}{CH}-O\right]_b\left[CH_2-CH_2-O\right]_a-\overset{\overset{\displaystyle O}{\|}}{C}-O-\!\!\bigcirc\!\!-NO_2$$

$NH_2CH_2CH_2NH_2$

$$HO-\left[CH_2-CH_2-O\right]_a\left[CH_2-\overset{\overset{\displaystyle CH_3}{|}}{CH}-O\right]_b\left[CH_2-CH_2-O\right]_a-\overset{\overset{\displaystyle O}{\|}}{C}-NHCH_2CH_2NH_2$$

Amino - terminated **Pluronic®**

Figure 1.17 Schematic synthesis of amino-terminated Pluronic® derivitive.

Table 1.2 Preparation of graft copolymers of Pluronic®-g-PAAc*

Code	Pluronic®	$W^®_{Pluronic}/W_{PAAc}$ in feed	Yield (wt%)	Pluronic® in copolymers** (wt%)
L61-1	L61	10/90	79	15
L61-2	L61	20/80	55	24
L61-3	L61	30/70	86	34
L92-3	L92	30/70	94	32
L122-1	L122	10/90	80	15
L122-2	L122	20/80	80	18
L122-3	L122	30/70	68	21
L122-4	L122	40/60	78	27
L122-5	L122	50/50	65	41

* Solvent: methanol, 100 ml; reaction temperature and time: room temperature, 24 h; molar ratios of DDC to Pluronic®, 2 : 1. The polymers were recovered by precipitation using THF or diethylether.
** The composition of Pluronic® in the graft copolymers were determined by back titration.

Formation of the graft copolymer was verified by gel permeation chromatography (GPC). As illustrated in Figure 1.18, the Pluronic® L-122 peak at 37.8 min can hardly be seen in the graft copolymers, but it is clearly seen in the physical mixture of PAAc with L-122 (30% L-122), indicating that almost no Pluronic® was physically-entrapped in the graft copolymers. In addition, the graft copolymers elute later than the 'parent' homoPAAc, indicating that they have smaller hydrodynamic radii than

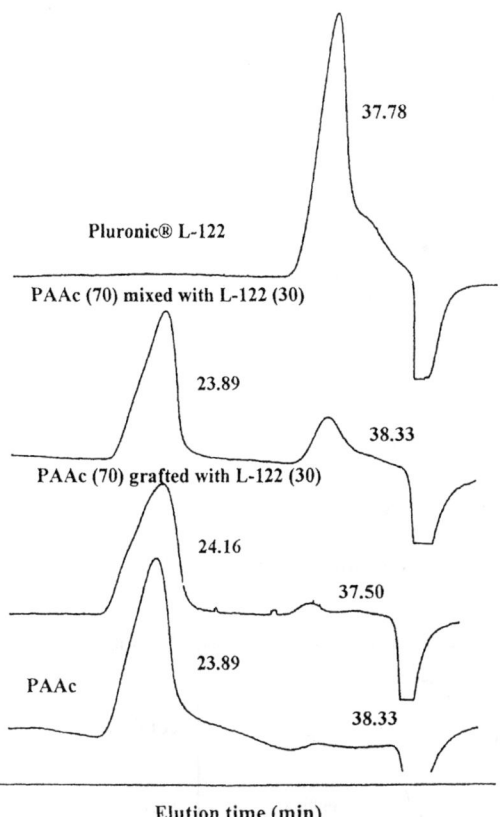

37.78

Pluronic® L-122

PAAc (70) mixed with L-122 (30)

23.89

38.33

PAAc (70) grafted with L-122 (30)

24.16

37.50

23.89

PAAc

38.33

Elution time (min)

Figure 1.18 GPC diagrams of (a) free Pluronic® L-122; (b) physical mixture of Pluronic® L122 with PAAc (30/70 by weight); (c) graft copolymer of Pluronic® L122-g PAAc (30/70 by weight); (d) parent homopolymer of PAAc. Eluent, DMF; temperature, 40 °C; elution rate, 0.7 ml/min.

PAAc, as one would expect due to the intramolecular interactions of the PPO block segments.

We have noted above that solutions of graft copolymers of PNIPAAm-g-PAAc or P(NIPAAm-co-BMA)-g-PAAc exhibited temperature-induced phase transitions at pH 7.4 despite the ionization of the backbone PAAc at that pH. The graft copolymer phase transitions also occurred at temperatures similar to the LCSTs of the grafted side chains (Chen and Hoffman, 1995a–d). In contrast with this, however, solutions of the Pluronic®-g-PAAc graft copolymers showed unique temperature-induced gelation behavior. Figure 1.19 illustrates the rapid rise in viscosities of the graft copolymer solutions vs. temperature, an opposite trend to viscosity/temperature behavior of most polymer solutions. In the vicinity of phase transition temperature (ca. 28–32 °C), the viscosity increased dramatically, and a gel formed. In contrast, Pluronic® by itself or physically mixed with PAAc, precipitates as its phase transition temperature is reached. It is worth noting that the graft copolymers have a higher phase transition temperature than the same free Pluronic® due to the concentration dependence of their phase transition behaviors (BASF Corporation, 1989). Normally, the higher the

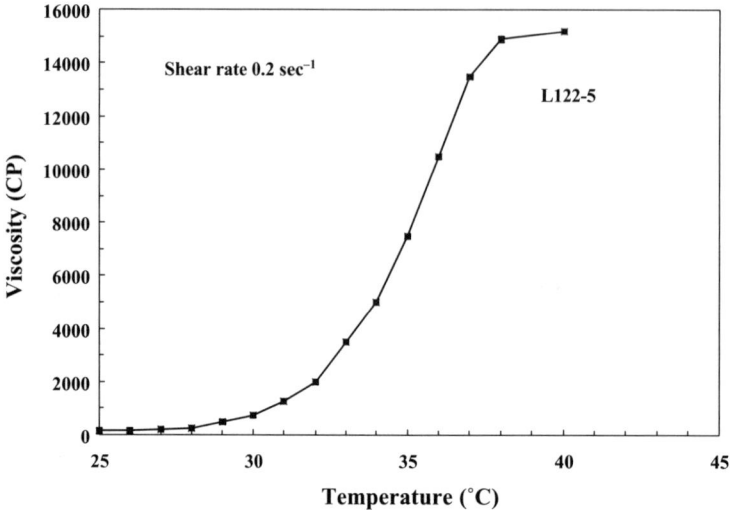

Figure 1.19 Viscosity of graft copolymer soultion (2.5 wt%) of Pluronic® L122-g-PAAc (L122-5) at pH 7.2 vs. temperature.

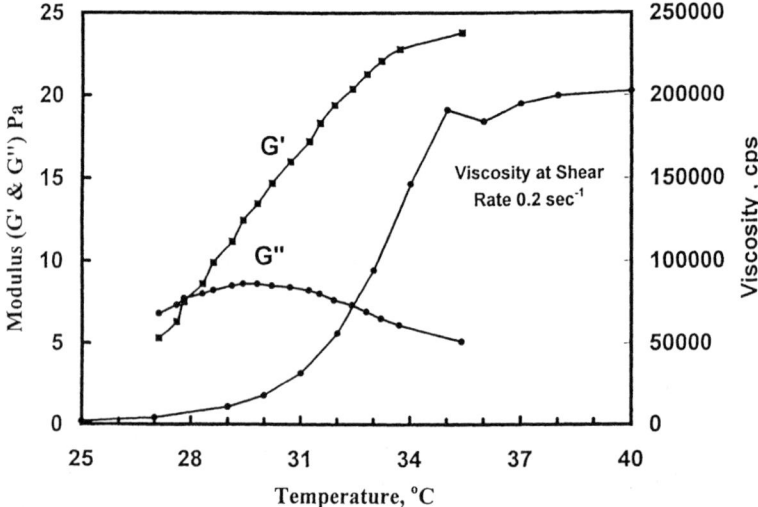

Figure 1.20 Elasticity and viscosity of Pluronic®-g-PAAc solution vs. temperature (with 41% of Pluronic® L122 in the graft chains). G' is the shear modulus and G'' is elastic modulus, respectively, of the graft copolymer solution.

concentration of the polymers, the lower the phase transition temperature; this should render the Pluronic®-g-PAAc graft copolymers more erodible as dilution increases during the drug delivery process. The shear modulus of the graft copolymer solution also increases with temperature, and the elastic modulus shows a maximum around 28–32 °C, corresponding to the gel point (Figure 1.20).

This temperature-induced gelation can be explained by the hydrophobic interactions of the polypropylene oxide (PPO) block segments of the grafted Pluronic®

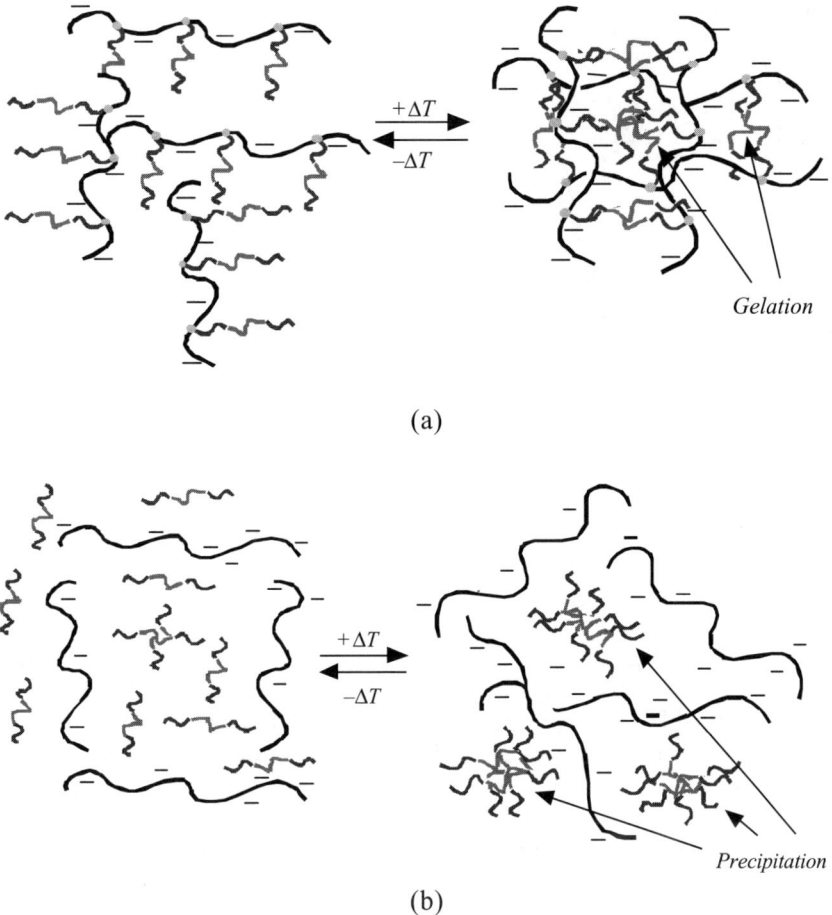

(a)

Gelation

(b)

Precipitation

Figure 1.21 (a) Sketch for the thermally-sensitive behavior of the graft copolymers of Pluronic®-g-PAAc at pH 7.4; (b) Sketch for the thermally-sensitive behaviour of the physical mixtures of Pluronic® with PAAc at pH 7.4.

chains. When the temperature is raised above the phase transition temperature of the Pluronic®, the PPO segments on different grafted chains should aggregate together, linking the various grafted chains into a physical network (Figure 1.21a), leading to the observed gelation. In contrast, free Pluronic® by itself or physically mixed with PAAc, both form low viscosity cloudy precipitates, and do not gel as temperature increases (Figure 1.21b).

This physical crosslinking associated with the hydrophobic aggregation of the PPO segments of the grafted chains is further evidenced by the fact that the viscosity increase can be reduced or even reversed with further increase of temperature or by increase of shear rate. As can be seen in Figure 1.22, the viscosity increase with increase of temperature, reaches a maximum and then decreases with further increase of temperature. It was also found that the maximum viscosity and the temperature at which maximum viscosity is observed decreases with increase of shear rate. This suggests that the aggregation of the PPO segments of the grafted Pluronic® chains with increase

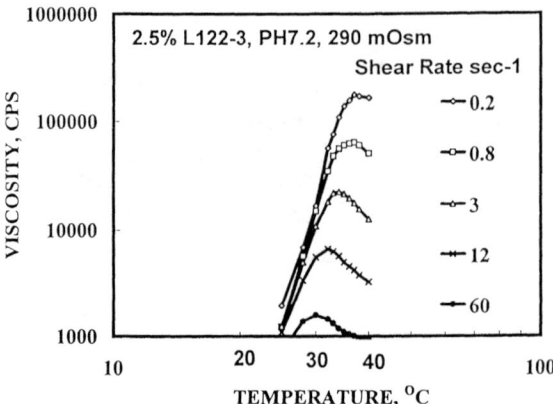

Figure 1.22 Viscosity of Pluronic®-g-PAAc solution vs. temperature. Effect of shear rate on the maximum viscosity.

of temperature has to compete with the increase in molecular motions at higher temperatures and/or mechanical shear. Hence, initially the viscosity increases with temperature, as hydrophobic aggregation of PPO segments on the Pluronic® grafted chains is the dominant action, but as temperature continues to increase, eventually the molecular motions resulting from thermal energy dominate. Addition of mechanical energy in the form of shear also helps to break up the PPO aggregates. These results further suggest that the PPO–PPO physical crosslinks formed as temperature increases are not permanent but can be easily reversed not only by reducing temperature but also by applying shear. It was found that viscosity was regained instantaneously upon removal of the shear forces.

The bioadhesive behavior of the graft copolymers was evaluated by N. Peppas and C. Brazel, who compared adhesion to a crosslinked mucin gel of crosslinked hydro-gels of PAAc vs. Pluronic®-g-PAAc. It was found that the graft copolymer hydrogels showed almost as good bioadhesive properties as the well-known bioadhesive gel of PAAc. These results will be reported elsewhere (Chen *et al.* 1995e; Hoffman *et al.*, 1999).

Timolol was loaded to the Pluronic®-g-PAAc graft copolymer matrix by dissolving both the polymer and the drug in a neutral solvent (methanol) and a film was cast on a glass disk from the copolymer–drug mixture, as described in previous sections. The dried coated glass disks were suspended in PBS buffer (pH 7.4) at 34 °C, and the amount of the drug released from the film was measured as a function of time. Figure 1.25 shows the drug release profiles from different composition graft copoly-mers of Pluronic®-g-PAAc. For comparison, the drug release profiles from homoPAAc, as well as a physical mixture of 80% homoPAAc and 20% L-61, are also presented in Figure 1.23. It can be seen that the graft copolymers show significantly prolonged dura-tions of drug release compared to homoPAAc alone and also comparing its physical mixture with 20% L-61 to the equivalent composition graft copolymer. Complete drug release from the graft copolymer of Pluronic®-g-PAAc with 20% or more Pluronic® in the grafted chains was prolonged to over 1 h vs. 15 min from the homoPAAc formu-lation or 25 min from the 80% : 20% physical mixture. This prolonged duration of drug release is clearly due to the gelation of the graft copolymer at the conditions of the

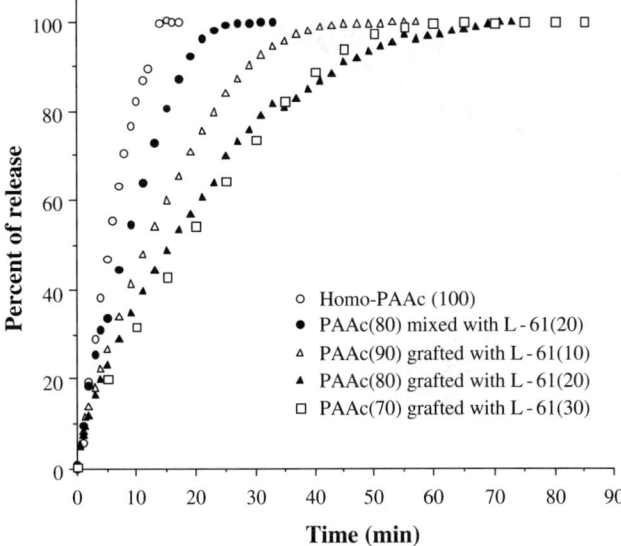

Figure 1.23 Percent of Timolol release from the Pluronic®-g-PAAc graft copolymer matrices (L61) as a function of time.

experiment (pH 7.4 and 34 °C). The gel formation also slowed the dissolution of the graft copolymer. In contrast, the drug formulation with the homopolymer swells and dissolves very rapidly, leading to fast drug release from the formulation. Furthermore, at the conditions of the release experiment, the Pluronic L-61 phase separates from the physical mixture of timolol plus 80% PAAc and 20% Pluronic® L-61. As a result, the drug in the PAAc phase is released rapidly due to the fast swelling and dissolution of PAAc, similar to the formulation of drug with homoPAAc. Evidently, the precipitated Pluronic® L-61 retarded the overall release rate a bit compared to the pure homoPAAc formulation where there is no precipitation.

Figure 1.24 illustrates the drug release profiles from the graft copolymers containing Pluronic® L-122, a more hydrophobic triblock copolymer with a longer PPO segment, as the grafted chains. It can be seen that the duration of drug release from those graft copolymer formulations is prolonged even further than the other graft systems. With 30% or more Pluronic® L-122 grafted to PAAc, the time for complete drug release was more than 4 h, compared to one hour when 30% Pluronic® L-61 was grafted. This indicates that longer hydrophobic PPO segments in the grafted chains might enhance the strength and extent of gelation, thus more effectively prolonging the drug release from the polymer matrix. This can be further evidenced from Figure 1.25 which compares the drug release profiles for three different triblock copolymers of Pluronic®, L-61, L-92 and L-122. The time for complete drug release from those three different polymer matrices are prolonged in the order L-122 > L-92 > L-61, while the length of PPO segments in those Pluronics® increases in the same order. This indicates that the longer hydrophobic segments in the grafted Pluronics® chains play an important role in prolonging the drug release due to their effect on the tightness of the gel network formed.

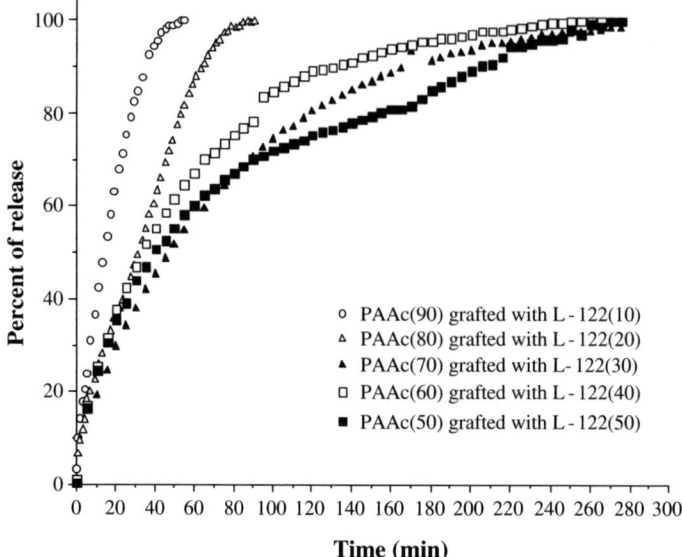

Figure 1.24 Percent of Timolol release from the Pluronic®-g-PAAc graft copolymer matrices (L122) as a function of time.

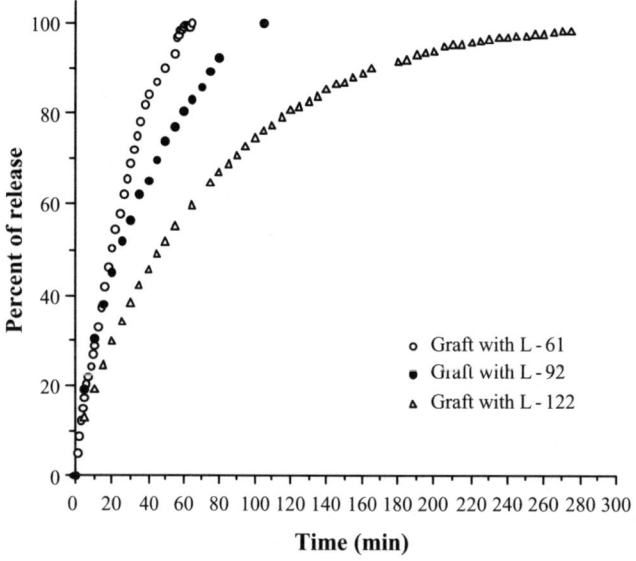

Figure 1.25 Kinetics of Timolol release from Pluronic®-g-PAAc graft copolymer matrices with different Pluronics® in the graft chains.

1.5. Conclusions

Graft copolymers of PNIPAAm-g-PAAc or P(NIPAAm-BMA)-g-PAAc exhibit temperature-induced phase transitions independent of the pH-responsive character of the PAAc back bone. However, when formulated with timolol maleate, these systems do not exhibit an increase in viscosity with temperature, and thus do not prolong significantly the duration of drug release. In contrast, graft copolymers of PAAc synthesized with grafts of thermally-sensitive triblock copolymers of PEO/PPO/PEO (Pluronic®) show a unique temperature-induced gelation. When used in drug delivery formulations with timolol maleate, complete drug release from the matrices is significantly prolonged to over four hours, compared to ca. 15 min from homoPAAc or less than half an hour for physical mixtures of homomoPAAc and Pluronic®. The Pluronic®-g-PAAc graft copolymers should be useful in mucoadhesive formulations for prolonged duration of drug delivery to mucosal surfaces.

1.6. Appendix

1.6.1. Synthesis of Graft Copolymer of PNIPAAm-g-PAAc

1.6.1.1. Synthesis of OligoNIPAAm with a Terminal Amino Group

The oligomer of N-isopropylacrylamide (oligoNIPAAm) with a terminal amino group was synthesized by free radical polymerization of NIPAAm in methanol at 60 °C for 20 h using 2,2′-azobisbutyronitrile (AIBN) and 2-aminoethanethiol hydrochloride (AET · HCl) as initiator and chain transfer reagent, respectively (Figure 1.26a). The ratio of monomer to chain transfer agent was 100 : 5. The oligomer was collected by precipitation into diethyl ether. In order to convert the amine hydrochloride end group

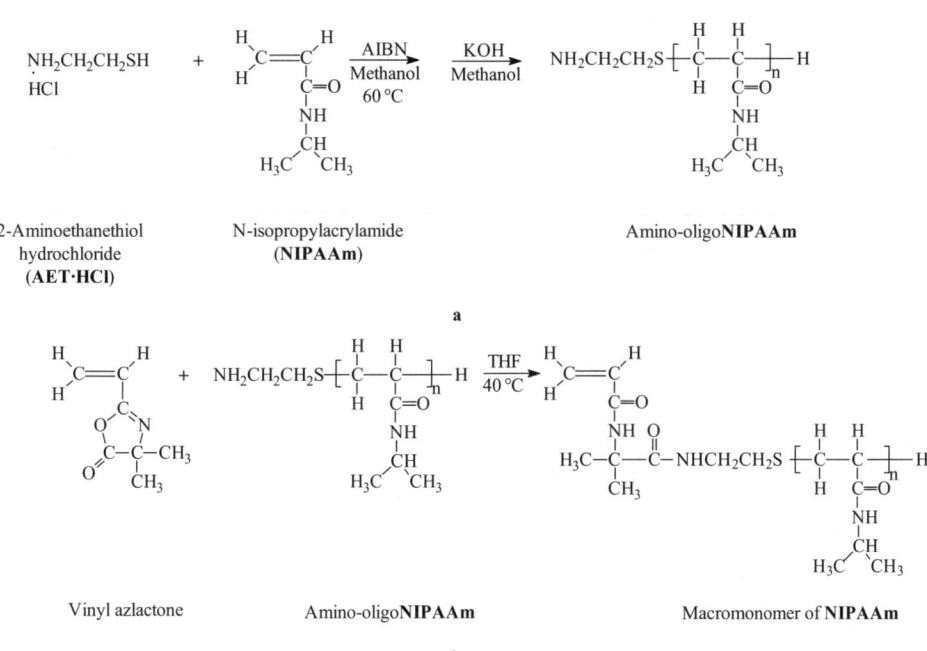

Vinyl azlactone   Amino-oligo**NIPAAm**   Macromonomer of **NIPAAm**

b

Figure 1.26 Schematic syntheses of (a) amino-terminated oligoNIPAAm and (b) macromonomer of NIPAAm

into the free amino group, the oligomer was re-dissolved in chloroform to which a certain amount of potassium hydroxide solution in methanol was added. The precipitated salt was removed by filtration and the oligomer was recovered by precipitation into diethyl ether. The oligomer was further purified by re-precipitation from chloroform into ether. The number average molecular weight of the oligomer was determined by titration to be 2,200.

1.6.1.2. Synthesis of the Macromonomer of NIPAAm

The macromonomer of NIPAAm was prepared by reaction of amino-terminated oligoNIPAAm with vinyl azlactone in dry tetrahydrofuran (THF) at 40 °C for 16 h (Figure 1.26b). The macromonomer was recovered by precipitating the reaction mixture into diethyl ether.

1.6.1.3. Synthesis of Graft Copolymer of PNIPAAm-g-PAAc

The graft copolymers with PNIPAAm as the grafted side chains and polyacrylic acid (PAAc) as the backbone were prepared in two ways: (1) by copolymerization of the macromonomer of NIPAAm with AAc (Figure 1.1); or (2) by coupling oligoNIPAAm onto the PAAc backbone through the reaction of the amino terminal group on the oligoNIPAAm chain with the carboxyl groups on the PAAc backbone (Figure 1.2).

For a typical preparation using the first method, the macromonomer of NIPAAm was copolymerized with acrylic acid (AAc) in methanol at 60 °C for 1.5 h using AIBN as initiator, using varying ratios of the two monomers and at a total monomer concentration of 10% w/v. After the copolymerization, the reaction solution was dropped into methyl ethyl ketone (MEK), a solvent for the macromonomer and AAc, to precipitate the copolymer. Any unreacted macromonomer and residual AAc were removed from the supernatant. The polymers were further purified by redissolving in methanol and reprecipitating into THF.

In the second method, graft copolymers of PNIPAAm-g-PAAc were synthesized in methanol by reaction of the terminal amino group of oligoNIPAAm (MW, 2,200) with the carboxyl group of PAAc (MW, 250,000) using N, N'-dicyclohexylcarbodiimide (DCC) as an activation reagent. PAAc (MW = 250,000) and the amino-terminated oligo(NIPAAm) were dissolved in methanol (total polymer concentration = 3.5% w/v). The solution was then cooled to 4 °C. To this a certain amount of DCC was added (50 mole% excess with respect to the amount of oligo[NIPAAm] added to the reaction solution), and the reaction was kept at room temperature for 24 h. After reaction, the mixture was dropped into MEK to precipitate the copolymer and to remove any unreacted oligo(NIPAAm) from the supernatant. The graft copolymer was further purified by redissolving in methanol and reprecipitating into THF.

For comparison, random copolymers of NIPAAm-AAc with various compositions were also prepared. NIPAAm and AAc were copolymerized in methanol at 60 °C using AIBN as initiator (Figure 1.5). After polymerization, the polymers were collected by precipitation into diethyl ether.

The compositions of the copolymers were determined by back titration. The copolymers were dissolved in an excess of 0.1 N NaOH to neutralize the carboxyl group and the excess of NaOH was back-titrated with 0.1 N HCl. The weight percentage of AAc in the copolymer was calculated based on the assumption that all the AAc carboxyl groups were neutralized by the NaOH.

1.6.2. *Synthesis of P(NIPAAm-BMA)-g-PAAc Graft Copolymer*

1.6.2.1. *Synthesis of Amino-terminated Oligo(NIPAAm-co-BMA)*

The amino-terminated oligo([NIPAAm]-co-butyl methacrylate [BMA]) was prepared by radical copolymerization of NIPAAm with BMA (3 mole% BMA and 97 mole% NIPAAm in the feed) in dimethyl formamide (DMF) at 60 °C for 1.5 h using AIBN and AET·HCl as initiator and chain transfer agent, respectively (Figure 1.11). The ratios of monomer to chain transfer agent was 100 : 5. The co-oligomer was collected by precipitation into diethyl ether. In order to convert the amine hydrochloride end group into the free amino group, the co-oligomer was redissolved in chloroform to which a certain amount of potassium hydroxide solution in methanol was added. The precipitated salt was removed by filtration and the co-oligomer was recovered by precipitation into diethylether. The co-oligomer was further purified by reprecipitation from chloroform into ether. The number average molecular weight (M_n) of the co-oligomer was determined by both vapor pressure osmometer (VPO, Knaur, German), and by titration with 0.1 N NaOH, which detects the amino group at one end of the oligomer. The BMA composition in the co-oligomer was determined to be 4 mole% by ^{1}H–NMR.

1.6.2.2. *Synthesis of Graft Copolymers of P(NIPAAm-BMA)-g-PAAc*

The graft copolymers of oligo(NIPAAm-co-BMA)-g-PAAc were synthesized by coupling oligo(NIPAAm-co-BMA) onto the PAAc backbone through the reaction of the amino terminal group in the co-oligomer with a carboxyl group on the PAAc backbone, using DCC as activation reagent (Figure 1.3). PAAc (MW = 250,000) and the amino-terminated co-oligomer were dissolved in methanol (total polymer concentration, 3.5% w/v). The solution was cooled to 4 °C. To this a certain amount of DCC (50 mole% excess with respect to the amount of co-oligomer added to the reaction mixture) was added, and the reaction was kept at room temperature for 24 h. After reaction, the mixture was dropped into MEK to precipitate the copolymer and to remove any un-reacted co-oligomer with the supernatant. The polymers were further purified by redissolving in methanol and reprecipitating into THF.

1.6.3. *Synthesis of Graft Copolymers of Pluronic®-g-PAAc*

1.6.3.1. *Synthesis of an Amino-terminated Tri-block Copolymer of PEO-PPO-PEO (Pluronic®)*

Triblock PEO–PPO–PEO copolymers with a reactive amino-terminal were prepared by a two step reaction. First, the triblock copolymer was reacted with 25 mole% access of 4-nitrophenyl chloroformate in methylene chloride in the presence of triethyleneamine at room temperature for 4 h, to yield a 4-nitrophenyl formate-derivatized intermediate (Figure 1.17). This intermediate was recovered by extraction using petroleum ether (grade of 35–60 °C) for three times, resulting in product with a yield of 70–80% by weight. In the second step, the intermediate was reacted with diaminoethylene (at a molar ratio of 3:1 with respect to the derivatized block copolymer) in methylene chloride at room temperature overnight (Figure 1.17). After reaction, the mixture was extracted with petroleum ether (grade of 35–60 °C) for three times, and dialyzed against distilled water for three days at 4 °C to remove any unreacted block copolymer (using a membrane with a molecular weight cut-off of 1,000 or 3,500 depending on the molecular weight of the Pluronic® used). The dialysed solution of the amino-terminated Pluronic® derivative was then lyophilized, with a product yield of 85% to 90% by weight. Functionality of the amino-terminated derivative was determined by back titration. The derivative was dissolved in an excess of 0.01 N HCl to neutralize the amino terminal group and the excess HCl was back-titrated with 0.01 N NaOH.

1.6.3.2. Synthesis of Graft Copolymers of Pluronic®-g-PAAc

Graft copolymers of Pluronic®-g-PAAc were synthesized by coupling the derivatized PEO–PPO–PEO block copolymers onto the PAAc backbone through the reaction of the terminal amino group on the Pluronic® molecule and the pendant carboxyl group on the PAAc backbone. Specifically, reaction between the amino group and carboxyl group was carried out with addition of the activation reagent, dicyclohexyl carbodiimide (DCC), as described above, and resulted in amide bond formation (Figure 1.4). The reaction was carried out in methanol at room temperature for 24 h, with a mole ratio of PEO–PPO–PEO to DCC of 2 : 1. The resulting graft copolymers were recovered from the reaction solution by precipitation into THF or diethylether.

1.7. Abbreviations

Aac	acrylic acid
AET HCl	2-aminoethanethiol hydrochloride
AIBN	2,2′-azobisbutyronitrile
BMA	butyl methacrylate
CP	cloud point
DCC	N,N′-dicyclohexylcarbodiimide
DMF	dimethyl formamide
GPC	gel permeation chromatography
GRAS	generally regarded as safe
LCST	lower critical solution temperature
MAAc	methacrylic acid
MEK	methyl ethyl ketone
NIPAAm	N-isopropylacrylamide
OligoNIPAAm	oligomer of N-isopropylacrylamide
PAAc	polyacrylic acid
PAAm	polyacrylamide
PEG	poly(ethylene glycol)
PEGM	PEG methacrylate
PEO	poly(ethylene oxide)
PMAAc	poly(methacrylic acid)
PNIPAAm	poly(N-isopropylacrylamide)

PPO   poly(propylene oxide)

THF   tetrahydrofuran

VPO   vapor pressure osmometer.

1.8. References

Bae, Y.H., Okano, T., Hsu, R., and Kim, S.W. (1987) Thermo-sensitive polymers as on-off switches for drug release, *Makromol. Chem., Rapid Commun.*, 8, 481–485.

BASF brochure (1989).

Chen, G.H. and Hoffman, A.S. (1993) Preparation and properties of thermo-reversible, phase-separating enzyme-oligo(N-isopropylacrylamide) conjugates, *Bioconjugate Chem.*, 4, 509–514.

Chen, G.H. and Hoffman, A.S. (1995a) Temperature induced phase transition behaviors of random vs. graft copolymers of N-isopropylacrylamide and acrylic acid, *Macromol. Rapid Commun.*, 16, 175–182.

Chen, G.H. and Hoffman, A.S. (1995b) Temperature- and pH-sensitive graft copolymers that exhibit temperature-induced phase transitions over a wide range of pHs, *Nature (London).*, 373, 49–52.

Chen, G.H. and Hoffman, A.S. (1995c) Temperature- and pH-sensitive random and graft copolymers for drug delivery, *21st Annual Meeting of the SOCIETY FOR BIOMATERIALS*, San Francisco, CA, March 18–22, 139–140.

Chen, G.H. and Hoffman, A.S. (1995d) Novel graft copolymers of a temperature-sensitive copolymer grafted to a bioadhesive polyelectrolyte for controlled drug delivery to mucosal surfaces, *7th International Symposium on Recent Advances in Drug Delivery Systems*, Salt Lake City, Utah, February 27–March 2, 139–140.

Chen, G.H., Hoffman, A.S., and Ron, E. (1995e) Novel hydrogels of a temperature-sensitive Pluronic® grafted to a bioadhesive polyacrylic acid backbone for vaginal drug delivery, *Proc. Intern. Symp. Contr. Rel. Bioact. Mater.*, 22, 167–168.

Dong, L.C. and Hoffman, A.S. (1991) A novel approach for preparation of pH-sensitive hydrogels for enteric drug delivery, *J. Controlled Release*, 15, 141–152.

Gu, J.M., Robinson, J.R., and Leung, S.-H.S. (1988) Binding of acrylic polymers to mucin/epithelial surface: structure-property relationships, *Crit. Rev. Therap. Drug Carrier Sys.*, 5, 21–67.

Hoffman, A.S. (1987) Applications of thermally reversible polymers and hydrogels in therapeutics and diagnostics, *J. Controlled Release*, 6, 297–305.

Hoffman, A.S., Chen, G.H., Wu, X.D., Ding, Z.L., Matsuura, J.E., and Gombotz, W.R. (1999) Stimuli-responsive polymers grafted onto polyacrylic acid and chitosan backbones as bioadhesive carriers for mucosal drug delivery, *Frontiers in Biomedical Polymer Applications*, Ottenbrite, R.M. Ed., 17–29.

Katono, J., Maruyama, A., Sanui, K., Ogata, N., Okano, T., and Sakurai, Y. (1991) Thermo-responsive swelling and drug release switching of interpenetrating polymer networks composed of poly(acrylamide-co-butyl methacrylate) and poly(acrylic acid), *J. Controlled Release*, 16, 215–228.

Kishi, R., Hara, M., Sawahata, K., and Osada, Y. (1991) *In "Polymer Gels"*, D. DeRossi *et al.*, Eds., Plenum, New York, pp. 205–220.

Klier, J., Scanton, A.B., and Peppas, N.A. (1990) Self-associating networks of poly(methacrylic acid-g-ethylene glycol), *Macromolecules*, 23, 4944–4949.

Kwon, I.C., Bae, Y.H., and Kim, S.W. (1991) Electrically erodible polymer gel for controlled release of drugs, *Nature (London)*, 354, 291–293.

Park, H. and Robinson, J.R. (1985) Physico-chemical properties of water insoluble polymers important to mucin/epithelial adhesion, *J. Controlled Release*, 2, 47–57.

Park, H. and Robinson, J.R. (1987) Mechanism of mucoadhesion of poly(acrylic acid) hydrogels, *Pharm. Res.*, **4**, 457–464.

Peppas, N.A. and Buri, P.A. (1985) Surface, interfacial and molecular aspects of polymer bioadhesion on soft tissues, *J. Controlled Release*, **2**, 257–275.

Priest, J.H., Murray, S.L., Nelson, J.R., and Hoffman, A.S. (1987) Lower critical solution temperatures of aqueous copolymers of N-isopropylacrylamide and other N-substituted acrylamides, *ACS Symp. Ser.*, **350**, 255–264.

Siegel, R.A. and Firestone, B.A. (1988) pH-dependent equilibrium swelling properties of hydrophobic polyelectrolyte copolymer gels, *Macromolecules*, **21**, 3254–3259.

Tanaka, T. (1979) Phase transitions in gels and a single polymer, *Polymer*, **20**, 1404–1412.

Yoshioka, H., Mikami, M., and Mori, Y. (1994) Preparation of poly(N-isopropylacrylamide)-b-poly(ethylene glycol) and calorimetric analysis of its aqueous solution, *J. Macromol. Sci., Pure Appl. Chem.*, **A31**, 109–112.

2 Synthesis of novel smart polymers for bioseparation and bioprocessing

Mitsuru Akashi and Akio Kishida

2.1. Introduction

It is known that water soluble polymers with hydrophobic moieties generally have their cloud points in aqueous solutions (Sakurada *et al.*, 1957; Baily and Callard., 1959; Drechsel, 1957; Heskins and Guillet, 1968; Horne *et al.*, 1971; Sarkar, 1979). Scheme 1 shows typical examples of such amphiphilic polymers. Among these, much attention has been particularly focused on poly(*N*-isopropylacrylamide) (poly(NIPAAm)) and its derivatives (Okahata *et al.*, 1986; Priest *et al.*, 1987; Hirotsu *et al.*, 1987; Kwon *et al.*, 1990; Fujishige *et al.*, 1989; Ito, 1989; Schild and Tirrel, 1990; Okano *et al.*, 1990; Winnik, 1990; Hoffman, 1995; Maeda *et al.*, 1994; Yoshida *et al.*, 1995; Shibayama *et al.*, 1996; Wu and Zhou, 1996; Chen *et al.*, 1996; Chen and Akashi, 1997a and b; Sakuma *et al.*, 1997a, b and c; Wang *et al.*, 1998; Seto *et al.*, 1998; Chen *et al.*, 1998a and b; Chen *et al.*, 1999a and b; Chiklis and Grasshoff, 1970; Chen and Hoffman, 1993; Feil *et al.*, 1992; Chen and Hoffman, 1994; Chen and Hoffman, 1995a and b). The coil-globule transition (Fujishige *et al.*, 1989) takes place for poly(NIPAAm) at the cloud point (32 °C). Lower critical solution temperature (LCST) behavior, such as precipitation of poly(NIPAAm) aggregates and shrinkage of crosslinked poly(NIPAAm) hydrogels, is observed with the rise in temperature above LCST. The phenomenon of the thermosensitivity of water-soluble polymers with appropriate hydrophilic part can be observed by reversible changes in light transmission of the aqueous solution of the polymer in response to an external temperature change. When the temperature of an aqueous solution is raised above

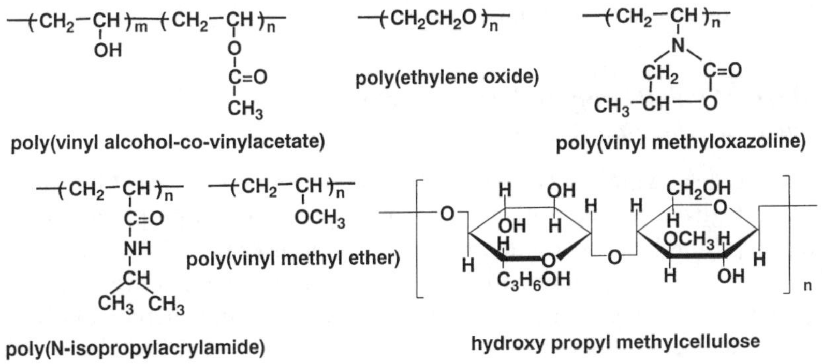

Scheme 1 Structures of some thermosensitive polymers.

the LCST, the hydratization structure of the polymers is broken up and they form insoluble aggregates. At temperatures below the LCST, the aggregates dissolve again to give a transparent polymer solution. The driving force for the phase transition of polymers, however, has not yet been fully elucidated. In addition, the relationship between the thermosensitive properties and the chemical structure of the polymers, has not yet been investigated at length.

A large number of so-called functional polymers have been designed and synthesized. Most of these polymers are soluble or are compatible in an organic medium. The number of water-soluble synthetic functional polymers is unfortunately limited, although they are important and indispensable in the fields of biorelated polymer science such as bioseparation and bioprocessing. Recently, we have began to call certain polymers 'smart' or 'intelligent', when they react with dramatic changes in physical properties to changes in certain environmental conditions. In this chapter, we will introduce our study of N-vinylamide polymers, particularly the N-vinylisobutyramide (NVIBA) polymer, which is quite similar to poly(NIPAAm) structurally and shows a thermosensitive property in aqueous solution (Akashi et al., 1996; Suwa et al., 1997a, b and c; Kishida et al., 1998; Kishida et al., 1999). This study is believed to contribute to both the understanding of LCST behavior of so-called thermosensitive polymers and the widening of their applications.

2.2. Synthesis of Poly(N-vinylisobutyramide) and its LCST Behavior

Aqueous solutions of poly(NIPAAm) show a LCST at 32 °C (Heskins and Guillet, 1968; Fujishige et al., 1989). Generally, the solubility-temperature relationship of water soluble polymers having alkyl groups on the side chains (Ito, 1989) or on the end (Horne et al., 1971) of the polymer chain depends on the hydrophilic–hydrophobic balance in the polymer structure, and the strength and the structure of hydration of the polymers play a dominant role in the phase transition (Drechsel, 1957). Not only synthetic polymers but also naturally occurring polymers such as elastin (Urry et al., 1988) exhibit LCST phenomena in aqueous solution. Therefore, studies on the interaction between water and synthetic polymers, particularly polymers as a protein model, have attracted much attention.

We demonstrated the synthesis and some properties of linear and crosslinked poly(N-vinylacetamide) (poly(NVA)), respectively (Akashi et al., 1990; Akashi et al., 1993). As poly(NVA) has an amide group on the side chain, it may be considered as a kind of polypeptide model compound.

2.2.1. Synthesis of Poly(N-vinylisobutyramide) by
Polymer Modification Reaction

Poly(NVA) is soluble in water and organic solvents such as alcohols, DMSO and DMF, because the amide group in the polymer side chain possesses amphiphilic properties. Similarly to poly(N-methylacrylamide), aqueous solutions of poly(NVA) do not show any LCST below 100 °C. To prepare novel thermosensitive polymers, we tried to introduce alkyl groups in the side chain of poly(VAm). As shown in Scheme 2, poly(VAm) was prepared by acid hydrolysis of poly(NVA). As poly(VAm) is not soluble in organic solvents, amidation of poly(VAm) was carried out in aqueous solution using a water-soluble carbodiimide. The products were identified by IR and NMR spectra. The

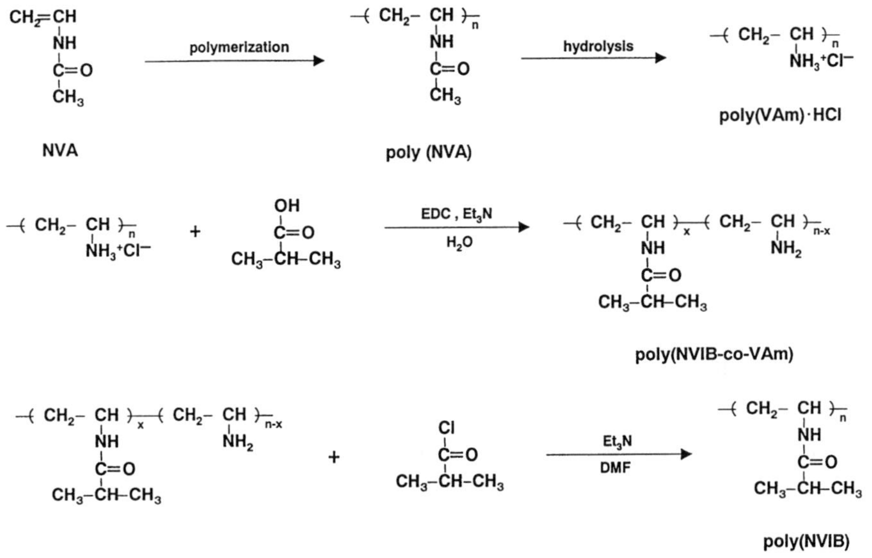

$CH_2=CH$
NH
$C=O$
CH_3

NVA

→ polymerization →

$-(CH_2-CH)_n$
NH
$C=O$
CH_3

poly (NVA)

→ hydrolysis →

$-(CH_2-CH)_n$
$NH_3{}^+Cl^-$

poly(VAm)·HCl

$-(CH_2-CH)_n$
$NH_3{}^+Cl^-$

+

OH
$C=O$
$CH_3-CH-CH_3$

→ EDC, Et$_3$N / H$_2$O →

$-(CH_2-CH)_x-(CH_2-CH)_{n-x}$
NH ... NH_2
$C=O$
$CH_3-CH-CH_3$

poly(NVIB-co-VAm)

$-(CH_2-CH)_x-(CH_2-CH)_{n-x}$
NH ... NH_2
$C=O$
$CH_3-CH-CH_3$

+

Cl
$C=O$
$CH_3-CH-CH_3$

→ Et$_3$N / DMF →

$-(CH_2-CH)_n$
NH
$C=O$
$CH_3-CH-CH_3$

poly(NVIB)

Scheme 2 Synthesis of poly(NVIBA).

intensity of absorption bands at 1505, 1610, 1950, and 2800 cm^{-1} in poly(VAm) decreased and new bands due to amide groups on the polymer side chain appeared at 1550 and 1650 cm^{-1}. From the NMR spectra, new peaks at 1.0 and 2.5 ppm were assigned to the protons of the isobutyl group. The degree of substitution was calculated from the intensity ratio of the methylene protons of poly(NVIBA-co-VAm) at 3.0–3.6 ppm and methine proton of poly(NVIBA) at 3.8–4.2 ppm. A maximum degree of amidation was 80% in this polymer modification reaction. The obtained copolymer consisting of NVIBA and VAm was soluble in both water and some organic solvents, when it contained a high amount of NVIBA. The aqueous solutions of poly(NVIBA-co-VAm), however, did not show any LCST below 100 °C.

As highly amidated copolymers were soluble in DMF, isobutyryl chloride could be reacted with remaining amino groups. NMR analysis suggested that the amino groups were reacted nearly quantitatively as a result of the polymer reaction. The resulting polymer was found to show a LCST. The transmittance change was traced by monitoring the transmittance of a 500 nm light beam at different temperatures. The temperature dependence of the transmittance of aqueous poly(NVIBA) solution (5 mg/mL) is shown in Figure 2.1. The rates of heating and cooling of the sample cells were adjusted to 1 °C/ min. As shown in Figure 2.1, the phase transition of the solution took place, when the temperature reached to 35 °C on heating and 38 °C on cooling. The reason why the transition is not sharp for poly(NVIBA) may be the non-substituted amino groups remained in the polymer side chain because the polymer was prepared by polymer modification reaction.

2.2.2. Synthesis of Poly(N-vinylisobutyramide) by Free Radical Polymerization

An alternative way to obtain poly(NVIBA) is radical polymerization. NVIBA was obtained (Wada *et al.*, 1993) by the pyrolysis of (*N*-α-isopropoxyethyl)isobutyramide

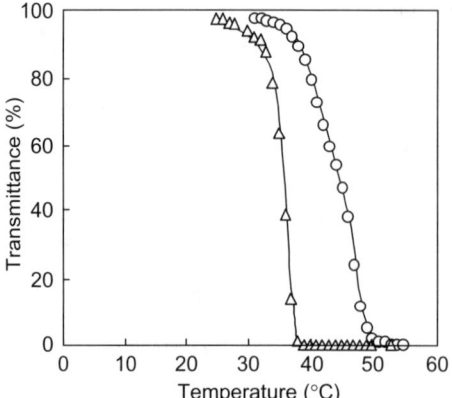

Figure 2.1 Effect of temperature on transmittance at 500 nm of the aqueous solution of poly (NVIBA). Circles – heating up, triangles – cooling down.

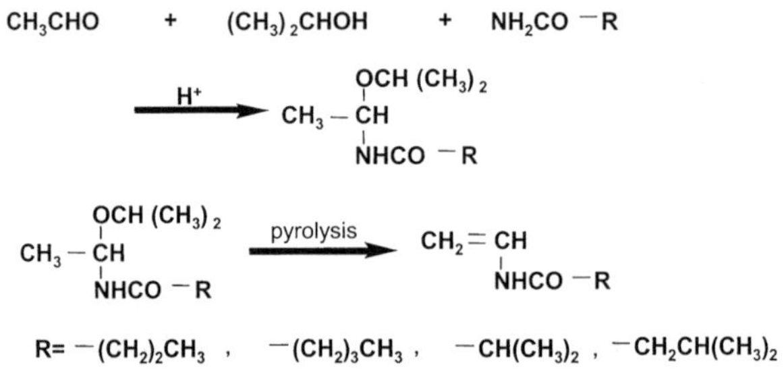

$$CH_3CHO \quad + \quad (CH_3)_2CHOH \quad + \quad NH_2CO-R$$

$$\xrightarrow{H^+} \quad CH_3-\underset{\underset{NHCO-R}{|}}{\overset{\overset{OCH(CH_3)_2}{|}}{CH}}$$

$$CH_3-\underset{\underset{NHCO-R}{|}}{\overset{\overset{OCH(CH_3)_2}{|}}{CH}} \quad \xrightarrow{pyrolysis} \quad CH_2=\underset{\underset{NHCO-R}{|}}{CH}$$

$$R= -(CH_2)_2CH_3 \ , \quad -(CH_2)_3CH_3 \ , \quad -CH(CH_3)_2 \ , \quad -CH_2CH(CH_3)_2$$

Scheme 3 Synthesis of N-vinylalkylamides by pyrolysis.

in a way similar to the synthetic method of NVA (Akashi *et al.*, 1990) (Scheme 3). Synthesis of NVIBA was also reported (Archibald and Fleming, 1993) by the nucleophilic opening reaction of N-(dimethylpropionyl)oxazoline-2-one, which is obtained in a high yield from the reaction of oxazolidin-2-one with butyllithium. In the latter case, the yield of NVIBA was considerably higher (46%) (Archibald and Fleming, 1993), although the treatment of chemicals such as lithium diisopropylamide, which is indispensable for this reaction, must be done carefully. On the other hand, however, pyrolysis of N-(α-alkoxyethyl)alkylamide is suitable for a large scale synthesis of N-vinylalkylamide. Actually, NVA is now produced in a manufacturing plant using the pyrolytic procedure, and N-vinylformamide is also prepared by pyrolysis industrially (Murao *et al.*, 1986) For synthesis of NVIBA, isobutyramide was used as a starting material, and the reaction with 2-propanol and acetaldehyde in the presence of concentrated sulfuric acid gave (N-α-isopropoxyethyl)isobutyramide in the first step. After distillation, pyrolysis was done in a pyrolysis quartz tube at 500 °C under reduced pressure. The condensed mixture was purified by vacuum distillation at 60 °C/1 mm Hg. NVIBA was recrystallised from n-hexane as white needles. The final yield was 20% based on isobutyramide. NVIBA was soluble in both polar solvents such as methanol, ethanol, and nonpolar solvents such as cyclohexane. It is also expected that other vinylalkylamides can be prepared by the synthetic route shown in Scheme 3. This

Table 2.1 Polymerization of NVIBA at 60 °C for 12 h[a]

Run	Solvent	Initiator	Temp. (°C)	Yield (%)	$Mn^b \times 10^{-4}$
1	H$_2$O	K$_2$S$_2$O$_8$	4	2.0	2.10
2	H$_2$O	VA-044	60	58.9	1.31
3	Ethanol	BPO	60	9.8	0.23
4	Ethanol	AIBN	60	84.0	0.70
5	DMF	AIBN	60	61.0	0.37
6	Dioxane	AIBN	60	61.0	1.00
7	Benzene	AIBN	60	85.0	1.82

a 0.88 mmol of NVIBA was used in 3 ml of solvent. [Initiator] : [NVIBA] = 1 : 100 (mol%).
b Estimated by GPC using PEGs as standards.

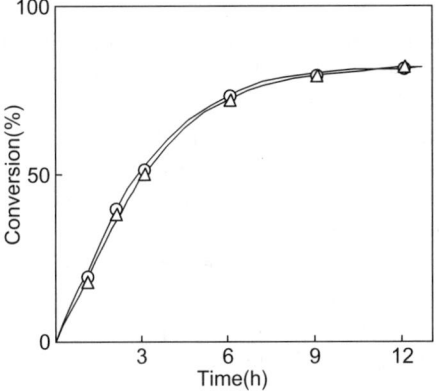

Figure 2.2 Time-conversion curves for the polymerization of NVIBA. Solvent: ethanol (circles), benzene (triangles).

makes it possible to design a new type of amphiphilic polymers that have hydrophilic and hydrophobic parts.

The free radical polymerization of NVIBA was carried out in different solvents in the presence of various initiators (Table 2.1). Figure 2.2 shows the time dependence of the polymerization of NVIBA. The resulting polymer, poly(NVIBA), was soluble in water, methanol, ethanol, DMSO, DMF, and was insoluble in benzene, and n-hexane similarly to poly(NVA). The results of the free radical polymerization of NVIBA suggest that the polymerizability of NVIBA as characterized by conversion and molecular weight in both water and organic solvents was fairly good. However, the conversion was extremely low, when peroxides were used as initiators. Consequently, the polymerizability of NVIBA was similar to that of NVA (Akashi *et al.*, 1990). The time dependence for production of poly(NVIBA) in benzene and ethanol suggest that the polymerization of NVIBA proceeds slowly and quantitatively. No difference in the time dependence of the polymerization in the solvents was observed, although it was found that the polymerization of NVA proceeded faster in benzene than in ethanol (Akashi *et al.*, 1990). One explanation for the difference may be ascribed to the fact that poly(NVIBA) is more hydrophobic than poly(NVA) due to the bulky structure of the isopropyl group of the side chain in contrast to the methyl group of NVA. It seems to be reasonable to believe that the isopropyl group in benzene contributes to the weakening of the intermolecular aggregate of NVIBA by hydrogen bonds in polymerization.

2.2.3. *Thermosensitive Behavior of Poly(N-vinylisobutyramide) and Its Copolymers*

Figure 2.3 shows the effect of temperature on the transmittance of an aqueous solution of poly(NVIBA) for both the heating and cooling processes as compared to an aqueous solution of poly(NIPAAm). The turbidity change of the poly(NVIBA) solution occurred reversibly and sharply, when the temperature reached near 40 °C during heating and near 39 °C on cooling. The reason why the transition of poly(NVIBA) that was prepared from poly(VAm) by polymer modification reaction, was not sharp enough was due to unreacted amino-groups, which prevented the aggregation of polymer molecules.

It is well known that in aqueous solution of poly(NIPAAm), hydrophobic groups in the polymer form insoluble aggregates above LCST (Heskins and Guillet, 1968; Fujishige *et al.*, 1989; Schild, 1992), because the hydratization of the polymer is broken above the LCST. The hydrophobic and the hydrophilic balance of the water soluble polymers has thus far been believed to be important. The LCSTs of poly(NVIBA) and poly(NIPAAm) are 39 and 32 °C, respectively. Although these chemical structures are different from each other, the total hydrophobic and hydrophilic balance of the polymers is exactly the same. Consequently, the present results suggest that the microstructure of the hydratization of the polymers is quite important. In other words, the hydrogen bonding interaction between amide groups of polymer side chains and water, and hydratization of an isopropyl group of a polymer side chain, must be important factors in phase transition.

To further clarify the thermosensitive phenomena of poly(NVIBA), a DSC study was performed (Figures 2.4 and 2.5). The temperatures that were obtained from the intersection point of the base line and the edge of the endotherm were almost the same as

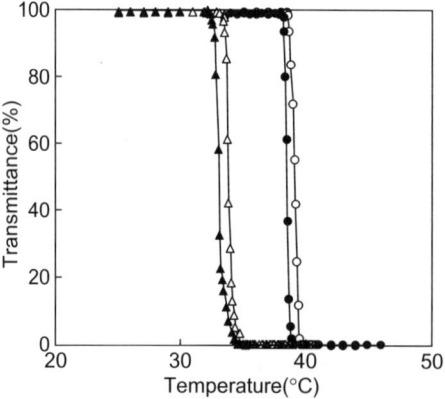

Figure 2.3 Effect of temperature on transmittance of the aqueous solution of poly(NVIBA) and poly(NIPAAm). Heating up (open circles) cooling down and (closed circles) for poly(NVIBA) (Mn = 1.0 × 10⁴, Mw/Mn = 2.3), heating up (open triangles) and cooling down (closed triangles) for poly(NIPAAm) (Mn = 7.5 × 10³, Mw/Mn = 2.3).

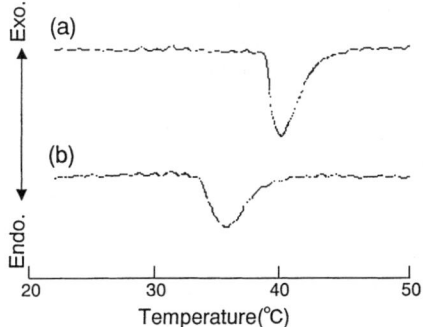

Figure 2.4 DSC thermograms of aqueous polymer solutions: (a) poly(NVIBA) (Mn = 1.1 × 10^4, Mw/Mn = 1.4), (b) poly(NIPAAm) (Mn = 1.0 × 10^4, Mw/Mn = 1.4).

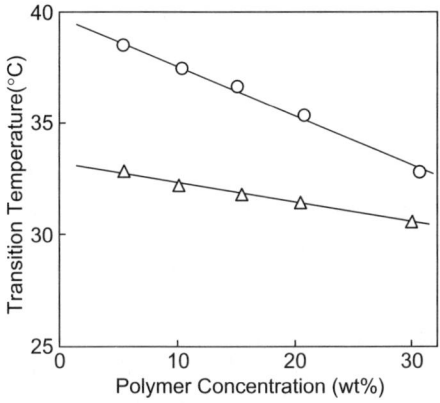

Figure 2.5 Transition temperature of aqueous polymer solutions plotted as a function of polymer concentration: poly(NVIBA) (Mn = 1.1 × 10^4, Mw/Mn = 1.4) (open circles), poly(NIPAAm) (Mn = 1.0 × 10^4, Mw/Mn = 1.4) (open triangles).

the cloud points that were determined by the measurement of turbidity. The transition temperature decreased with an increase in polymer concentration. This phenomenon has already been reported for poly(NIPAAm) aqueous solutions, but the tendency was more pronounced for poly(NVIBA). It can be assumed that intermolecular aggregation between water-soluble polymers occurs, particularly in the poly(NVIBA) solution, and influences the LCST.

2.3. Thermosensitive Properties of Vinylamide Copolymers

A series of polyacrylamide derivatives having different alkyl groups in their side chains was synthesized and their LCST behaviors were found to depend mainly on polymer side chains (Ito, 1989). LCSTs of copolymers of NIPAAm with other N-alkylacrylamides, such as acrylamide (AAm), N-methylacrylamide

(NMAAm), N-ethylacrylamide (NEAAm), or N-butylacrylamide (NBAAm), N-tert-butylacrylamide (NTBAAm) have also been reported (Priest et al., 1987; Priest et al., 1986). It is clear that their transition temperatures depend on the hydrophobic-hydrophilic balance of the polymers (Priest et al., 1987; Feil et al., 1993).

On the other hand, the solution properties of N-vinyl-n-butyramide (NVBA), N-vinylisovaleramide (NVIVA), N-vinyl-n-valeramide (NVVA), etc., which are NVA derivatives in a polymeric form, have never been examined, although their LCST behavior is of great scientific interest. So far, a few studies on NVA (Akashi et al., 1990) and NVIBA (Akashi et al., 1996; Suwa et al., 1997a, b and c) have been performed. The fact that the LCSTs of a poly(NIPAAm) can be increased or decreased by combining hydrophilic or hydrophobic units into the polymer chain (Priest et al., 1987; Feil et al., 1993; Chen and Hoffman, 1995a and b) encouraged us to try to prepare a series of copolymers consisting of NVIBA and other N-vinylalkylamides. The copolymers exhibit LCSTs that are higher or lower in accordance with their comonomer content. Moreover, when these monomers with various alkyl groups are used as a starting material, thermosensitive microspheres (Chen et al., 1996), stimuli-responsive polymer systems in a solution as well as thermosensitive hydrogels can be obtained.

2.3.1. Synthesis of N-vinylalkylamides

As shown in Scheme 3, N-vinylalkylamides such as NVBA, NVVA, and NVIVA were synthesized by the two-step reaction process that is similar to the NVIBA synthetic method as shown in the previous section. When n-butyramide, isovaleramide, and n-valeramide were used as starting materials, the reaction with 2-propanol and acetaldehyde in the presence of concentrated sulfuric acid followed by pyrolysis results in NVBA, NVIVA, and NVVA, respectively. For NVIVA synthesis, IVA was synthesized in advance from isovaleryl chloride in the presence of aqueous ammonia (Kent and McElvain, 1955), because it is not commercially available. Each of the condensed mixtures, which were obtained in the process of each intermediate pyrolysis, was purified by vacuum distillation. They were then purified by repeated crystallization from n-hexane. As a result, white needles of NVVA were obtained, while NVBA and NVIVA were obtained only as oily substances. The final yields of NVBA, NVIVA, and NVVA were 4.2, 21.5, and 13.4% based on n-butyramide, isovaleramide, and n-valeramide, respectively. Due to the difficulties in separation of NVBA and NVIVA from by-products of pyrolysis, the yields were low and it therefore was impossible to measure the melting points of NVBA and NVIVA. To characterize NVBA and NVIVA, however, an elemental analysis of the resulting polymers was performed. The results of the elemental analysis of poly(NVBA) and poly(NVIVA) as well as poly(NVIBA) and poly(NVVA)agreed with the theoretical value. Detailed synthetic procedures are shown in the Appendix.

2.3.2. Polymerization and Copolymerization of N-vinylalkylamides and Thermosensitive Properties of the Polymers

The free radical polymerization of NVIBA, NVBA, NVIVA, and NVVA was carried out in ethanol in the presence of AIBN as initiator (Table 2.2). The resulting polymer, poly(NVBA), which has n-propyl groups on its side chains, was soluble in cold water, methanol, ethanol, DMSO, DMF, and was insoluble in benzene, and

Table 2.2 Polymerization of N-vinylalkylamides at 60 °C for 24 h[a]

Abbreviation	R	Initiator	Temp. (°C)	Yield (%)	$Mn^b \times 10^{-4}$	Mw/Mn
NVIBA	$-CH(CH_3)_2$	AIBN	60	86.0	1.00	2.30
NVB	$-(CH_2)_2CH_3$	AIBN	60	90.0	0.84	1.34
NVIV	$-CH_2-CH(CH_3)_2$	AIBN	60	24.0	1.70	1.80
NVV	$-(CH_2)_3CH_3$	AIBN	60	56.0	2.90	1.95

a Monomers: NVIBA (0.88 mmol), NVBA (0.88 mmol), NVIVA (0.79 mmol), NVVA (0.79 mmol), solvent; ethanol (3 ml), [Initiator] : [Monomer] = 1 : 100 (mol%).
b Estimated by GPC using PEGs as standards.

n-hexane, as similar to poly(NVIBA) and poly(NVA). With an increase in the chain length of alkyl groups in the side chains, the polymers became more hydrophobic, and finally they were insoluble even in cold water. All polymers were soluble in methanol, ethanol, DMSO, DMF and were almost insoluble in benzene and in n-hexane, though poly(NVIVA) and poly(NVVA) swell in these solvents.

The results of the free radical polymerization of NVIVA and NVVA suggest that their polymerizability was relatively low, although the conversion of NVBA was fairly good similarly to NVIBA. NVIBA (M_1) and NVA (M_2) were copolymerized in ethanol at 60 °C. The composition ratios of the copolymers were determined by ^{1}H-NMR (400 MHz, D_2O) measurement. The signal intensities of methyne protons (3.7 ppm) of the main chain in NVIBA and NVA units were compared with those of methyne protons (2.4 ppm) of side chains in NVIBA units. The monomer reactivity ratios r_1 and r_2 values (M_1 = NVIBA, M_2 = NVA) were estimated by the curve-fitting method based on a nonlinear least squares procedure.

Using the curve-fitting method based on a nonlinear least-square procedure (Yamada *et al.*, 1978), the apparent monomer reactivity ratios were estimated to be $r_1 = 0.94$ and $r_2 = 0.99$. The reactivity of NVIBA vinyl groups was nearly the same as that of NVA. These results suggest that NVIBA like NVA could be classified as vinyl monomer of a non-conjugated type. Therefore, the copolymerization of NVIBA with NVA is regarded as an ideal copolymerization. The copolymers of NVIBA-NVA, NBIBA-NVVA, NVVA-NVA, and NIPAAm-NVA were also prepared by the solution polymerization described in the Appendix. The resulting polymers and copolymers were dialyzed against regularly exchanged distilled water to remove unreacted monomers, and the isolated polymer and copolymers were dried *in vacuo*. The conversion of NIPAAm was *ca.* 90 wt%. The conversions of NVIBA-NVA, NVIBA-NVVA, NVVA-NVA, and NIPAAm-NVA were 87, 63, 86, and 92 wt%, respectively.

Figure 2.6 shows the temperature dependence of the light transmittance of aqueous solutions of poly(NVIBA) and poly(NVBA) on both the heating and cooling processes as compared to poly(NIPAAm). When the N-alkyl substituent is changed from isopropyl to n-propyl, the LCST shifted from 39 to 32 °C without any drop in the sharpness of profile. The homopolymers of NVIVA or NVVA that have isobutyl groups or n-butyl groups on their side chains were insoluble in cold water.

The LCST for a given copolymer is expected to vary, depending on the hydrophobic–hydrophilic balance of the copolymers. The LCSTs of copolymers of NVIBA and NVA increased with an increase in NVA content while the LCSTs of copolymers of NVIBA with NVVA decreased with an increase in NVVA content. Copolymers of NVVA with

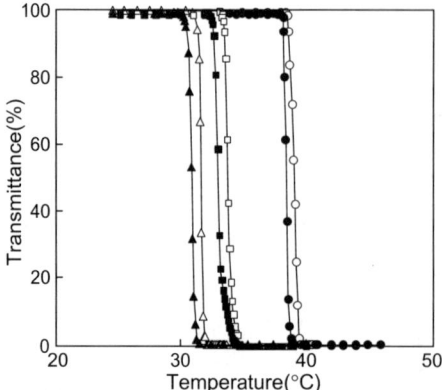

Figure 2.6 Temperature dependence of the light transmittance of the aqueous solution of poly(NVIBA), poly(NVBA), and poly(NIPAAm). Heating up (open circles) and cooling down (closed circles) for poly(NVIBA) (Mn = 1.0 × 10⁴, Mw/Mn = 2.3). Heating up (open squares) and cooling down (closed squares) for poly(NIPAAm) (Mn = 7.5 × 10³, Mw/Mn = 2.3). Heating up (open triangles) and cooling down (closed triangles) for poly(NVBA) (Mn = 8.4 × 10³, Mw/Mn = 1.3).

NVA exhibited LCST sharply at 70 °C. The incorporation of hydrophilic comonomers such as NVA led to a higher LCST while hydrophobic comonomers caused a decrease in LCST without any precipitation of copolymers coming out above the LCSTs of their copolymers. Moreover, the profiles of the transition were sharp and the cloud points were clear between *ca.* 10 and 90 °C in accordance with their comonomer content. This phenomenon was also observed in random copolymers of NIPAAm with other *N*-alkylacrylamides systems with the LCSTs varying between 0 and 65 °C. LCST of copolymers of NIPAAm with NVA, increased with an increase in NVA content from *ca.* 32 to 53 °C with some broadening of the response profile (Figure 2.7).

2.4. Crosslinked Poly(N-vinylisobutyramide) Thermosensitive Hydrogels

Crosslinked poly(NIPAAm) forms hydrogels, which undergo a volume phase transition in water in response to temperature changes (Heskins and Guillet, 1968; Hoffman, 1995; Wu *et al.*, 1992). Numerous studies have been performed on the thermosensitive properties of poly(NIPAAm) gels (Shibayama *et al.*, 1996; Miura *et al.*, 1994; Hoffman *et al.*, 1993; Hoffman *et al.*, 1997; Inomata *et al.*, 1995; Bae *et al.*, 1990; Inomata *et al.*, 1990). Recently, poly(NIPAAm) gels with rapid deswelling volume changes in response to temperature changes have also been examined (Yoshida *et al.*, 1995; Kaneko *et al.*, 1995). Little attention, however, has been given to NVA and its derivatives with the reverse amide structure as compared to acrylamide (AAm) and its derivatives. NVA was copolymerized with divinyl derivatives of NVA to give stable hydrogels with a high yield (Akashi *et al.*, 1993). NVIBA, and other *N*-vinylalkylamides are expected to give hydrogels with a high yield and to show amphiphilic properties similar to crosslinked poly(NVA). NVIBA can be regarded as a novel compound to produce thermosensitive hydrogels, which differ from NIPAAm.

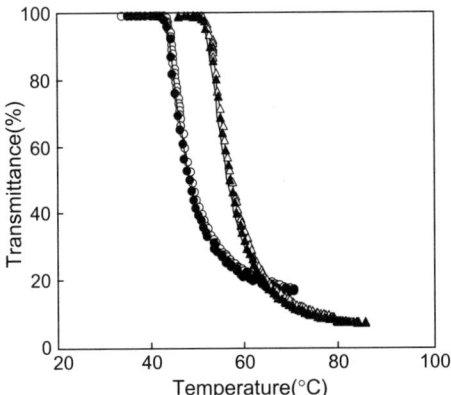

Figure 2.7 Temperature dependence of the light transmittance of the aqueous solution of NIPAAm-NVA copolymers with varying composition. Heating up (open circles) and cooling down (closed circles) for poly(NIPAAm-co-NVA, 90/10) (Mn = 7.5 × 10³, Mw/Mn = 2.3), heating up (open triangles) and cooling down (closed triangles) for poly(NIPAAm-co-NVA, 80/20) (Mn = 1.5 × 10⁴, Mw/Mn = 3.1).

Poly(NVIBA) and poly(NIPAAm) gels with different butylene-bis-N-vinylacetamide (B-BNVA) and N,N'-methylene-bis-acrylamide (Bis-A) contents were prepared by radical polymerization at 60 °C for 24 h (Scheme 4). The gels were transparent when ethanol had been used as a solvent. Figure 2.8 demonstrates the time dependence of polymerization for poly(NVIBA) and poly(NIPAAm) gels in ethanol at 60 °C, and Table 2.3 shows the yield of crosslinked polymers together with the percentage of crosslinker. The conversions for each gel reached about 100% after 24 h. The initial conversion for poly(NVIBA) gel was lower than that for poly(NIPAAm) gel up to 15 h. This may be due to the difference in polymerizability between NVIBA and NIPAAm.

Swelling ratios of poly(NVIBA) and poly(NIPAAm) gels in organic solvents are shown in Table 2.4. It is clear that poly(NVIBA) gels are amphiphilic and swell well in organic solvents, similar to poly(NVA) gels (Akashi *et al.*, 1993). They even swell in hydrophobic solvents such as benzene and n-hexane. As for poly(NVA) gels, even if they were prepared with a 10 mol% of a crosslinker, they absorbed large amounts of water. When they were compared with the same crosslinker concentration (10 mol%), the swelling ratios of poly(NVA) gel were much higher than those of the poly(NVIBA) gel. By comparing them, it is obvious that poly(NVIBA) gel is more hydrophobic than the poly(NVA) gels due to the isopropyl groups of the side chains. The poly(NVIBA) gels showed a lower swelling ratio than poly(NIPAAm) gels in hydrophobic solvents such as benzene and n-hexane.

The swelling ratios of poly(NVIBA) and poly(NIPAAm) gels at 23 °C decreased with an increase in crosslinker concentration. There were no significant differences in the swelling ratio between poly(NVIBA) and poly(NIPAAm) gels due to variation in crosslinker concentration.

Water-soluble poly(NVIBA) and poly(NIPAAm) show LCST sharply at 39 and 32 °C in water, respectively. As the phase transition behavior of crosslinked poly(NIPAAm) hydrogel occurs clearly above its LCST, a similar phenomenon should take place

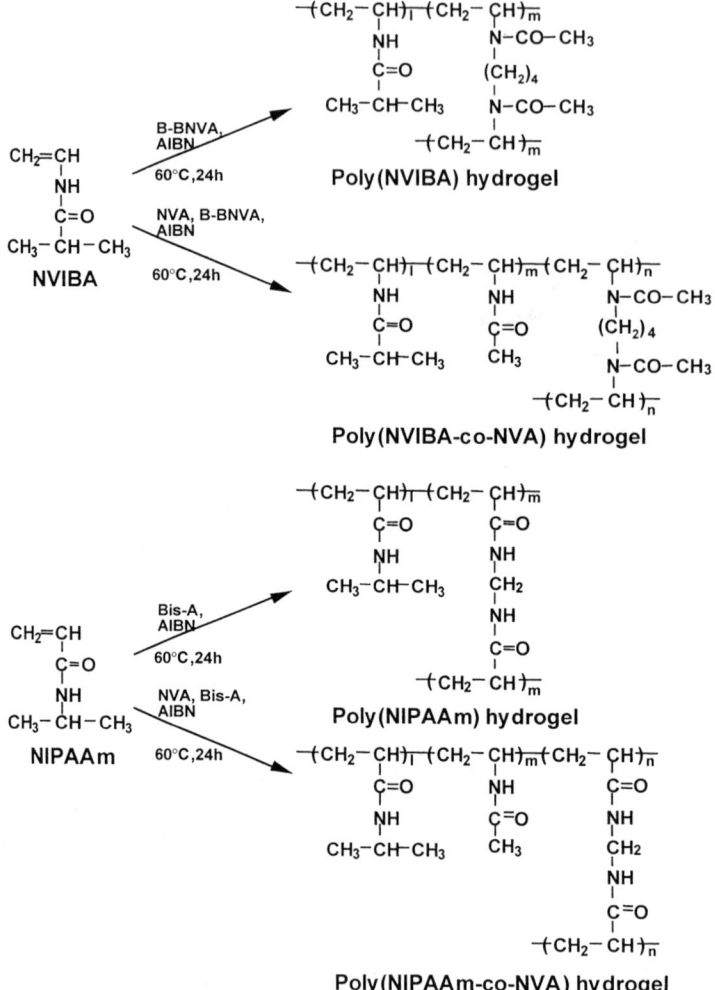

Scheme 4 Preparation of some thermosensitive hydrogels.

for crosslinked poly(NVIBA). Actually, when the transparent poly(NVIBA) gel was warmed in hot water, it was found to shrink to become a white pellet. Changes in the swelling ratios of poly(NVIBA) and poly(NIPAAm) gels with different molar ratios of crosslinkers are shown in Figure 2.9a and b as a function of temperature, respectively. Transition temperature, T_ts, for the poly(NVIBA) gel was about 41 °C, while it was *ca.* 33 °C for the poly(NIPAAm) gel, independent of crosslinker concentration. The T_t values of both gels nearly agreed with LCSTs of poly(NVIBA) and poly(NIPAAm).

However, shrunken states above T_t, were slightly different for these two gels. The poly(NVIBA) gels shrank almost completely and desorbed 98% of water above T_t, whereas the poly(NIPAAm) gel desorbed only about 90% of water. Poly(NIPAAm) gel is reported after increasing temperature to shrink slowly due to the skin layer of the collapsed polymer on the gel surface. Shrunken state of poly(NIPAAm) gels above

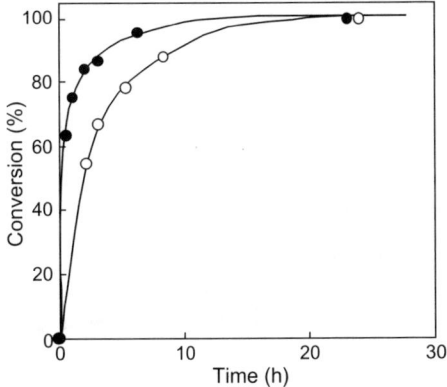

Figure 2.8 Time dependence of polymerization of poly(NVIBA) and poly(NIPAAm) gels in ethanol at 60 °C. NVIBA/B-BNV A = 100/1 (open circles) and NIPAAm/Bis-A = 100/l (closed circles).

Table 2.3 Preparation of hydrogels by copolymerization of NVIBA or NIPAAm with divinyl monomers as crosslinker at 60 °C for 24 h[a]

Run	Monomer (mmol)		Percent of crosslinker	Yield[b] (%)
1	NVIBA (4)	B-BisNVA (0.04)	1	92
2	NVIBA (4)	B-BisNVA (0.20)	5	93
3	NVIBA (4)	B-BisNVA (0.40)	10	92
4	NVIBA (4)	B-BisNVA (0.80)	20	92
5	NIPAAm (4)	BisA (0.04)	1	91
6	NIPAAm (4)	BisA (0.20)	5	93
7	NIPAAm (4)	BisA (0.40)	10	90
8	NIPAAm (4)	BisA (0.80)	20	90

a Polymerization was carried out in ethanol (2 ml, total concentration of monomers 4 mM; [Initiator] : [Monomer] = 1 : 100 (mol%).
b Calculated by the weight of the dried gel.

Table 2.4 Gel swelling ratio of polyNVIBA and polyNIPAAm gels in various solvents at room temperature

Gel	H_2O	Methanol	Ethanol	Iso-propanol	DMF	Benzene	n-Hexane
PolyNVIBA[a]	25.0	12.2	19.0	24.0	16.0	4.12	4.38
PolyNIPAAm[b]	18.0	16.0	17.0	17.6	14.7	10.0	9.30

a Run 1 in Table 2.3.
b Run 5 in Table 2.3.

the T_t may be affected by the skin layer formation. In the case of the poly(NVIBA) gels, however, the skin layer formation was not observed. Moreover poly(NVIBA) gels swelled more than poly(NIPAAm) gels at temperatures below T_t. We assume that the amount of the hydratization of poly(NVIBA) and poly(NIPAAm) must be different from each other, which leads to the difference in the swelling ratio of their

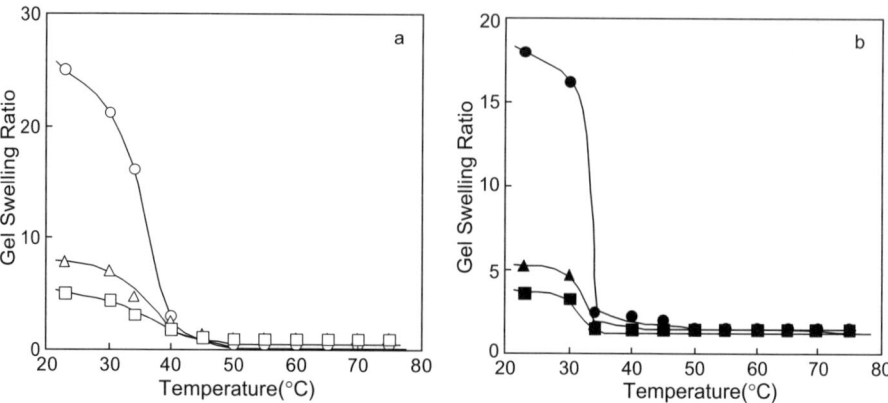

Figure 2.9 Swelling ratios of poly(NVIBA) gels (a) and poly(NIPAAm) gels (b) as a function of temperature, (a): NVIBA/B-BNVA = 100/1 (open circles), NVIBA/B-BNVA = 100/5 (open triangles), NVIBA/B-BNVA = 100/10 (open squares), (b): NIPAAm/Bis-A = 100/1 (closed circles), NIPAAm/Bis-A = 100/5 (closed triangles), NIPAAm/Bis-A = 100/10 (closed squares). The swelling ratio is defined as W/W_d, where W is the swollen weight of a gel in pure water and W_d is the dry weight of a gel that has been dried *in vacuo*. Three samples were used for each data. The transition temperature was determined from the intersection point of the baseline in the shrunken states and the edge of the swelling transition.

gels. Currently, the reason why poly(NVIBA) gels swelled more than poly(NIPAAm) gels at lower temperature is not clear.

The hydrogels of crosslinked copolymers of NVIBA and NVA were prepared and compared with crosslinked copolymers of NIPAAm and NVA. The T_t and the conversions for both kinds of gels, poly(NVIBA-co-NVA) and poly(NIPAAm-co-NVA) gels, are shown in Table 2.5. The conversion of both gels was over 90%. T_t for poly(NVIBA-co-NVA) and poly(NIPAAm-co-NVA) gels (molar ratios, 80/20, and 60/40) increased with an increase in the NVA content; however, the swelling ratio below T_t decreased (Figure 2.10a and b). The T_t, of these copolymer gels was almost the same as LCSTs of copolymers of NVIBA with NVA and NIPAAm with NVA.

For the deswelling behavior of the gels, there was a little difference. Poly(NVIBA-co-NVA) gels desorbed almost all water above T_t and poly(NIPAAm-co-NVA) gels desorbed up to 90% of water. These differences were considered as follows. Poly(NVIBA-co-NVA) gels are supposed to be composed of crosslinked ideal random copolymers of NVIBA with NVA, while poly(NIPAAm-co-NVA) gels are composed of the heterogeneous structure of NIPAAm with NVA due to the different polymerizability of the different monomers. It can be assumed that the heterogeneous structure of poly(NIPAAm-co-NVA) inhibits the formation of polymer network, even if the NIPAAm chains start to become less hydratized and become hydrophobic above T_t. As a result, poly(NIPAAm-co-NVA) gels retain more water above their T_t. On the other hand, in the case of poly(NIPAAm-co-NVA) gels, an increase in NVA content increased the T_t, however, the degree of swelling was not influenced.

In conclusion, the poly(NVIBA-co-NVA) show a swelling transition in water at a temperature higher than 40 °C, depending on NVA content. These features may

Table 2.5 Poly(NVIBA-co-NVA) and poly(NIPAAm-co-NVA) gels[a]

(mol%)			Yield (%)	Tt^b (°C)	$LCST^c$ (°C)
NVIBA	NVA	B-BNVA			
100	0	1	92	41	39
80	20	1	92	55	53
60	40	1	92	70	70
(mol%)			Yield (%)	Tt^b (°C)	$LCST^c$ (°C)
NIPAAm	NVA	Bis-A			
100	0	1	91	33	32
80	20	1	90	43	42
60	40	1	90	53	53

a Polymerization was carried out in ethanol (2 ml) total concentration monomers 4 mM; [Initiator] : [Monomer] = 1 : 100 (mol%),
b T_t: transition temperature of the gels.
c LCSTs of linear copolymers of NVIBA-NVA and NIPAAm-NVA were determined from transmittance plots at 500 nm.

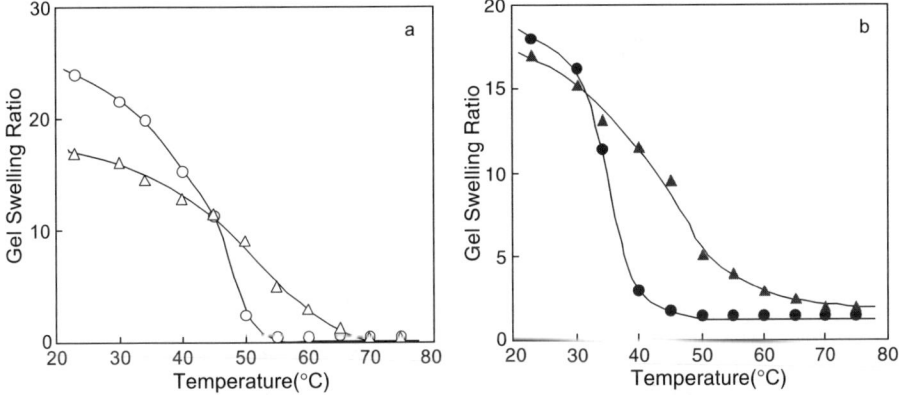

Figure 2.10 Swelling ratios of poly(NVIBA-co-NVA) gels (a) and poly(NIPAAm-co-NVA) gels (b) as a function of temperature, (a): NVIBA/NVA/B-BNVA = 80/20/1 (open circles), NVIBA/NVA/B-BNVA = 60/40/1 (open triangles). (b): NIPAAm/NVA/Bis-A = 80/20/1 (closed circles), NIPAAm/NVA/Bis-A = 60/40/1) (closed triangles).

be very useful in designing gels with various characteristics. The poly(NVIBA) and poly(NVIBA-co-NVA) gels shrank almost completely releasing about 98% of absorbed water.

2.5. Two-phase Separation Using N-vinylamide Polymers

Aqueous two-phase systems, which were introduced as a separation method for biomolecules in the middle of 1950s by Albertsson (1956, 1958), are extensively studied in these days (Walter *et al.*, 1985; Albertsson, 1986; Walter and Johansson, 1994; Dissing and Mattiasson, 1993; Bergfeldt *et al.*, 1995; Svensson *et al.*, 1995). The most commonly used polymer systems are based on poly(ethylene glycol)(PEG) and dextran. In some applications, the polymers have been modified with covalently bound groups, for example hydrophobic ligands and biospecific ligands for affinity partitioning (Walter *et al.*, 1991; Johansson and Shanbhag, 1984; Johansson and Joelsson, 1987; Sivars *et al.*, 1996; Birkenmeier and Kunath, 1996). It is easy to modify dextran because it has many hydroxyl groups, however, as PEG has few functional groups, it is very difficult to attach the affinity ligands to the polymer. It is of interest to introduce a novel polymer which has the ability to form two-phase systems and has many functional groups, which could be used for modification e.g., incorporation of affinity ligands.

Poly(NVA) forms aqueous two phase system with dextran similar to PEG–dextran system with poly(NVA)-rich top phase and dextran-rich bottom phase. Figure 2.11 shows the phase-diagrams of poly(NVA)–dextran and PEG–dextran two-phase systems for polymers with different degree of polymerization. In the poly(NVA)–dextran system, the phase separation is obtained at lower concentrations compared to the PEG–dextran system. The phase separation at such low polymer concentrations is attractive from the industrial point of view.

The molecular weights of the monomer unit are different for PEG (molecular weight monomer unit: 44) and poly(NVA)(molecular weight of monomer unit: 85). If the polymers are used in the same weight concentration, the total number of the poly(NVA) monomer unit will be the approximately half of that of PEG. For comparing the repulsion of monomer–monomer interaction of poly(NVA)–dextran and PEG–dextran systems, the polymer concentrations when phase separation take place, were compared in the monomer unit concentration (Tables 2.6 and 2.7). The phase

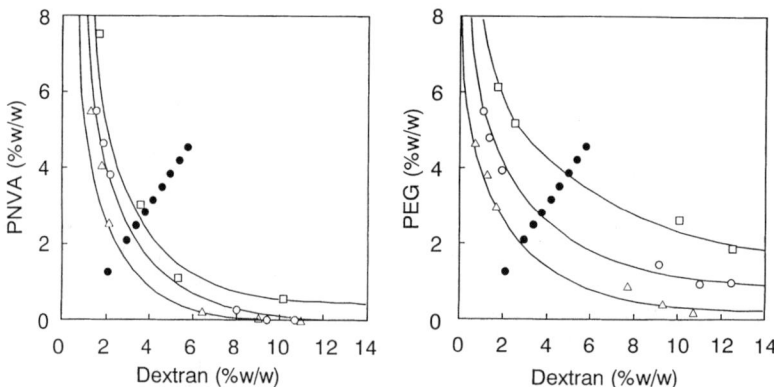

Figure 2.11 Phase diagrams of aqueous two-phase systems of different degree of polymerizations (DP). (a) PNVA–dextran system. DP of PNVA: 240 (open squares); 590 (open circles); 1410 (open triangles). (b) PEG–dextran system. DP of PEG: 140 (open squares); 450 (open circles); 1600 (open triangles). Filled circles are the concentrations in feed. The compositions of the phases in the mixtures were determined as following. The systems were mixed and left to separate overnight at 4 °C. The phases were collected separately using a syringe and diluted as appropriate for the concentration determination. The dextran concentration were determined by polarimetry. The total concentration of polymers in the different phases was determined by gravimetry after freeze-drying of the solutions.

Table 2.6 Composition of PNVA–dextran aqueous two-phase system

Total composition (mmol/ml)		Top phase (wt%)		Bottom phase (wt%)	
Poly(NVA)	Dextran	Poly(NVA)	Dextran	Poly(NVA)	Dextran
3.5	1.2	3.0	3.5	1.1	5.3
6.0	1.7	7.5	1.7	0.5	10.2
8.1	2.0	9.9	1.4	0.0	12.7
10.5	3.3	13.7	1.1	0.0	18.5

Table 2.7 Composition of PEG–dextran aqueous two-phase system

Total composition (mmol/ml)		Top phase (wt%)		Bottom phase (wt%)	
PEG	dextran	PEG	Dextran	PEG	Dextran
3.5	1.2	N.F.[a]	N.F.	N.F.	N.F.
6.0	1.7	N.F.	N.F.	N.F.	N.F.
8.1	2.0	N.F.	N.F.	N.F.	N.F.
10.5	3.3	7.8	0.8	0.5	17.0

a N.F.: not formed.

separation was observed at 3.5 mmol/ml for poly(NVA)–dextran system, whereas it needs higher concentration (10 mmol/ml) for PEG–dextran system. The repulsion between the monomer unit of PNVA and that of dextran seemed to be stronger than that of PEG and dextran, probably due to higher hydrophobicity of poly(NVA) as compared to PEG.

For PEG–dextran system, the critical concentration of phase separation was strongly affected by the degree of polymerization of the polymer whereas for poly(NVA)–dextran system the dependence was much less pronounced (Figure 2.11). One of the reasons could be the difference in the structure of water molecules in vicinity of poly(NVA) and PEG. PEG is reported to form helix-like structures in water. So when the degree of polymerization of PEG become high, the length and content of helix-like structure increase. It affects the viscosity, motility and hydration state, and hence the change of degree of polymerization affects the phase separation.

The partition coefficients (K) for myoglobin were smaller than 1 in all cases studied (Figure 2.12). Myoglobin is a hydrophilic protein and it partitions preferentially to the hydrophilic dextran phase. The preferential partitioning of myoglobin to the dextran phase increased with increasing the polymer concentration. The partition coefficient of myoglobin in poly(NVA)–dextran system is smaller than that in PEG–dextran system indicating that myoglobin partitioned more to the bottom phase in poly(NVA)–dextran system than in PEG–dextran system. This is due to the higher hydrophobicity of poly(NVA) than PEG.

Most of the recent research on two-phase systems that are based on polymer/salt or polymer/polymer has focused on their practical industrial use. One main goal of researching the two-phase system is the development of a low-cost two-phase system, and the other is to design polymers that can be easily recovered or that have a high

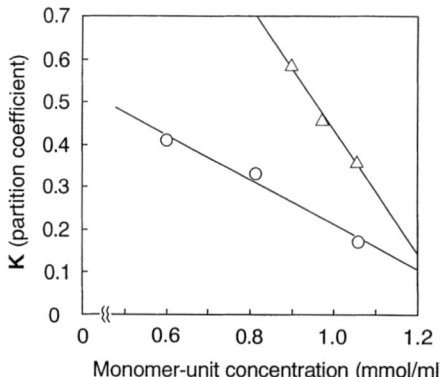

Figure 2.12 Partition coefficient, K, for myoglobin separation in PNVA–dextran system (open circles) and in PEG–dextran system (open triangles). K is defined as $K = C_T/C_B$ where C_T and C_B are the equilibrium concentration of the partitioned substance in the top (upper) and bottom (lower) phases, respectively. The protein concentration was determined by absorption at 280 nm. All results are average values after partition of protein in two equal systems with two repeated measurements of protein content.

level of separation performance. Several studies have been done on phase-separation phenomena with the above two goals (Walter *et al.*, 1991; Johansson and Shanbhag, 1984; Johansson and Joelsson, 1987; Sivars *et al.*, 1996; Birkenmeier and Kunath, 1996; Lu *et al.*, 1994; Carlsson *et al.*, 1993; Li *et al.*, 1997; Johansson *et al.*, 1996; Miyazaki and Kataoka, 1996).

Both poly(NVIBA) and poly(NIPAAm) form aqueous two-phase systems with dextran with synthetic polymers, (poly(NVIBA) or poly(NIPAAm) forming top phase and dextran–rich bottom phase (Figure 2.13). In both systems, the critical concentrations of phase separation were affected by the molecular weight of the polymers. When polymers with a higher molecular weight were used, a two-phase system was formed at the lower polymer concentrations. This was due to the enlargement of the repulsion force between the synthetic polymer and the dextran with increasing molecular weight. There was no apparent difference in the phase diagrams between the poly(NVIBA)- and poly(NIPAAm)–dextran systems indicating that the hydrophilic/hydrophobic balance of the two polymers is almost equivalent.

The partitioning of myoglobin in poly(NVIBA)– and poly(NIPAAm)–dextran systems was very similar to that in poly(NVA)–dextran system: partition coefficients were less than 1 in all systems and preferential partitioning of myoglobin to the dextran phase increased with increasing the polymer concentration. A comparison of the partition coefficient between poly(NVIBA)–dextran, poly(NIPAAm)–dextran and PEG–dextran shows that both the partition coefficients in the poly(NVIBA)–dextran and poly(NIPAAm)–dextran systems are smaller than that in the PEG–dextran system. This is caused by the higher hydrophobicity of poly(NVIBA) and poly(NIPAAm) (as compared to PEG), i.e., both polymers have a strong repulsion to myoglobin, which is a hydrophilic protein.

As the LCST of poly(NVIBA)(39 °C) was higher than that of poly(NIPAAm)(32 °C), poly(NVIBA) seems to have higher hydrophilicity than poly(NIPAAm). By comparing

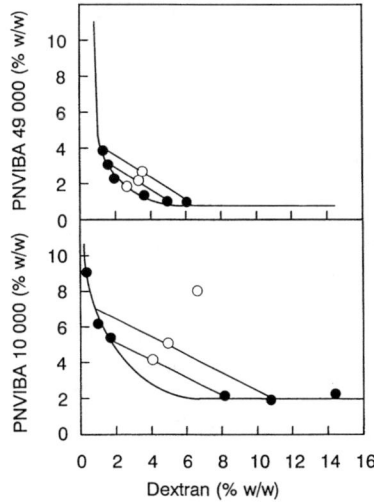

Figure 2.13 Phase diagram of PNVIBA–dextran system with various molecular weight of PNVIBA. Open circles are the concentrations in feed, and filled circles are experimental data of upper phase and lower phase.

the partition coefficients at the same concentration, it is clear that myoglobin (hydrophilic protein) partitioning was more effective in the poly(NIPAAm)–dextran system than in the poly(NVIBA)–dextran system. These results show that it may be possible to determine the hydrophilicity series of water-soluble polymers using aqueous two-phase partitioning.

The above results show that poly(NVIBA) and poly(NIPAAm) are useful as phase-forming polymers in aqueous two-phase systems. For a two-phase system with thermosensitive polymers, Miyazaki and Kataoka (1996) found that the copolymer of N-hydroxymethylacrylamide and N-phenylacrylamide formed a two-phase system by coacervation. They noted that partition coefficient (K) drastically changed across the LCST. Although their system is slightly different from that in this study, it was expected that poly(NVIBA) and/or poly(NIPAAm)–dextran systems have the thermosensitivity in regard to protein separation. In fact, when a poly(NVIBA) or a poly(NIPAAm) rich phase was heated to above its LCST, insoluble aggregates of thermosensitive polymers were observed. Further research including recovering and recycling of the poly(NVIBA) and poly(NIPAAm) is required.

2.6. Conclusion

New thermosensitive N-vinylamide-based polymers have been developed. Poly(NVIBA) has a LCST of 39 °C, and the LCSTs of copolymers consisting of N-vinylamides could be controlled between the freezing and boiling points by varying copolymer composition. Poly(NVIBA) hydrogel also clearly showed a swelling transition in water at *ca.* 41 °C. In this article, we demonstrated that thermosensitive polymers are useful as polymers forming aqueous two-phase systems. Moreover, N-vinylamide polymers stable enough and predicted to be useful for bioseparation or bioprocessing. Future development on novel thermosensitive N-vinylamide polymers is expected.

2.7. Appendix

2.7.1. Synthesis of Poly(N-vinylisobutyramide) by Polymer Modification Reaction

Poly(VAm) was first prepared. NVA (17.0 g, 0.2 mol) was dissolved in 150 ml of distilled water, and polymerized at 60 °C for 12 h using 2,2'-azobis(N,N''-dimethylene isobutyramidine)dihydrochloride as free-radical initiator under nitrogen atmosphere. The resulting polymer was purified by dialysis and freeze-drying to collect 12.0 g (90%) of poly(NVA) (Akashi et al., 1990). From gel permeation chromatographic (GPC) analysis, M_n and M_w/M_n were determined as 1.53×10^5, respectively. Then, 1.6 g of poly(NVA) was hydrolyzed in 10 ml of 2N HCl solution at 120 °C for 12 h in a sealed tube after degassing. The reaction mixture was poured into acetone to collect 1.2 g (80%) of poly(VAm). After reprecipitation twice, poly(VAM) was identified by NMR (Akashi et al., 1993). In the next step, isobutyric acid was reacted with the amino group of poly(VAm). Triethylamine was added to 20 ml of aqueous solution containing 0.4 g (5 mmol) of isobutyric acid to adjust the pH to 5.0. 1-Ethyl-3-[3-(dimethylamino)propyl]carbodiimide hydrochloride (EDC) (1.0 g, 5 mmol) was added to the solution with stirring for 6 h at 0 °C) pH 7.0). After activation with EDC for 5 min, 20 ml of aqueous solution containing 0.1 g (1.3 mmol) of poly(VAm) was added to the solution with stirring for 6 h at 0 °C. The resulting polymer, poly(NVIBA-co-VAm) was purified by dialysis and freezedrying. In 15 ml of N,N'-dimethylformamide (DMF), 0.1 g of poly(NVIBA-co-VAm) and 0.7 ml of triethylamine were dissolved. Then, 0.6 ml (0.3 mmol) of isobutyryl chloride in 5 ml of DMF was added to the solution, and stirred at −30 °C for 1 h under nitrogen atmosphere. The mixture was allowed to stand at room temperature for another 24 h, and precipitated triethylamine hydrochloride was isolated by filtration. DMF was removed by evaporation under reduced pressure, and the product was purified by dialysis and freeze-drying to collect 10 mg (10%) of poly(NVIBA).

2.7.2. Synthesis of Poly (N-vinylisobutyramide) by the Free Radical Polymerization

First, N-(α-isopropoxyethyl) isobutyramide (IPEIBA) was synthesized. A 1 l three-neck flask that was equipped with a thermometer, a Dimroth condenser, a dry-ice condenser, and a dropping funnel was charged with 50 g (0.574 mol) of isobutyramide, 440 ml (5.75 mol) of 2-propanol and 3.06 ml (0.574 mol) of concentrated sulfuric acid. Acetaldehyde (320 ml, 5.74 mol) was added sequentially under stirring, and the reaction vessel was heated at 90 °C for 2 h. After neutralisation of the reaction mixture with 1 N NaOH, the unreacted acetaldehyde and 2-propanol were evaporated. The condensed mixture was extracted with chloroform and water. The aqueous layer was extracted by chloroform several times. The organic layers were combined, dehydrated by the addition of Na_2SO_4, and evaporated. Then the mixture was purified by a distillation method under reduced pressure of 1 mm Hg at 70 °C. IPEIBA was identified by ^{1}H-NMR (400 MHz, $CDCl_3$). Totally, 50 g of IPEIBA was obtained. The yield was

50% based on isobutyramide. Then, NVIBA was synthesized in a process similar to that for NVA. A 200 ml three-neck flask was equipped with a dropping funnel and was connected with a quartz 40 × 2 cm pyrolysis tube. The tube was filled with a quartz fiber, and another side of the tube was connected with two traps. The first trap was immersed in dry ice/methanol (−50 °C) and the second trap that was connected to the pump was in liquid nitrogen. The tube was heated to 500 °C with an electric furnace and the internal pressure was controlled to 1 mm Hg. Then the flask was heated to 170–180 °C in an oil bath. A solution of 50 g of IPEIBA in 300 ml of 2-propanol was dropped into the flask at rates of about 1 ml/min. The condense of the first trap was recovered, and 2-propanol was evaporated. The condensed mixture was purified by vacuum distillation at 60 °C/1 mm Hg and by repeated crystallization from n-hexane. NVIBA was identified by ^{1}H-NMR (400 MHz, CDCl$_3$). Totally, 13 g of NVIBA were obtained as white needles, so that the total yield from the isobutyramide was 20% mp 58–59 °C (lit. −58–59 °C).

The homopolymers of NVIB were prepared by the solution polymerization of NVIBA (0.1 g; 0.88 mmol) in a variety of solvents (3 ml) and initiators (1 mol% to monomer) at 60 °C in a sealed tube under shaking in an incubator. Postassium persulfate (KPS), 2,2-azobis(N,N-dimethylene-isobutyramidine) dihydrochloride, benzoylperoxide (BPO), or 2, 2′-azobisisobutyronitrile (AIBN) were used as an initiator. When using a redox initiator, KPS-TEMED, 2 mg of KPS and 20 μl of TEMED were added to the mixture. After the polymerization, the resulting polymers were dialyzed against regularly exchanged distilled water to remove unreacted monomer and the isolated polymer was dried *in vacuo*. The homopolymer of NIPAAm was also prepared by the solution polymerization of NIPAAm (0.1 g; 0.88 mmol) in ethanol (3 ml) at 60 °C using AIBN as initiator.

2.7.3. *Synthesis of* N-*vinylalkylamides*

2.7.3.1. N-(α-*isopropoxyethyl*)-*n*-*butyramide* (IPEBA)

In a 1 l four neck flask that was equipped with a sealed stirrer, two Dimroth condensers through which coolant −25 °C circulated, and a thermometer were placed in 50.0 g (574 mmol) of n-butyramide, 440 ml (5.74 mol) of 2-propanol, and 3.06 ml (57.4 mmol) of concentrated sulfuric acid. To this, 320 ml (5.74 mol) of acetaldehyde was added successively. After this addition the reaction mixture was heated for 2 h at 90–120 °C. The reaction mixture was neutralized with 1 N NaOH and was evaporated and condensed. The condensed mixture was poured into 200 ml of water and the layers were separated. The aqueous layer was extracted with chloroform (3 × 100 ml), and all organic layers were combined. The organic layer was dried with sodium sulfate and after filtration the filtrate was evaporated. The resulting product, IPEBA, was purified by distillation under reduced pressure. IPEBA was identified by ^{1}H-NMR spectroscopy. Yield: 38.0 g (0.220 mol) (38% based on n-butyramide); b.p. 107–108 °C (7.0 mm Hg).

2.7.3.2. N-*vinyl*-*n*-*butyramide* (NVBA)

A 200 ml three-necked flask that was equipped with a dropping funnel was heated to 170 °C in an oil bath and was conneted with a pyrolysis tube (quartz, i.d.3 × 30 cm), which was filled with quartz fibers and was heated to 500 °C with an electric furnace. Another side of the tube was connected with two traps: the first one was cooled to –30 °C and the second one, which was connected with a vacuum pump, was in liquid nitrogen. The internal pressure of this system was reduced to 0.1–0.5 mm Hg. A solution of 26.0 g of IPEBA in 155 ml of 2-propanol was added into this flask drop by drop at a rate of 1–1.5 ml/min. The condensate of the first trap was evaporated, and the resulting product, NVBA, was purified by distillation under reduced pressure. NVBA was identified by ^{1}H-NMR. Yield: 1.87 g (16.5 mmol) (11% based on IPEBA); b.p. 101–102 °C (6.0 mm Hg).

2.7.3.3. Isovaleramide (IVA)

Cold concentrated aqueous ammonia (900 ml, *ca.* 15 mol) was placed in a flask that was equipped with a 500 ml dropping funnel and a thermometer and surrounded by an ice-salt freezing mixture. To this, 248 g (2.05 mol) of isovaleryl chloride was added drop by drop with rapid stirring at such a rate that the temperature of the reaction mixture did not rise above 10 °C. Stirring was continued for 1.5 h after the addition of the acid chloride. The reaction mixture was evaporated to dryness under reduced pressure. This dry residue of ammonium chloride and isovaleramide was boiled for 10 min with 1.5 l of dry ethyl acetate and the boiling solution was quickly filtered. The residue on the filter was extracted with 500 ml of dry ethyl acetate. The combined ethyl acetate extracts were cooled to 0 °C and the crystalline amide that separated was removed by filtration. The filtrate was concentrated to about 700 ml and chilled and a second crop of amide was collected. The two crops of amide were combined and dried in a vacuum desiccator. Glistening white needle-shaped crystals of IBA were obtained. IBA was identified by ^{1}H-NMR spectroscopy. Yield: 113 g (1.12 mol) (55% based on isovaleryl chloride).

2.7.3.4. N-(α-isopropoxyethyl)isovaleramide (IPEIVA)

IPEIVA was obtained under the same conditions as IPEBA. The reagents for this synthesis were as follows: isovaleramide (50.0 g, 494 mmol), 2-propanol (378 ml, 4.94 mmol), concentrated sulfuric acid (2.63 ml, 49.4 mmol), acetaldehyde (276 ml, 4.94 mmol). IPEIVA was identified by ^{1}H-NMR spectroscopy. Yield: 45.9 g (245 mmol) (50% based on isovaleramide); b.p. 112–114 °C (8.0 mm Hg).

2.7.3.5. N-(α-isopropoxethyl)-n-valeramide (IPEVA)

IPEVA was obtained under the same conditions as IPEBA. The reagents for this synthesis were as follows: n-valeramide (25.0 g, 247 mmol), 2-propanol (189 ml, 2.47 mol), concentrated sulfuric acid (1.32 ml, 24.7 mmol), acetaldehyde (138 ml, 2.47 mol). IPEVA was identified by ^{1}H-NMR spectroscopy. Yield: 19.4 g (104 mmol) (42% based on n-valeramide), b.p. 108–109 °C (9.0 mm Hg).

2.7.3.6. N-vinylisovaleramide (NVIVA)

NVIVA was obtained under the same conditions as NVBA. A solution of 45.9 g of IPEIVA in 275 ml of 2-propanol produced crude NVIVA and it was then purified by distillation under reduced pressue. NVIVA was identified by ^{1}H-NMR spectroscopy. Yield: 13.3 g (105 mmol) (43% based on IPEIVA); b.p. 81–82 °C (2.0 mm Hg).

2.7.3.7. N-*vinyl-n-valeramide (NVVA)*

NVVA was obtained under the same conditions as NVBA. A solution of 52.6 g of IPEVA in 300 ml of 2-propanol produced crude NVVA. The crude product was purified by disitillation under reduced pressure and was then recrystallized from n-hexane. NVVA was identified by ^{1}H-NMR spectroscopy. Yield: 11.4 g (89.8 mmol) (32% based on IPEVA); b.p. 86–87 °C (2.0 mm Hg); m.p. 46.5–48.5 °C.

2.7.4. Synthesis of Crosslinked Poly(N-*vinylisobutyramide) Hydrogels*

First butylene-bis-N-vinylacetamide (B-BNVA) as a crosslinker, was synthesized by the reaction of NVA (118 mmol) and 1,4-dibromobutane (60.2 mmol) in the presence of the reducing agent sodium hydride (118 mmol) at room temperature in DMF (120 ml). Ten grams of NVA were dissolved in 60 ml of DMF and added to the solution of 2.82 g of NaH in 60 ml of DMF in a 200 ml three-necked flask Then, 13.0 g of 1,4-dibromobutane were dropped sequentially into the flask with stirring under N$_2$ atmosphere. After the reaction mixture was stirred at room temperature for 3 h under N$_2$ atmosphere, the DMF was evaporated. The condensed mixture was extracted with 100 ml of chloroform and 30 ml of water. The aqueous layer was extracted three times by chloroform. The organic layers were combined, washed with saturated sodium chloride solution, dehydrated by the addition of Na$_2$SO$_4$, and evaporated. After the mixture was purified by column chromatography on silica gel in a solution of n-hexane and ethylacetate mixture (volume ratio: 1 : 4), the resulting product was isolated in a solution of chloroform and 2-propanol mixture (volume ratio: 15 : 1). The product was purified by repeated crystallization from n-hexane. Yield: 5.6 g (43%); (white needles, m.p. 43–44 °C). Purified B-BNVA was identified by ^{1}H-NMR spectroscopy. Then, poly(NVIBA) and poly(NIPPAm) gels with different B-BNVA and N,N'-methylene-bis-acrylamide (Bis-A) contents were prepared by radical polymerization at 60 °C for 24h; ethanol as solvent, 2 ml; both kinds of monomers, 4 mM; both kinds of crosslinkers, 0.04, 0.2, 0.4, 0.8 mM, respectively; AIBN as an initiator, 0.04 mM. Poly(NVIBA-co-NVA) and poly(NIPPAAm-co-NVA) gels were also prepared by radical polymerization as before; ethanol as solvent, 2 ml; monomer ratios of NVIBA/NVA and NIPAAm/NVA, 100 : 0, 80 : 20, 60 : 40, respectively; total amount of the monomers, 4 mmol; both B-BNVA and Bis-A as a crosslinking agent, 0.04 mmol; AIBN as an initiator, 0.04 mmol. After the solutions were bubbled with dry nitrogen gas for 20 min, the solutions were injected into glass molds that were separated by a Tenon gasket (6 × 8 cm) (2.0 mm thickness). Solutions were polymerized at 60 °C for 24 h. After polymerization, the gel membranes were separated from the glass molds and were cut into disks (10 mm diameter) using a cork borer. These gel membranes were then immersed in methanol to remove unreacted compounds. The methanol was replaced every day. The disk samples that had been swollen in methanol were then soaked in distilled water for 1 week at room temperature and the water was changed daily. Samples were finally freeze dried.

2.8. Abbreviations

Aam	acrylamide
AIBN	2,2'-azobisisobutyronitrile
B-BNVA	butylene-bis-N-vinylacetamide
Bis-A	N,N'-methylene-bis-acrylamide
BPO	benzoylperoxide
DMF	N,N-dimethylformamide
EDC	ethyl-3-[3-(dimethylamino)propyl]carbodiimide hydrochloride
IPEBA	N-(α-isopropoxyethyl)-n-butyramide
IPEIBA	N-(α-isopropoxyethyl)isobutyramide
IPEIVA	N-(α-isopropoxyethyl)isovaleramide
IPEVA	N-(α-isopropoxyethyl)-n-valeramide
IVA	isovaleramide
KPS	potassium persulfate
LCST	lower critical solution temperature
NBAAm	N-butylacrylamide
NEAAm	N-ethylacrylamide
NIPAAm	N-isopropylacrylamide
NMAAm	N-methylacrylamide
NTBAAm	N-tert-butylacrylamide
NVA	N-vinylacetamide
NVBA	N-vinyl-n-butyramide
NVIBA	N-vinylisobutyramide
NVIVA	N-vinylisovaleramide
NVVA	N-vinyl-n-valeramide
poly(VAm)	poly(vinylamine)
VA-044	2,2-azobis(N,N'-dimethylene-isobutylamidine)dihydrochloride

2.9. References

Akashi, M., Yashima, E., Yamashita, T., Miyauchi, N., Sugita, S., and Marumo, K. (1990) A novel synthetic procedure of vinylacetamide and its free radical polymerization, *J. Polym. Sci., Part A, Polym. Chem. Ed.*, **28**, 3487–3497.

Akashi, M., Saihata, S., Yashima, E., Sugita, S., and Marumo, K. (1993) Novel nonionic and cationic hydrogels prepared from N-vinylacetamide, *J. Polym. Sci., Part A, Polym. Chem. Ed.*, **31**, 1153–1160.

Akashi, M., Nakano, S., and Kishida, A. (1996) Synthesis of poly(N-vinylisobutyramide) from poly(N-vinylacetamide) and its thermosensitive property, *J. Polym. Sci., Part A, Polym. Chem.*, **34**, 301–303.

Albertsson, P.-Å. (1956) Chromatography and partition of cells and cell fragments, *Nature*, **177**, 771–772.

Albertsson, P.-Å. (1958) Particle fractionation in liquid two-phase systems. The composition of some phase systems and the behavior of some model particles in them. Application to the isolation of cell walls from microorganisms, *Biochim. Biophys. Acta*, **27**, 378–388.

Albertsson, P.-Å. (1986) *Partition of Cell Particles and Macromolecules*, 3rd Ed., Wiley, New York.

Archibald, S.C., and Fleming, I. (1993) Two unexpected but understandable reactions with lithium diisopropylamide (LDA), *J. Chem. Soc., Perkin Trans.*, **1**, 751–758.

Bae, Y.H., Okano, T. and Kim, S.W. (1990) Temperature dependence of swelling of crosslinked poly(N,N-alkylsubstituted acrylamides) in water, *J. Polym. Sci., Part B: Polym. Phys.*, **28**, 923–929.

Baily, F.E. Jr. and Callard, R.W. (1959) Thermodynamic parameters of poly(ethylene oxide) in aqueous solution, *J. Appl. Polym. Sci.*, **1**, 56–64.

Bergfeldt, K., Piculell, L., and Tjerneld, F. (1995) Phase separation phenomena and viscosity enhancements in aqueous mixtures of poly(styrenesulfonate) with poly(acrylic acid) at different degrees of neutralization, *Macromolecules*, **28**, 3360–3370.

Birkenmeier, G. and Kunath, M. (1996) Ligand interaction of human α2-macroglobulin–α2-macroglobulin receptor studied by partitioning in aqueous two-phase systems, *J. Chromatogr., B*, **680**, 97–103.

Carlsson, M., Linse, P., and Tjerneld, F. (1993) Temperature-dependent protein partitioning in two-phase aqueous polymer systems, *Macromolecules*, **26**, 1546–1554.

Chen, C.-W. and Akashi, M. (1997a) Preparation of ultrafine platinum particles protected by poly(*n*-isopropylacrylamide) and their catalytic activity in the hydrogenation of allyl alcohol, *J. Polym. Sci. Part A: Polym. Chem.*, **35**, 1329–1332.

Chen, C.-W. and Akashi, M. (1997b) Synthesis, characterization, and catalytic properties of colloidal platinum nanoparticles protected by poly(N-isopropylacrylamide), *Langmuir*, **13**, 24, 6465–6472.

Chen, C.-W., Chen, M.-Q., Serizawa, T., and Akashi, M. (1998a) *In situ* synthesis and the catalytic properties of platinum colloids on polystyrene microspheres with surface-grafted poly(N-isopropylacrylamide), *J. Chem. Soc., Chem. Com.*, 831–832.

Chen, C.-W., Chen, M.-Q., Serizawa, T., and Akashi, M. (1998b) *In situ* formation of silver nanoparticles on poly(N-isopropylacrylamide)-coated polystyrene microspheres, *Advanced Materials*, **10**, 14, 1122–1126.

Chen, C.-W., Serizawa, T., and Akashi, M. (1999a) Preparation of platinum colloids on polystyrene nanospheres and their catalytic properties in hydrogenation, *Chem. Mater.*, **11**, 1381–1389.

Chen, C.-W., Serizawa, T., and Akashi, M. (1999b) Synthesis and characterization of poly(N-isopropylacrylamide)-coated polystyrene microspheres with silver nanoparticles on their surfaces, *Langmuir*, **15**, 23, 7998–8006.

Chen, G.H. and Hoffman, A.S. (1993) Preparation and properties of thermoreversible, phase-separating enzyme-oligo(N-isopropylacrylamide) conjugates, *Bioconjugate Chem.*, **4**, 509–514.

Chen, G.H. and Hoffman, A.S. (1994) Synthesis of carboxylated poly(NIPAAm) oligomers and their application to form thermo-reversible polymer-enzyme conjugates, *J. Biomater. Sci. Polym. Ed.*, **5**, 371–382.

Chen, G.H. and Hoffman, A.S. (1995a) Graft co-polymers that exhibit temperature-induced phase transitions over a wide range of pH, *Nature*, **373**, 49–52.

Chen, G. and Hoffman, A.S. (1995b) A new temperature- and pH-responsive copolymer for possible use in protein conjugation, *Macromol. Chem. Phys.*, **196**, 1251–1259.

Chen, M.-Q., Kishida, A., and Akashi, M. (1996) Graft co-polymers having hydrophobic backbone and hydrophilic branches. XI. Preparation and thermosensitive properties of polystyrene microspheres having poly(N-isopropylacrylamide) branches on their surfaces, *J. Polym. Sci. Part A: Polym. Chem.*, **34**, 2213–2220.

Chiklis, C.K. and Grasshoff, J.M. (1970) Swelling of thin films. I. Acrylamide-N-isopropylacrylamide co-polymers in water, *J. Polym. Sci.*, **8**, 1617–1626.

Dissing, U. and Mattiasson, B. (1993) Poly(ethyleneimine) as a phase-forming polymer in aqueous two-phase systems, *Biotechnol. Appl. Biochem.*, **17**, 15–21.

Drechsel, E.K. (1957) N-Vinyl-2-oxazolidinone and its polymers, *US Patent* 2,818,362 and 2,818,399 (*Chem. Abstracts* 52:5882c)

Feil, H., Bae, Y.H., Feijen, J., and Kim, S.W. (1992) Mutual influence of pH and temperature on the swelling of ionizable and thermosensitive hydrogels, *Macromolecules*, **25**, 5528–5530.

Feil, H., Bae, Y.H., Feijen, J., and Kim, S.W. (1993) Effect of co-monomer hydrophilicity and ionization on the lower critical solution temperature of N-isopropylacrylamide co-polymers, *Macromolecules*, **26**, 2496–2500.

Fujishige, S., Kubota, K., and Ando, I. (1989) Phase transition of aqueous solutions of poly(N-isopropylacrylamide) and poly(N-isopropylmethacrylamide), *J. Phys. Chem.*, **93**, 3311–3313.

Heskins, M. and Guillet, J.E. (1968) Solution properties of poly(N-isopropylacrylamide), *J. Macromol. Sci. Chem.*, **A2**, 1441–1455.

Hirotsu, S., Hirokawa, Y., and Tanaka, T. (1987) Volume-phase transitions of ionized N-isopropylacrylamide gels, *J. Chem. Phys.*, **87**, 1392–1395.

Hoffman, A.S. (1995) "Intelligent" polymers in medicine and biotechnology, *Artif. Organs*, **19**, 458–467.

Hoffman, A.S., Antonsen, K.P., Ashida, T., Bohnert, J.L., Dong, L.C, Nabeshima, Y., Nagamatsu, S., Park, T.G., and Sheu, M.S. (1993) Characterizing pore sizes and water "structure" in stimuli-responsive hydrogels, *Polym. Prepr. (Am. Chem. Soc., Div. Polym. Chem.)*, **34**, 1, 828.

Hoffman, A.S., Stayton, P.S., Ding, Z., Bulmus, V., Hayashi, Y., Furuzono, T., Saito, H., Long, L.C., Wu, G., Chen, X., Matsuura, J.E., Gombotz, W.R. (1997) Graft copolymers of stimuli-responsive polymers on biomolecule backbones synthesis and biomedical applications, *Polym. Prepr. (Am. Chem. Soc., Div. Polym. Chem.)*, **38**, 2, 532–533.

Horne, R.A., Almeida, J.P., Day, A.F., and Yu, N.T. (1971) Macromolecule hydration and the effect of solutes on the cloud point of aqueous solutions of poly(vinyl methyl ether): possible model for protein denaturation and temperature control in homeothermic animals, *J. Colloid Interface Sci.*, **35**, 77–84.

Inomata, H., Goto, S., and Saito, S. (1990) Phase transition of n-substituted acrylamide gels, *Macromolecules*, **25**, 4887–4888.

Inomata, H., Wada, N., Yagi, Y., Goto, S., and Saito, S. (1995) Swelling behaviors of N-alkylacrylamide gels in water: effects of co-polymerization and crosslinking density, *Polymer*, **36**, 875–877.

Ito, S. (1989) Phase transition of aqueous solution of poly(N-alkylacrylamide) derivatives. Effects of side chain structure, *Kobunshi Ronbunshu*, **46**, 437–443.

Johansson, G. and Shanbhag, V.S. (1984) Affinity partitioning of proteins in aqueous two-phase systems containing polymer-bound fatty acids. I. Effect of polyethylene glycol palmitate on the partition of human serum albumin and α-lactalbumin, *J. Chromatogr.*, **284**, 63–72.

Johansson, G. and Joelsson, M. (1987) Affinity partitioning of enzymes using dextran-bound Procion Yellow HE-3G, *J. Chromatogr.*, **393**, 195–208.

Johansson, H.-O., Lundh, G., Karlström, G., and Tjerneld, F. (1996) Effects of ions on partitioning of serum albumin and lysozyme in aqueous two-phase systems containing ethylene oxide/propylene oxide co-polymers, *Biochim. Biophys. Acta*, **1290**, 289–298.

Kaneko, Y., Sakai, K., Kikuchi, A., Yoshida, R., Sakurai, Y., and Okano, T. (1995) Influence of freely mobile grafted chain length on dynamic properties of comb-type grafted poly(N-isopropylacrylamide) hydrogels, *Macromolecules*, **28**, 7717–7723.

Kent, R.E. and McElvain, S.M. (1955) Isobutyramide, *Org. Synth.*, **111**, 490.

Kishida, A., Nakano, S., Kikunaga, Y., and Akashi M. (1998) Synthesis and functionalities of poly(N-vinylalkylamide) VII. A novel aqueous two-phase systems based on poly(n-vinylacetamide) and dextran, *J. Appl. Polym. Sci.*, **67**, 255–258.

Kishida, A., Kikunaga, Y., and Akashi, M. (1999) Synthesis and functionalities of poly(N-vinylalkylamide). X. A novel aqueous two-phase systems based on thermosensitive polymers and dextran, *J. Appl. Polym. Sci.*, **73**, 2545–2548.

Kwon, I.C., Bae, Y.H., Okano, T., Berner, B., and Kim., S.W. (1990) Stimuli sensitive polymers for drug delivery systems, *Makromol. Chem., Macromol. Symp.*, 33, 265–277.

Li, M., Zhu, Z.-Q., and Mei, L.-H. (1997) Partitioning of amino acids by aqueous two-phase systems combined with temperature-induced phase formation, *Biotechnol. Prog.*, 13, 105–108.

Lu, M., Albertsson, P.-Å., Johansson, G., and Tjerneld, F. (1994) Partitioning of proteins and thylakoid membrane vesicles in aqueous two-phase systems with hydrophobically modified dextran, *J. Chromatogr. A.*, 668, 215–228.

Maeda, M., Nishimura, C., Umeno, D., and Takagi, M. (1994) Psoralen-containing vinyl monomer for conjugation of double-helical DNA with vinyl polymers, *Bioconjugate Chem.*, 5, 527–531.

Miura, M., Cole, C.A., Monji, N., and Hoffman, A.S. (1994) Temperature-dependent adsorption/desorption behavior of lower critical solution temperature (LCST) polymers on various substrates, *J. Biomater. Sci. Polym. Ed.*, 5, 6, 555–568.

Miyazaki, H. and Kataoka, K. (1996) Preparation of polyacrylamide derivatives showing thermo-reversible coacervate formation and their potential application to two-phase separation processes, *Polymer*, 37, 681–685.

Murao, Y., Sawayama, S., and Sato, K. (1986) Manufacture of N-vinylformamide polymers, *Jp. Pat.*, 61-97309 (*Chem. Abstracts* 105:227548v).

Okahata, Y., Noguchi, H., and Seki, T. (1986) Thermoselective permeation from a polymer-grafted capsule membrane, *Macromolecules*, 19, 494–495.

Okano, T., Bae, Y.H., Jacobs, H., and Kim, S.W. (1990) Thermally on-off switching polymers for drug permeation and release, *J. Control. Release*, 11, 255–265.

Priest, J.H., Murray, S.L., and Nelson, R.J. (1986) Lower Critical Solution Temperatures of some co-polymers of N-isopropyl acrylamide with other N-substituted acrylamides, *Polym. Prep. (Am. Chem. Soc., Div. Polym. Chem.)*, 27, 239–240.

Priest, J.H., Murray, S.L., Nelson, R.J., and Hoffman, A.S. (1987) Lower Critical Solution Temperatures of aqueous co-polymers of N-isopropylacrylamide and other N-substituted acrylamides, *ACS Symp. Ser.*, 350, 255–264.

Sakuma, S., Ishida, Y., Sudo, R., Suzuki, N., Kikuchi, H., Hiwatari, K., Kishida, A., Akashi, M., and Hayashi, M. (1997a) Stabilization of salmon calcitonin by polystyrene nanoparticles having surface hydrophilic polymeric chains, against enzymatic degradation, *Int. J. Pharm.*, 159, 181–189.

Sakuma, S., Suzuki, N., Kikuchi, H., Hiwatari, K., Arikawa, K., Kishida, A., and Akashi, M. (1997b) Oral peptide delivery using nanoparticles composed of novel graft copolymers having hydrophobic backbone and hydrophilic branches, *Int. J. Pharm.*, 149, 93–106.

Sakuma, S., Suzuki, N., Kikuchi, H., Hiwatari, K., Arikawa, K., Kishida, A., and Akashi, M. (1997c) Absorption enhancement of orally administered salmon calcitonin by polystyrene nanoparticles having poly(N-isopropylacrylamide) branches on their surfaces, *Int. J. Pharm.*, 158, 69–78.

Sakurada, I., Sakaguchi, Y., and Ito, Y. (1957) Phase equilibrium of partially acetylated polyvinyl alcohols, *Kobunshi kagaku*, 14, 41–48.

Sarkar, N. (1979) Thermal gelation properties of methyl and hydroxypropyl methylcellulose, *J. Appl. Polym. Sci.*, 24, 1073–1087.

Schild, H.G. (1992) Poly(N-isopropylacrylamide): experiment, theory and application, *Prog. Polym. Sci.*, 17, 163–249.

Schild, H.G. and Tirrel, D.A. (1990) Microcalorimetric detection of lower critical solution temperatures in aqueous polymer solutions, *J. Phys. Chem.*, 94, 4352–4356.

Seto, F., Fukuyama, K., Muraoka, Y., Kishida, A., and Akashi, M. (1998) Thermosensitive surface properties of polyethylene film with poly(N-isopropylacrylamide) chains prepared by corona discharge induced grafting, *J. Appl. Polym. Sci.*, 68, 1773–1779.

Shibayama, M., Mizutani, S., and Nomura, S. (1996) Thermal properties of co-polymer gels containing N-isopropylacrylamide, *Macromolecules*, **29**, 2019–2024.

Sivars, U., Bergfeldt, K., Piculell, L., and Tjerneld, F. (1996) Protein partitioning in weakly charged polymer-surfactant aqueous two-phase systems, *J. Chromatogr., B*, **680**, 43–53.

Suwa, K., Morishita, K., Kishida, A., and Akashi, M. (1997a) Synthesis and functionalities of poly(N-vinylalkylamide) V. Control of a Lower Critical Solution Temperature of poly(N-vinylalkylamide), *J. Polym. Sci. Part A: Polym. Chem.*, **35**, 3087–3094.

Suwa, K., Wada, Y., Kishida, A., and Akashi, M. (1997b) Synthesis and functionalities of poly(N-vinylalkylamide) VI. A novel thermosensitive hydrogel crosslinked poly(N-vinylisobutyramide), *J. Polym. Sci. Part A: Polym. Chem.*, **35**, 3377–3384.

Suwa, K., Wada, Y., Kikunaga, Y., Morishita, K., Kishida, A., and Akashi, M. (1997c) Synthesis and functionalities of poly(N-vinylalkylamide). IV. Synthesis and free radical polymerization of N-vinylisobutyramide and thermosensitive properties of the polymer, *J. Polym. Sci. Part A: Polym. Chem.*, **35**, 1763–1768.

Svensson, M., Linse, P., and Tjerneld, F. (1995) Phase behavior in aqueous two-phase systems containing micelle-forming block co-polymers, *Macromolecules*, **28**, 3597–3603.

Urry, D.W., Harris, R.D., and Prasad, K.U. (1988) Chemical potential driven contraction and relaxation by ionic strength modulation of an inverse temperature transition, *J. Am. Chem. Soc.*, **110**, 3303–3305.

Wada, Y., Nakano, S., Kishida, A., and Akashi, M. (1993) Synthesis of a novel thermosensitive polymer, *Polym. Prep. Jpn.*, **42**, 2742.

Walter, H., Brooks, D.E., and Fischer, D. (1985) *Partitioning in Aqueous-Two Phase Systems*, Academic Press, London.

Walter, H. and Johansson, G. (1994) Aqueous two-phase systems, *Methods Enzymol.*, Academic Press, San Diego, USA.

Walter, H., Johansson, G., and Brooks, D.E. (1991) Partitioning in aqueous two-phase systems: recent results, *Anal. Biochem.*, **197**, 1–18.

Wang, X., Qiu, X., and Wu, C. (1998) Comparison of the coil-to-globule and the globule-to-coil transitions of a single poly(N-isopropylacrylamide) homopolymer chain in water, *Macromolecules*, **31**, 2972–2976.

Winnik, F.M. (1990) Fluorescence studies of aqueous solutions of poly(N-isopropylacrylamide) below and above their LCST, *Macromolecules*, **23**, 233–242.

Wu, C. and Zhou, S. (1996) Internal motions of both poly(N-isopropylacrylamide) linear chains and spherical microgel particles in water, *Macromolecules*, **29**, 1574–1578.

Wu, X.S., Hoffman, A.S., and Yager, P. (1992) Synthesis and characterization of thermally reversible macroporous poly(N-isopropylacrylamide) hydrogels, *J. Polym. Sci. Part A: Polym. Chem. Ed.*, **30**, 2121–2129.

Yamada, B., Itahashi, M., and Otsu, T. (1978) Estimation of monomer reactivity ratios by nonlinear least-squares procedure with consideration of the weight of experimental data. *J. Polym. Sci., Polym. Chem. Ed.*, **16**, 1719–1733.

Yoshida, R., Uchida, K., Kaneko, Y., Sakai, K., Kikuchi, A., Sakurai, Y. and Okano, T. (1995) Comb-type grafted hydrogels with rapid de-swelling response to temperature changes, *Nature*, **374**, 240–242.

3 Affinity precipitation of proteins using smart polymers

Igor Yu. Galaev and Bo Mattiasson

3.1. Introduction

All bioseparation processes include three stages: (1) the preferential partitioning of target substance and impurities between two phases (liquid–liquid or liquid–solid); (2) the mechanical separation of the phases (e.g. separation of the stationary and mobile phase in a chromatographic column); and (3) recovery of the target substance from the enriched phase. Selective precipitation of a target protein from a crude mixture is very attractive as a separate protein enriched phase forms from a homogeneous solution during the process and the mechanical separation of protein enriched phase (pellet) from protein depleted phase (supernatant) is easily achieved by well established techniques like filtration or centrifugation, eliminating the need for sophisticated and expensive equipment e.g. columns and adsorbents used in chromatography. Moreover, in precipitated form proteins are usually more stable which is important when dealing with crude extracts containing proteases. Another attractive feature of protein precipitation is the possibility to concentrate the target protein. The protein precipitated from a large volume of the crude extract could be dissolved afterwards in a small buffer volume.

Traditionally precipitation of the target protein is achieved by the addition of large amounts of salts like ammonium sulfate, polymers, like polyethylene glycol (PEG) or organic solvents miscible with water like acetone or ethanol (Scopes, 1994). The precipitation of the target protein occurs because changing bulk parameters of the medium are needed, the driving force being integral physico-chemical and especially surface properties of the protein macromolecule. To change bulk parameters of the medium, hundreds of grams per liter of ammonium sulfate, tens of grams per liter of PEG or hundreds of ml per liter of organic solvents are added. The macromolecular nature of protein molecules combined with a general principle of their folding, hydrophilic amino acids at the surface, hydrophobic – inside the core, make proteins rather similar in their surface properties. One could not expect high selectivity to be achieved by traditional precipitation techniques as the selectivity of precipitation is limited to the differences in integral surface properties of protein molecules. The low selectivity of traditional precipitation techniques is considered to be their main drawback. Nevertheless, the techniques are widely used as the first step of protein purification combining capture of the target protein with some purification. The introduction of higher selectivity to precipitation techniques is certainly of great importance. The present tendency in down stream processing of proteins is the introduction of highly selective affinity steps at the early stage of purification protocol when using robust, biologically and chemically stable ligands (Kaul and Mattiasson, 1992). Affinity precipitation, which is discussed in the present chapter, is an attempt to introduce the highly selective affinity technique at the very beginning of protein purification

3.2. Homo- and Heterobifunctional Mode of Affinity Precipitation

The basic idea of affinity precipitation is the use of a macroligand, i.e. ligand with two or more affinity sites. The macroligands could be synthesized either by covalent linking of two low-molecular-weight ligands (directly or through a short spacer) or by covalent binding of several ligand molecules to a water-soluble polymer.

If the protein has a few ligand-binding sites, complexation with a macroligand results in a formation of poorly soluble large aggregates (Figure 3.1) which could be separated by filtration or centrifugation from the supernatant containing soluble impurities. Alternatively, the polymer backbone of the macroligand could be used as a driving force for precipitation, provided the solubility of the polymer changes drastically when small changes in the environment (pH, temperature, ionic strength, presence of specific substances) take place (Gupta and Mattiasson, 1994). The polymers with reversed solubility are often combined under the names 'smart polymers', 'intelligent polymers' or 'stimuli responsive polymers'. The smart polymers are critical for the development of affinity precipitation of protein in heterobifunctional mode.

The homobifunctional mode of affinity precipitation was introduced by Larsson and Mosbach who performed quantitative precipitation to purify lactate dehydrogenase

Homobifunctional affinity precipitation

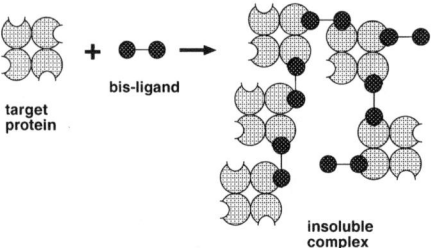

Heterobifunctional affinity precipitation

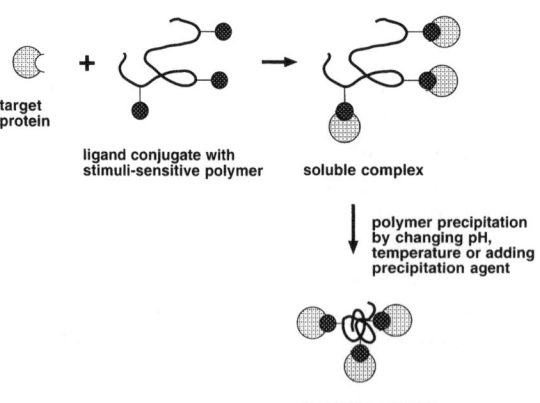

Figure 3.1 Homo- and heterobifunctional mode of affinity precipitation.

(tetrameric enzyme) and glutamate dehydrogenase (hexamer) using bis-NAD ($N^2,N^{2'}$-adipohydrazidobis(N^6carbonylmethyl)-NAD) (Larsson *et al.*, 1984; Larsson and Mosbach, 1979). The precipitate was easily soluble in the presence of NADH that forms stronger bonds with lactate dehydrogenase and replaces the enzyme from the complex with bis-NAD.

As the homobifunctional mode of affinity precipitation does not exploit the smart polymers, it falls out beyond the scope of this chapter. The reader is addressed to reviews of (Galaev and Mattiasson, 1997; Gupta and Mattiasson, 1994; Kaul and Mattiasson, 1992; Mattiasson and Kaul, 1993) for additional information.

Affinity precipitation in a heterobifunctional mode is based on using a conjugate of smart polymers with convalently coupled ligands specific for the target proteins. The conjugate first forms a complex with the target protein and phase separation of the complex is triggered by small changes in the environment resulting in transition of polymer backbone into an insoluble state. The target protein is then either eluted from the insoluble macroligand–protein complex or the precipitate is dissolved, the protein dissociated from the macroligand and the ligand–polymer conjugate reprecipitaed without the protein which remains in the supernatant in a purified form. Various ligands, such as triazine dyes, sugars, protease inhibitors, antibodies and nucleotides, have been successfully used for affinity precipitation (Gupta and Mattiasson, 1994; Galaev *et al.*, 1996).

3.3. Smart Polymers Used in Affinity Precipitation of Proteins

Smart or intelligent materials have the capability to sense changes in their environment and respond to the changes in a pre-programmed and pronounced way (Gisser *et al.*, 1994). Referring to water soluble polymers and hydrogels this definition can be formulated as follows. Smart polymers undergo fast and reversible changes in microstructure triggered by small changes of medium property (pH, temperature, ionic strength, presence of specific chemicals, light, electric or magnetic field). These microscopic changes of polymer microstructure manifest themselves at the macroscopic level as a precipitate formation in a solution or as manyfold decrease/increase of the hydrogel size and hence of water content. Macroscopic changes in the system with smart polymer/hydrogel are reversible and elimination of the trigger changes in the environment returns the system to its initial state.

3.3.1. pH-sensitive Smart Polymers

In general, all smart polymer/hydrogel systems can be divided into three groups (Figure 3.2). The first group consists of the polymers for which poor solvent conditions are created by decreasing net charge of the polymer/hydrogel. The net charge can be decreased by changing pH to neutralize the charges on the macromolecule and hence to reduce the repulsion between polymer segments. For instance, copolymers of methylmethacrylate and methacrylic acid precipitate from aqueous solutions on acidification to pH around 5 while copolymers of methyl methacrylate with dimethylaminethyl methacrylate are soluble at acidic conditions but precipitate at slightly alkaline conditions (Röhm Pharma, 1993). The reversible solubility of the Eudragit is used for enteric coating. Pills coated with a polymer film are insoluble in the acidic environment of the stomach and hence the active component of the pill is protected

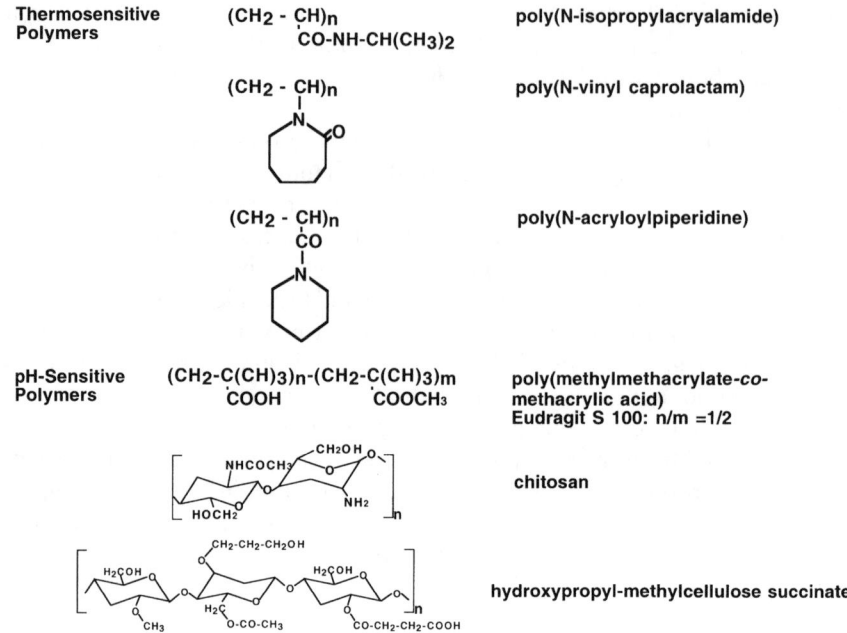

Figure 3.2 Structures of smart polymers frequently used in affinity precipitation.

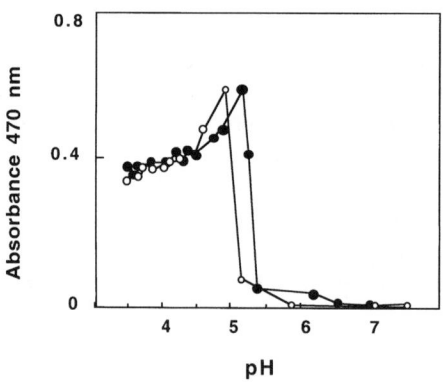

Figure 3.3 Precipitation curves of Eudragit S 100 (open circles) and *p*-amino-phenyl-
α-D-glucopyranoside-modified Eudragit S 100 (closed circles) measured as
turbidity at 470 nm. Some decrease in turbidity at lower pH values is caused
by flocculation and sedimentation of polymer precipitate. Reproduced from
(Linné-Larsson and Mattiasson, 1994) with permission.

against the harmful action of the stomach content. When coming into the intestine
with its neutral pH, the polymer coating dissolves and the active component of the pill
is released.

Figure 3.3 presents a pH-dependent precipitation curve of a random copolymer
of methacrylic acid and methyl methacrylate (commercialized as Eudragit S 100
by Röhm Pharma GMBH, Weiterstadt, Germany) as well as *p*-amino-phenyl-α-D-
glucopyranoside-modified Eudragit S 100. Replacement of some carboxy-groups with

a non-charged sugar moiety increased the hydrophobicity of the copolymer resulting in precipitation at higher pH (Linné-Larsson and Mattiasson, 1994).

The charges on the macromolecule can be neutralized also by addition of an efficient counterion, e.g., low molecular weight counter ion or a polymer molecule with the opposite charges. The latter systems are combined under the name of polycomplexes. The cooperative nature of interaction between two polymers with the opposite charges makes polycomplexes very sensitive to the changes in pH or ionic strength (Izumrudov *et al.*, 1999, see also Chapter 5 in this volume). The complex formed by poly(methacrylic acid) (polyanion) and poly(N-ethyl-4-vinyl pyridinium bromide) (polycation) undergoes reversible precipitation from aqueous solution at any desired pH-value in the range 4.5–6.5 depending on the ionic strength and polycation/polyanion ratio in the complex (Figure 3.4) (Dainiak *et al.*, 1999). Polyelectrolyte complexes formed by poly(ethyleneimine) and poly(acrylic acid) undergo soluble-insoluble transition in an even broader pH range, from pH 3 to 11 (Dissing and Mattiasson, 1996).

When antigen, inactivated glyceraldehyde-3-phosphate dehydrogenase from rabbit, was covalently coupled to a polycation, the resulting complex was used for purification of monoclonal antibodies from the 6G7 clone specific towards inactivated

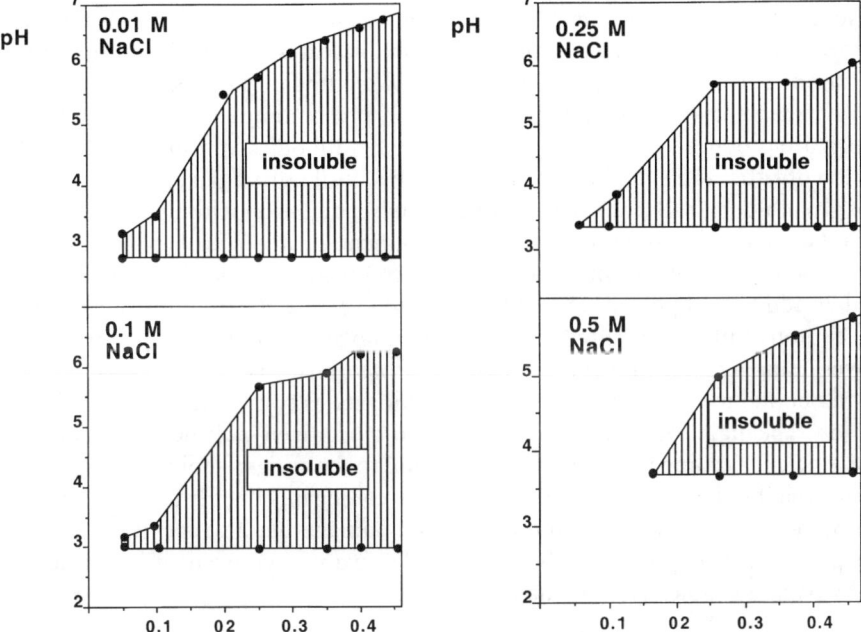

Poly(N-ethyl-4-vinyl pyridinium bromide)/Poly(methacrylic acid) Ratio

Figure 3.4 Phase diagram of the poly(N-ethyl-4-vinyl pyridinium bromide)–poly(methacrylic acid) system. The dots (present pH values at which the turbidity of the polymer solutions was first observed at 470 nm. Ionic strength was 0.01, 0.1, 0.25 and 0.5 M NaCl. Dashed area represents pH/composition range where the complex is insoluble. Reproduced from (Dainiak *et al.*, 1999) with permission.

Table 3.1 Antibody purification by affinity precipitation using glyceraldehyde-
3-phosphate dehydrogenase bound to polyelectrolyte complex

Polycation/polyanion ratio	0.45	0.35	0.2	0.15
pH of incubation	7.3	6.5	5.5	4.6
pH of precipitation	6.5	6.0	5.3	4.5
Recovery, %	98	96	60	60
Purification factor (SDS-PAGE data)	1.8	1.4	1.4	1.4
Purification factor (ELISA data)	1.5	1.5	1.0	—

glyceraldehyde-3-phosphate dehydrogenase bound to a polyelectrolyte complex and the precipitation of the latter was carried at 0.01 M NaCl and pH 4.5, 5.3, 6.0 and 6.5 using complexes with polycation/polyanion ratio of 0.45, 0.3, 0.2 and 0.15, respectively. Purified antibodies were eluted at pH 4.0 where polyelectrolyte complexes of all compositions used were insoluble. Quantitative recoveries were achieved under optimal conditions (Table 3.1). Relatively low values of purification factors were due to the relative purity of the preparation of antibodies used in this work. Precipitation of polyelectrolyte complexes is accompanied only by small nonspecific co-precipitation of proteins. Precipitated polyelectrolyte complexes could be dissolved at pH 7.3 and used repeatedly (Dainiak et al., 1999).

The successful affinity precipitation of antibodies using glyceraldehyde-3-phosphate dehydrogenase bound to polyelectrolyte complex indicates that the ligand is exposed into solution. This fact was used to develop a new method of the production of monovalent Fab fragments of antibodies containing only one binding site. Traditionally, Fab fragments are produced by proteolytic digestion of antibodies in soultion followed by separation of Fab fragments. In case of monoclonal antibodies against inactivated subunits of glyceraldehyde-3-phosphate dehydrogenase, digestion with papain resulted in significant damage of binding sites of the Fab fragment. Proteolysis of monoclonal antibodies in the presence of the antigen–polycation conjugate followed by (i) precipitation induced by addition of polyanion, poly(methacrylic) acid and pH-shift from 7.3 to 6.5 and (ii) elution at pH 3.0 resulted in 90% immunologically competent Fab fragments. Moreover, papain concentration required for proteolysis was 10 times less in case of antibodies bound to the antigen–polycation conjugate as compared to free antibodies in soultion. The digestion of antibodies bound to the antigen–polyelectrolyte complex was less efficient as compared to antibodies bound to the antigen–polycation conjugate suggesting that binding to the antigen–polycation conjugate not only protected binding sites of monoclonal antibodies from proteolytic damage but also facilitated the proteolysis probably by exposing antibody molecules in a way convenient for proteolytic attack by papain (Table 3.2) (Dainiak et al., 2000).

3.3.2. Thermosensitive Smart Polymers

The second group of smart polymers consists of uncharged polymers soluble in water due to the hydrogen bonding with water molecules. Changes in hydrophobic–hydrophilic balance are induced by increasing temperature or ionic strength. The efficiency of hydrogen bonding is reduced on raising temperature. There is a critical temperature for some polymers at which the efficiency of hydrogen bonding becomes insufficient for solubility of macromolecule and the phase separation of polymer takes place.

Table 3.2 Yield of Fab fragments after digestion of antibodies at different conditions

	Free antibodies		Antibodies coupled to polycation		Antibodies coupled to nonstoichiometric polyelectrolyte
	Low papain	High papain	Low papain	High papain	Low papain
Antibodies[a], %	90%	0%	10%	0%	50%
Fab-fragments (active)[a], %	0%	0%*	90%	70%*	50%

a initial concentration of immunologically competent antibodies was taken as 100%.
* We observed that the part or all (in the case of free antibodies) Fab fragments lost affinity to antigen after digestion at high concentration of papain.

On raising the temperature of aqueous solutions of smart polymers to a point higher than the critical temperature (lower critical solution temperature, LCST, or 'cloud point'), separation into two phases takes place. A polymer enriched phase and an aqueous phase containing practically no polymer are formed. Both phases can be easily separated. This phase separation is completely reversible and the smart polymer dissolves in water on cooling. Three thermosensitive smart polymers are most widely studied and used, poly(N-isopropyl acrylamide) (poly(NIPAM)) with LCST 32–34 °C poly(vinyl methyl ether) with LCST 34 °C, and poly(N-vinyl caprolactam) (poly(VCL)) with LCST 32–40 °C. The modern polymer chemistry provides polymers with different transition temperatures from 4–5 °C for poly (N-vinyl piperidine) to 100 °C for poly(ethylene glycol) (Galaev and Mattiasson, 1993b).

Opposite to pH-sensitive smart polymers, which contain carboxy or amino groups which could be used for the covalent coupling of ligands, the thermosensitive polymers do not have inherent reactive groups. Thus copolymers containing reactive groups should be synthesized. N-hydroxyacrylsuccinimide (Liu *et al.*, 1995) or glycidyl methacrylate (Mori *et al.*, 1994) were used as active comonomers in copolymerization with NIPAM allowing further coupling of amino group-containing ligands to the synthesized copolymers. An alternative strategy is to modify the ligand with acryloyl group and then co-polymerize the modified ligand with NIPAM (Maeda *et al.*, 1993; Umeno *et al.*, 1998).

Incorporation of hydrophilic comonomers increases the transition temperature for the polymer. In a copolymer of NIPAM and vinyl imidazole (poly(VI/NIPAM)) the incorporation of relatively hydrophilic imidazole moieties hindered the hydrophobic interactions of the native poly-NIPAM and resulted in increase in the precipitation temperature (Figure 3.5). The effect was more evident at lower pH values where imidazole moieties were protonated and hence rendered the polymer more hydrophilic. Poly(VI/NIPAM) (15.6 mol% VI), for instance, did not precipitate at all on heating up to 70 °C at pH 4 and 6, whereas precipitation of homopolymer, poly(NIPAM) was independent of pH (Figure 3.5) (Galaev *et al.*, 1999).

Poly(VI/NIPAM) copolymers are capable of binding metal ions via imidazole moieties (Galaev *et al.*, 1999). When loaded with Cu(II), all the Cu-loaded copolymers did not precipitate on heating up to 70 °C. The increase in ionic strength dramatically facilitated precipitation of Cu(II)-poly(VI/NIPAM) (Figure 3.6). Sodium sulfate facilitated thermoprecipitation, as expected, more efficient than sodium chloride while

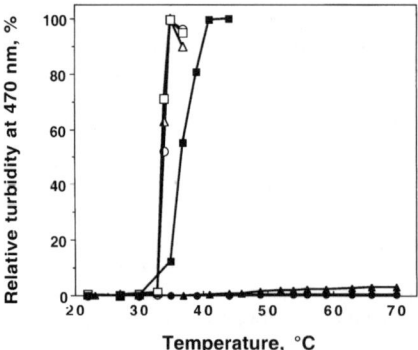

Figure 3.5 Thermoprecipitation of poly-NIPAM (open symbols) and (VI/NIPAM) (15.6 mol% VI) (closed symbols) from aqueous solution at pH 4.0 (circles), 6.0 (triangles) and 8.0 (squares) monitored as turbidity at 470 nm. Polymer concentration 1.0 mg/ml. Reproduced from (Galaev *et al.*, 1999) with permission.

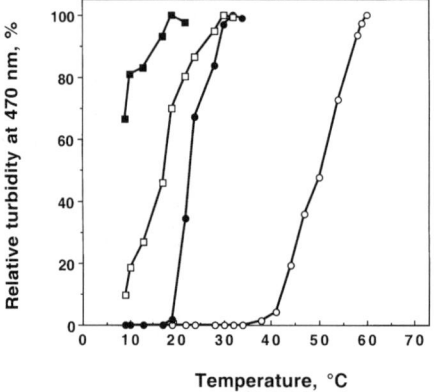

Figure 3.6 Thermoprecipitation of Cu(II)-poly(VI/NIPAM) (15.6 mol% VI) at 0.05 M (open circles), 0.1 M (closed circles), 0.2 M (open squares) and 0.4 M NaCl (closed squares) at pH 6.0 monitored as turbidity at 470 nm. Polymer concentration 1.0 mg/ml. Reproduced from (Galaev *et al.*, 1999) with permission.

ammonium sulfate was slightly less efficient than sodium sulfate. When NaCl was added up to a final concentration of 0.6 M at room temperature, all the polymer was instantaneously precipitated and flocculated in a clump, the remaining solution being transparent. The efficient precipitation of Cu(II)-poly(VI/NIPAM) by high salt concentrations at mild temperatures is very convenient for metal affinity precipitation. This allows its application for the purification of a wide range of proteins. High salt concentration does not interfere with protein–metal ion–chelate interaction (Porath and Olin, 1983) and on the other hand it reduces the possibility of nonspecific binding of foreign proteins to the polymer both in solution and after precipitation.

Cu(II)-ion can from a complex with up to four imidazoles in solution. When imidazole groups are coupled to the polymer, about two imidazole groups formed a complex with one Cu(II) ion (Kumar *et al.*, 1998a). When Cu(II) binds to imidazole in solution, log K (where K is association constant, M^{-1}) for each imidazole ligand decreases from log $K_1 = 3.76$ for binding the first imidazole ligand to log $K_4 = 2.66$ for binding the fourth imidazole ligand (Liu and Gregor, 1965). The binding of a single imidazole ligand to the Cu(II)-ion in solution is much weaker comparing to the binding of tridentate iminodiacetic acid (log $K = 11$ (Todd *et al.*, 1994)). On the other hand, when the Cu(II)-ion forms a complex with four imidazole ligands the combined binding constant log $K = \log K_1 + \log K_2 + \log K_3 + \log K_4 = 12.6 - 12.7$. The efficiency of this complex is close to that of Cu(II)-ion complex with poly (1-vinylimidazole) (poly-VI), log $K = 10.64 - 14.72$ (Gold and Gregor, 1960; Liu and Gregor, 1965) and comparable with the binding of tridentate ligand of iminodiacetic acid. With about two imidazole ligands bound to the Cu(II)-ion one could expect binding strength of log $K = 5.5 - 6.0$. When coupled to solid matrices, imidazole ligands are spatially separated and the proper orientation of the ligands to form a complex with the same Cu(II)-ion is unlikely and the imidazole ligands are not used for Immobilized Metal Affinity (IMA)-chromatography (Galaev *et al.*, 1997). In solution, the flexible polymer like poly (V/NIPAM) can adopt a soultion-phase conformation where two imidazole ligands are close enough to form a complex with the same Cu(II)-ion providing significant strength of interaction.

The flexibility of polymer backbone of the macroligand and hence the possibility for a few ligand molecules to interact simultaneously with the protein molecules increases dramatically the apparent efficiency of the protein binding to the macroligand. Even when the single ligand–protein interaction is relatively weak as in the case of imidazole-metal ion, the multiple nature of these interactions is well sufficient to ensure strong protein–macroligand binding. Hence, affinity precipitation allows in principle the use of weaker ligands as compared to affinity chromatography, provided multisite ligand–protein interactions are possible.

An interesting example of using poly(N-acryloylpiperidine) terminally modified with maltose for affinity precipitation was presented recently (Hoshino *et al.*, 1998). The use of the polymer with extermely low critical temperature (soluble below 4 °C and completely insoluble above 8 °C) made it possible to use the technique for purification of thermolabile α-glucosidase from cell-free extract of *Saccharomyces cerevisiae* achieving 206-fold purification with 68% recovery.

Thermosensitive polymers undergo phase transition because they become progressively more hydrophobic with the increase in temperature. At some point the hydrophobic interactions between polymer molecules become more favorable then polymer–water interactions and the polymer molecules aggregate forming a separate phase, a polymer precipitate. Hydrophobic interactions could be promoted by high salt concentrations, shifting cloud point to the temperatures below room temperature. For example, affinity precipitation of α-amylase inhibitor with the Cu-loaded poly(V/NIPAM) was carried out at room temperature by addition of 2 M NaCl (Kumar *et al.*, 1998a). This approach could be used only for the systems where ligand–protein interactions are not affected by high salt concentrations. Interaction of histidine residues at the surface of protein molecules with the metal ion, immobilized on the polymer is an example of such salt-independent interactions (Porath and Olin, 1983).

Recently, a new carrier, polymerized liposomes was developed for salt-induced affinity precipitation (Sun *et al.*, 1995). Polymerized liposomes, prepared using a synthesized phospholipid with a diacetylene moiety in the hydrophobic chain and an amino group in the hydrophilic head, showed a reversible precipitation on salt addition and removal. Polymerized liposomes with immobilized soybean trypsin inhibitor were used to resolve a model mixture of trypsin and chymotrypsin.

Thermosensitive latexes composed of thermosensitive polymers or with a layer of thermosensitive polymer at the surface present another example of insoluble but reversibly suspended system in response to increasing/decreasing temperature. In more detail these systems will be discussed in the chapter 'Smart Latexes for Bioseparation and Bioprocessing' by H. Kawaguchi and K. Fujimoto. Up till now thermosensitive latexes were used mainly as carriers for reversibly soluble biocatalyst (Kondo and Fukuda, 1997; Kondo et al., 1994; Okubo and Ahmad, 1998; Shiroya et al., 1995a and b). Their potential for affinity precipitation remains to be evaluated.

pH-sensitive latex based on the copolymer of poly(methacrylate) and poly(methylmethacrylic acid) was compared with a soluble polymer of the similar composition, Eudragit L 100 as carriers for precipitation of positively charged protein, lysozyme. Latex particles precipitated much faster but their capacity was about one order of magnitude less as compared to soluble polymer (Chern et al., 1996).

3.3.3. Reversibly Cross-linked Polymer Networks

The third group of smart polymers combines systems with reversible non-covalent cross linking of separate polymer molecules into insoluble polymer network. The most familiar systems of this group are Ca-alginate (Charles et al., 1974; Linné et al., 1992) and boric acid-polyols (Wu and Wisecarver, 1992; Kokufuta and Matsukawa, 1995; Kitano et al., 1993) or boric acid-polysaccharides (Bradshaw and Sturgeon, 1990). This type of polymers has found a limited application as carriers in affinity precipitation but they are more promising for the development of 'smart' drug delivery systems capable of releasing drugs in response to the signal, e.g. release of insulin when glucose concentration is increasing (Lee and Park, 1996, 1997).

3.4. Properties of the Polymer Precipitate

When designing a polymer for affinity precipitation one should consider also properties of polymer precipitate (or polymer enriched phase formed). When phase separation of the polymer takes place, three extreme types of the precipitate could be formed (Figure 3.7).

The first type of polymer enriched phase is usually formed after pH-induced precipitation of chitosan (Senstad and Mattiasson, 1989a and b; Tyagi et al., 1996) or Ca-induced precipitation of alginate (Linné et al., 1992). This type of gel has high water content and when formed, a lot of solution is entrapped. The lose structure of such gel makes it difficult to separate the gel from the supernatant and the entrapped solution containing impurities decreases the purification factor. The attractive feature of such polymers is their relatively low hydrophobicity. Large protein molecules and even cells can be processed without a risk of denaturation due to hydrophobic interactions with the polymer.

The second type of polymer phase with high polymer content occurs during thermoprecipitation of polymers like poly(N-vinyl caprolactam) (Galaev and Mattiasson, 1993a) or poly(ethylene glycol-co-propylene glycol) random and block co-polymers

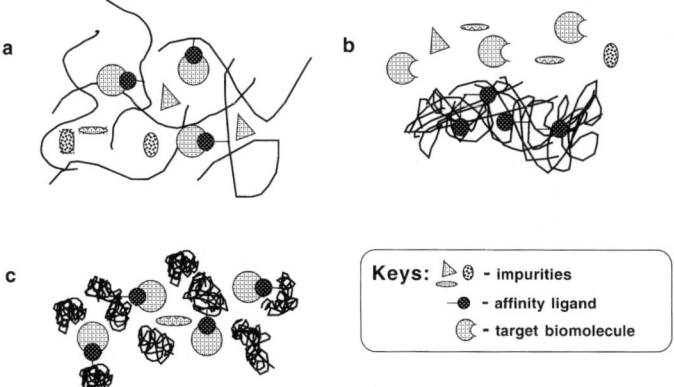

Figure 3.7 Types of the polymer enriched phases formed after the phase separation of the polymer during affinity precipitation. (a) Lose gel with high water content containing bound target protein and entrapped impurities, (b) compact hydrophobic phase with low water content which provides unfavorable environment for target protein, the latter remains in solution when polymer precipitates and (c) suspension of compact polymer aggregates with bound protein molecules exposed in solution.

(Lu *et al.*, 1996). The polymer phase is easily separated from supernatant with practically no entrapment of impurities from solution. High polymer content renders the polymer phase very hydrophobic and makes it an unfavorable environment for proteins. In fact, proteins are pushed off from the complex with the polymer, when the polymer phase is formed. High affinity ligands with strong binding towards target protein are required to achieve specific co-precipitation of the protein with the polymer (Galaev and Mattiasson, 1992; Galaev and Mattiasson, 1993a). On the other hand, this property disadvantageous for affinity precipitation could be quite useful in another mode of protein purification, namely, partitioning in aqueous two-phase polymer systems (Alred *et al.*, 1994). Some of the thermosensitive polymers are able to form aqueous two-phase systems with dextrans. When a specific ligand is coupled to the polymer, the target protein partitions preferentially into the aqueous phase formed by this polymer. After mechanical separation from the dextran phase, the target protein can be recovered by thermoprecipitation of the polymer. The latter forms a compact polymer enriched phase while the purified target protein remains in the supernatant. The high hydrophobicity of the polymer phase results in small co-precipitation of the target protein with the polymer especially, when the conditions (pH, ionic strength) are also changed to reduce specific protein–polymer interactions (Franko *et al.*, 1997; Harris *et al.*, 1991). The use of smart polymers for partitioning of biomolecules in aqueous two-phase polymer systems is discussed in detail in the chapter of this book 'Aqueous Two-phase Systems with Smart Polymers' by H.-O. Johansson *et al.*

The third case looks the most preferable for affinity precipitation. Compact particles of aggregated polymer are easily separated from the supernatant accompanied with only minimal entrapment of the supernatant and impurities. The ligands with bound target protein are exposed to the solution and hence are in the comfortable environment.

We have studied the rheological properties of the polymer phase formed after pH-induced precipitation of Eudragit S-100. The neutralization of the negatively charged carboxy groups results in an increase in hydrophobicity of the polymer molecules. The latter aggregate due to the hydrophobic interactions and form a three-dimensional network with high viscosity and strong shear-thinning behavior. Upon further decrease in pH and increase in hydrophobicity of the polymer molecules, the latter rearrange from three-dimensional network into a suspension of particles of aggregated polymer chains. This rearrangement is accompanied by a significant drop in viscosity and much less pronounced shear-thinning behavior indicating formation of the polymer precipitate of the third (Linné-Larsson *et al.*, 1996). It is not surprising that quite a few successful affinity precipitation procedures were reported when using this polymer (Guoqiang *et al.*, 1993, 1994a and b, 1995a and b; Gupta *et al.*, 1994; Kamihira *et al.*, 1992; Mattiasson and Kaul, 1994; Shu *et al.*, 1994) or the similar polymer Eudragit L 100 (Chern *et al.*, 1996a). An alternative pH-sensitive polymer successfully used for affinity precipitation is hydroxypropylmethylcellulose acetate succinate (Taniguchi *et al.*, 1989, 1990) which is also used for enteric coating like Eudragit polymers.

Affinity precipitation is readily combined with other protein isolation techniques e.g. partitioning in aqueous two-phase polymer systems (Chen and Jang, 1995; Kamihira *et al.*, 1992; Mattiasson and Kaul, 1994; Guoqiang *et al.*, 1994a). Partitioning of protein complex with ligand–polymer conjugate is usually directed to the upper hydrophobic phase of aqueous two-phase polymer systems formed by poly(ethylene glycol) and dextran or hydroxypropyl starch while most of proteins present in crude extracts or cell homogenates partition into the lower hydrophilic phase. Then the precipitation of the protein–polymer complex is promoted by changing pH. Trypsin was purified using a conjugate of soybean trypsin inhibitor with hydroxypropyl cellulose succinate acetate (Chen and Jang, 1995), lactate dehydrogenase and protein A – using conjugate of Eudragit S 100 with a triazine dye, Cibacron Blue (Guoqiang *et al.*, 1994b) and immunoglobulin G (Kamihira *et al.*, 1992; Mattiasson and Kaul, 1994), respectively. Combination of partitioning with affinity precipitation improves yield and purification factor.

The choice of the polymer carrier for affinity precipitation is decisive for the success of purification procedure. As a rule, increase in the molecular weight of the thermosensitive polymer reduces the precipitation temperature and makes the transition between soluble and insoluble state of the polymer more sharp (Galaev *et al.*, 1996a). On the other hand polymers with very high molecular weights from viscous solutions difficult for handling. The molecular weight distribution of the polymer is also important for the sharpness of soluble–insoluble transition. The broader the molecular weight distribution, the less sharp is the transition. Especially, the effect is pronounced for the polymers with a relatively low molecular weight.

Natural pH-sensitive smart polymers are usually polysaccharides e.g. hydrophobic cellulose derivatives bearing carboxy groups, alginate or chitosan. These polymers have a broad molecular weight distribution as well as a broad compositional distribution appeared as the result of the polymer isolation from the natural source and further processing. Thus, natural smart polymers do not have as a rule sharp transitions. An extensive change of the environmental parameter is often required to achieve a complete precipitation of these polymers. Fractionation by separation of the polymer precipitated after only a small change of the parameter improves significantly the sharpness of the soluble–insoluble transition (Linné *et al.*, 1992).

3.5. Ligands Used in Affinity Precipitation

Affinity precipitation contrary to affinity chromatography is a one plate process. A protein molecule that has dissociated from the ligand into solution will on precipitating the polymer have practically no chance to interact with the ligand again as the polymer is removed from solution when precipitating. Thus, to be successful, the affinity precipitation requires stronger protein–ligand interactions as compared to affinity chromatography, where the protein molecule dissociated into solution from the ligand has high chances to be bound again as it moves along the column matrix with its high concentration of the ligands available. The rough estimation is that the binding constant for a single ligand–protein interaction in affinity precipitation should be at least 10^{-5} M (Galaev and Mattiasson, 1993a). On the other hand, flexibility of the polymer chain allows interaction of a few ligands with multisubunit proteins hence increasing significantly the polymer–protein avidity. In case of two site ligand–protein interaction, even relatively weak interactions with binding constants of about 10^{-3} M provide sufficient macroligand-binding strength with binding constant being $10^{-3} \times 10^{-3} = 10^{-6}$ M, a value fully sufficient for the successive affinity precipitation. This fact allows exploiting relatively weak binding pairs as sugar–lectin for the efficient affinity precipitation of tetrameric lectin, concanavalin A using pH-sensitive conjugate of Eudragit S 100 with *p*-aminophenyl-N-acetyl-D-galactosamine (Linné-Larsson and Mattiasson, 1996) or *p*-aminophenyl-α-D-glucopyranoside (Linné-Larsson and Mattiasson, 1994).

The affinity precipitation is designed to be applied at the first stages of purification protocol dealing with crude unprocessed extracts. Hence, the ligands used should be robust to withstand the action of both harmful components present in the crude extracts and precipitating/eluting agents. Triazine dyes, which are robust affinity ligands for many nucleotide dependent enzymes, were successfully used in conjugates with Eudragit S to purify dehydrogenases from various sources by affinity precipitation (Guoqiang *et al.*, 1993, 1994a and b, 1995b; Shu *et al.*, 1994). Sugar ligands constitute another attractive alternative for being used in bioseparation of lectins (Hoshino *et al.*, 1998; Linné-Larsson and Mattiasson 1994, 1996) and ligands with immobilized metal ions and efficient for affinity precipitation of proteins having histidine residues at the surface (Galaev *et al.*, 1997, 1999; Kumar *et al.*, 1999, 1998a and b; Mattiasson *et al.*, 1998). *p*-Aminobenzamidine ligands were used for affinity precipitation of proteolytic enzymes (Pécs *et al.*, 1991; Nguyen and Luong, 1989).

More complex ligands and even protein ligands were used in affinity precipitation as well. Poly(NIPAM) conjugate with $(dT)_8$ was used for a model separation of a complementary oligonucleotide $(dA)_8$ from a mixture of $(dA)_8$ and $(dA)_3(dT)(dA)_4$. Affinity precipitation resulted in precipitation of 84% of $(dA)_8$ while 92% of $(dA)_3(dT)(dA)_4$ remained in solution (Umeno *et al.*, 1998). Restriction endonuclease Hind III was isolated using poly(NIPAM conjugate with λ phageDNA (Maeda *et al.*, 1993) and C-reactive protein was isolated using poly(NIPAM) conjugate with *p*-aminophenylphosphorylcholine (Mori *et al.*, 1994).

Protein ligands provide good selectivity and high binding strength needed for affinity precipitation. Provided protein ligands are stable enough under conditions of polymer precipitation and elution, they could be used in affinity precipitation. Such proteins include soybean trypsin inhibitor (Chen and Jang, 1995; Galaev and Mattiasson, 1992; Kumar and Gupta, 1994), immunoglobulin G (Kamihira

Table 3.3 Published protocols of protein isolation using affinity precipitation

Target protein	Polymer	Ligand	Method used to precipitate polymer	Protein recovery, %	Purification fold	Reference
Wheat germ agglutinin	Chitosan	*	ΔpH	55–70	10	(Senstad and Mattiasson, 1989b)
Lysozyme, lectins specific for N-acetylglucosamine	Chitosan	*	ΔpH	**	**	(Tyagi et al., 1996)
Trypsin	Chitosan	Soybean trypsin inhibitor	ΔpH	93	5.5	(Senstad and Mattiasson, 1989a)
Trypsin	Copolymer NIPAM and N-acrylhydroxysuccinimide	p-amino-benzamidine	ΔT	74–82	**	(Nguyen and Luong, 1989)
Trypsin	Alginite	Soybean trypsin inhibitor	addition of Ca^{2+}	30–36	**	(Linné et al.,1992)
Trypsin	Poly(N-vinyl caprolactam)	Soybean trypsin inhibitor	ΔT	54	28	(Galaev and Mattiasson, 1992)
Protein A	Hydroxypropyl-methylcellulose succinate	IgG	ΔpH	50–70 / 53–90	4.9–5.4 / 32	(Taniguchi et al., 1989, 1990)
Human IgG	Galactomannan	Protein A	addition of tetraborate	**	**	(Bradsaw and Sturgeon, 1990)
Alkaline protease	Poly-NIPAM	p-amino-benzamidine	ΔT	**	**	(Pécs et al., 1991)
β-Glucosidase	Chitosan	*	ΔpH	**	**	(Homma et al., 1993)
Restriction endonuclease Hind III	Poly-NIPAM	Phage λ DNA	ΔT	**	**	(Maeda et al., 1993)]
Xylanase from Trichoderma viridie	Eudragit S 100	*	ΔpH	70	4.6	(Gupta et al., 1994)
D-Lactate dehydrogenase from Lactobacillus bulgaricus	Eudragit S 100	*	ΔpH	85	2.4	(Guoqiang et al., 1993)
Endopoly-galacturonase from Aspergillus niger	Eudragit S 100	*	ΔpH	86	8.8	(Gupta et al., 1993)

Target protein	Polymer	Ligand	Method	Yield (%)	Purification factor	Reference
Trypsin	Eudragit S 100	Soybean trypsin inhibitor	ΔpH	74	1.8	(Kumar and Gupta, 1994)
Concanavalin A	Eudragit S 100	p-aminophenyl-α-D-glucopyranoside	ΔpH	83–91	**	(Linné-Larsson and Mattiasson, 1994)
D-Lactate dehydrogenase from *Leuconostoc mesenteroides* ssp. *cremoris*	Eudragit S 100	Cibacron Blue	ΔpH	56	4.3	(Shu et al., 1994)
Lactate dehydrogenase, and	Eudragit S 100	Cibacron Blue	ΔT	63	7.8	(Guoqiang et al., 1994b)
pyruvate kinase from porcine muscle				59	4.4	
Alcohol dehydrogenase from bakers yeast	Eudragit S 100	Cibacron Blue in the presence of Zn^{2+}	ΔT	40–42	8.2–9.5	(Guoqiang et al., 1995b)
β-Glucosidase from *Trichoderma longibrachiatum*	I. chitosan II. Eudragit S 100	*	ΔpH	82–86	10–14	(Agarwal and Gupta, 1996)
α-Amylase inhibitor from wheat	Copolymer of NIPAM and 1-vinyl imidazole	Immobilized Cu ions	Addition of salt	89	4	(Kumar et al., 1998a)
α-Amylase inhibitors I-1 and I-2 from seeds of ragi (Indian finger millet)	Copolymer of NIPAM and 1-vinyl imidazole	Immobilized Cu ions	Addition of salt	84 (I-1)	13 (I-1)	(Kumar et al., 1998b)
				89 (I-2)	4(I-2)	
Monoclonal antibodies	Polyelectrolyte complex formed by poly-(methacrylic acid) and poly(N-ethyl-4-vinylpyridinium bromide)	Antigen (inactivated glyceraldehyde-3-phosphate dehydrogenase)	ΔpH	96–98	1.4–1.8	(Dainiak et al., 1999)
Lactate dehydrogenase, from porcine muscle	Carboxymethyl cellulose	Cibacron Blue	Addition of Ca^{2+} and PEG	87	21	(Lali et al., 1999)
α-glucosidase from *Saccharomyces cerevisiae*	Poly(N-acryloylpiperidine)	Maltose	ΔT	54–68	170–200	(Hoshino et al., 1998)

* target protein has the affinity for the polymer itself.

** not presented.

et al., 1992; Mattiasson and Kaul, 1994), protein antigen (for binding monoclonal antibodies (Dainiak *et al.*, 1999)).

The selectivity of affinity precipitation as a protein purification technique depends on the selectivity of the ligands used as well as on the non-specific interactions of proteins with the polymer when the precipitate is formed. Precipitation of non-charged thermosensitive polymers is less complicated with non-specific co-precipitation of proteins as compared to pH-sensitive polymers where ionic interactions can contribute to the unwanted polymer–protein interactions. In the worst case, a nearly quantitative non-specific co-precipitation of proteins took place (Kumar *et al.*, 1994). When polyelectrolyte complexes precipitate, the charges of one polyanion are compensated by the charges of the oppositely charged polyanion. Strong cooperative interactions of oppositely charged polyions resulted in efficient displacement of non-specifically bound proteins, the amount of non-specifically bound protein does not exceed a few percents of that present in solution (Dainiak *et al.*, 1999). On the other hand, the affinity of proteins to non-modified polymer could be harnessed for purification of the latter. Endo-polygalacturonase was purified by co-precipitation with alginate (Gupta *et al.*, 1993); xylanase from *Trichoderma viride* – by co-precipitation with Eudragit S 100 (Gupta *et al.*, 1994); β-glucosidase from *Trichoderma longibrachiatum* – by sequential precipitation of chitosan (to remove co-precipitated cellulase activity) and co-preciptiation with Eudragit S 100 (Agarwal and Gupta, 1996); lysozyme and lectins specific for N-acetly glucosamine (e.g. lectins from rice, tomato, and potato) – by co-precipitation with chitosan (Tyagi *et al.*, 1996).

Table 3.3 summarizes published protocols of protein isolation using affinity purification technique.

3.6. Practical Affinity Precipitation

A general affinity precipitation protocol is presented in Figure 3.8. With a proper choice of affinity ligand and precipitation conditions, affinity precipitation procedure allows to capture about 90% of the target protein (Kumar *et al.*, 1998b; Guoqiang *et al.*, 1993; Senstad and Mattiasson, 1989b). An important issue in affinity precipitation protocol is co-precipitation of proteins nonspecifically bound to the polymer via electrostatic and/or hydrophobic interactions (Guoqiang *et al.*, 1995a). To reduce these interactions, some additives like detergents and salts are added to the buffer during precipitation or washing steps (Guoqiang *et al.*, 1993). The formation of weak complexes between the polymer and a low-molecular weight compound present in solution could also reduce the non-specific protein–polymer interactions and hence improve the purification efficiency. We have found that Tris (tris(hydroxymethyl)aminomethane) forms weak complexes with Eudragit S 100, probably due to the hydrogen bonding of hydroxy groups of Tris molecules with carboxy groups of the polymer and electrostatic interactions of protonated amino groups of Tris with carboxyl anions. These weak interactions decrease significantly non-specific co-precipitation of proteins by Eudragit S 100 when Tris-buffer is used as compared to the use of phosphate buffer (Arasaratnam *et al.*, 2000). Polyelectrolyte complexes present attractive carriers for pH-dependent affinity precipitation as the non-specific electrostatic interactions of the proteins with the polymer are relatively small. Oppositely charged polymers bind to each other more efficiently than proteins bind to the polymers due to the higher charge density of the polymer partners as compared to that of proteins. The non-specifically

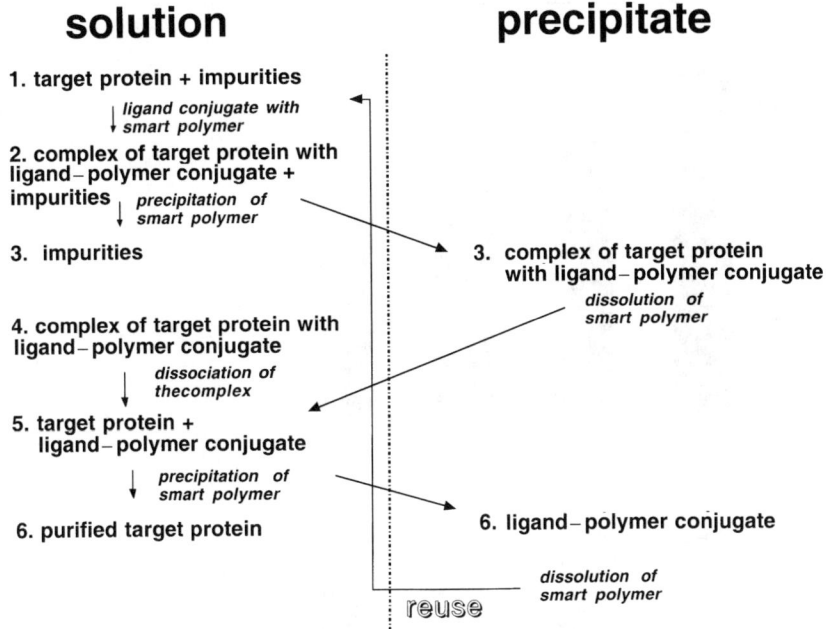

Figure 3.8 Affinity precipitation procedure.

bound proteins are essentially 'pushed out' by the polymer from the complex (Dainiak *et al.*, 1999).

Elution procedure is usually carried out by resolubilization of precipitated and washed polymer–protein complex. Then some agent which causes the dissociation of the polymer–protein complex is added. For example, addition of EDTA, which binds Cu(II) ions, destroys the complex poly(VI/NIPAM)-Cu-protein (Kumar *et al.*, 1998a and b). The polymer could be precipitated again with the purified target protein remaining in solution. Alternatively, the target protein could be eluted directly from the precipitate e.g. by increased salt concentrations (Gupta *et al.*, 1993; Dainiak *et al.*, 1999). Sequential increase in salt concentration of the eluent allows specific elution of two enzymes bound to the conjugate of Eudragit S 100 with Cibacron blue (Guoqiang *et al.*, 1994b). The affinity precipitation procedure is usually characterized by reasonably high yields around 70–90% and purification folds up to 100, depending on the content of target protein in the crude extract used (Table 3.3).

The polymer–ligand conjugate could be collected with 90–95% recovery after the protein purification and reused for the purification of the next portion of the protein. As a rule, performance of the macroligand deteriorate somewhat in the following cycles (Senstad and Mattiasson, 1989b; Kumar *et al.*, 1999; Dainiak *et al.*, 1999). For example, precipitation of wheat germ agglutinin from the crude extract decreased from the 98% in the first cycle to 71% in the third cycle. On the contrary, the recovery of bound agglutinin in the elution step improved from 71% in the first cycle to 103% in the third cycle (Senstad and Mattiasson, 1989b). An improvement in protein recovery in the second cycle was also noticed during protein A purification using conjugate of human IgG with pH-sensitive polymer, hydroxypropylmethylcellulose

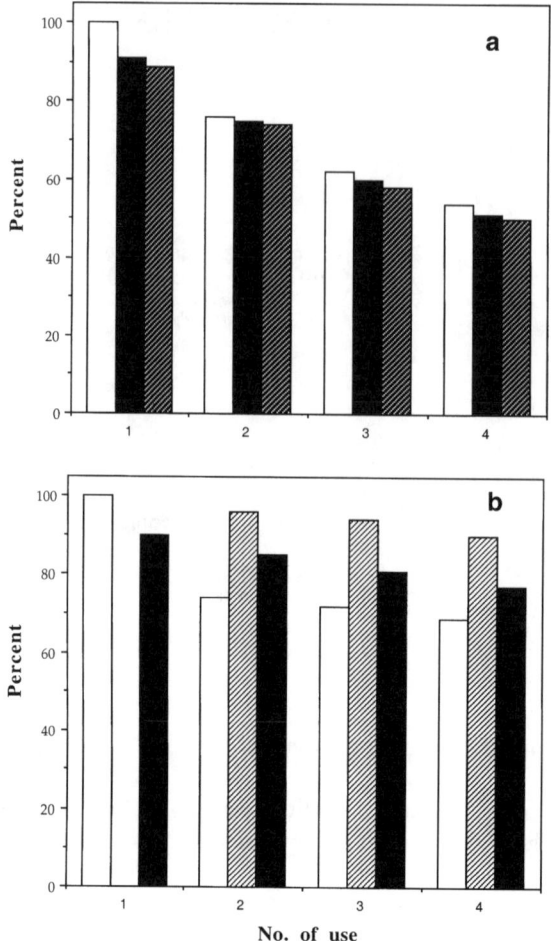

Figure 3.9 Precipitation and recovery of α-amylase inhibitor from wheat meal with Cu(II)-
poly(VI-NIPAM) after each recycle without (a) and with (b) reloading the
polymer with the same amount of Cu(II) lost in each recycle use. Cu(II) concentration on the polymer initially and after each recycle (open bars) and after
reloading (slightly dashed bars); precipitation of α-amylase inhibitor (closed
bars); recovery of α-amylase inhibitor (densely dashed bars).

acetate succinate (Taniguchi *et al.*, 1989). The deterioration of the efficiency of affinity precipitation with increasing cycle numbers when Cu(II)-loaded poly(VI-NIPAM)
was used (Figure 3.9) happened to be due to the stripping of Cu(II) ions from the
polymer in elution step. Recharging the polymer with Cu(II) ions after every reuse
improved significantly the performance of the system (Kumar *et al.*, 1999).

3.7. Conclusion

Affinity precipitation of proteins using smart polymers emerged in early eighties and
since then it matured to a technique capable of simple, fast and efficient purification of a

variety of proteins. To be successful affinity precipitation requires a robust ligand with affinity for a target protein, the ligand should be coupled to a polymer backbone which renders the conjugate a property of reversible solubility in response to small changes of pH, ionic strength or temperature. Commercial pH-sensitive polymers like different Eudragits are available. Thermosensitive polymers based on N-isopropylacrylamide could be easily synthesized in aqueous solutions following essentially the standard procedure for the production of electrophoretic polyacrylamide. Thus, protein purification using affinity precipitation constitutes a potentially very attractive mode of operation when designing protein purification protocols.

3.8. Abbreviations

NIPAM	N-isopropylaceylamide
PEG	poly(ethyleneglycol)
VCL	N-vinylcaprolactam
VI	1-vinyl imidazole

3.9. References

Agarwal, R. and Gupta, M.N. (1996) Sequential precipitation with reversibly soluble polymers as a bioseparation strategy: Purification of β-glucosidase from *Trichoderma longibrachiatum*, *Protein Expression Purification*, 7, 294–298.

Alred, P.A., Kozlowsi, A., Harris J.M., and Tjerneld F. (1994) Application of temperature-induced phase partitioning at ambient temperature for enzyme purification, *J. Chromatogr. A*, 659, 289–298.

Arasaratnam, V., Galaev, I.Y., and Mattiasson, B. (1999) Reversibly soluble biocatalyst: Optimization of trypsin coupling to Eudragit S-100 and biocatalyst activity in soluble and precipitated forms, *Enzyme and Microb. Technol.*, 27, 254–263.

Bradshaw, A.P. and Sturgeon, R.J. (1990) The synthesis of soluble polymer–ligand complexes for affinity precipitation studies, *Biotechnol. Techniq.*, 4, 67–71.

Charles, M., Coughlin, R.W., and Hasselberger, F.X. (1974) Soluble-insoluble enzyme catalysts, *Biotechnol. Bioeng.*, 16, 1553–1556.

Chen, J.-P. and Jang, F.-L. (1995) Purification of trypsin of affinity precipitation combining with aqueous two-phase extraction, *Biotechnol. Techniq.*, 9, 461–466.

Chern, C.S., Lee, C.K., Chen, C.Y., and Yeh, M.J. (1996) Characterization of pH-sensitive polymeric supports for selective precipitation of proteins, *Colloids and Surfaces B: Biointerfaces*, 6, 37–49.

Dainiak, M.B., Izumrudov, V.A., Muronetz, V.I., Galaev, I.Yu., and Mattiasson, B. (1999) Affinity precipitation of monoclonal antibodies by nonstoichiometric polyelectrolyte complexes, *Bioseparation*, 7, 231–240.

Dainiak, M.B., Izumrudov, V.A., Muronetz, V.I., Galaev, I.Yu., and Mattiasson, B. (2000) Production of Fab fragments of monoclonal antibodies using polyelectrolyte complexes, *Anal. Biochem.*, 277, 58–66.

Dissing, U. and Mattiasson, B. (1996) Polyelectrolyte complexes as vehicles for affinity precipitation of proteins, *J. Biotechnol.*, 52, 1–10.

Franko, T.T., Galaev, I.Yu., Hatti-Kaul, R., Holmberg, N., Bülow, L., and Mattiasson, B. (1997) Aqueous two-phase system formed by thermoreactive vinyl imidazole/vinyl caprolactam copolymer and dextran for partitioning of a protein with a polyhistidine tail, *Biotechnol. Techniq.*, 11, 231–235.

Galaev, I.Yu., Kumar, A., Agarwal, R., Gupta, M.N., and Mattiasson, B. (1997) Imidazole – a new ligand for metal affinity precipitation. Precipitation of Kunits soybean trypsin inhibitor using Cu(II)-loaded copolymers of 1-vinylimidazole with N-vinylcaprolactam and N-isopropylacrylamide, *Appl. Biochem. Biotechnol.*, **68**, 121–133.

Galaev, I.Yu., Gupta, M.N., and Mattiasson, B. (1996) Use smart polymers for bioseparations, *CHEMTECH*, #12, 19–25.

Galaev, I.Yu., Kumar, A., and Mattiasson, B. (1999) Metal-copolymer complexes of N-isopropylacrylamide for affinity precipitation of proteins, *J. Macromol. Sci.*, **36**, 1093–1105.

Galaev, I.Yu. and Mattiasson, B. (1992) Affinity thermoprecipitation of trypsin using soybean trypsin inhibitor conjugated with a thermo-reactive polymer, poly(N-vinyl caprolactam), *Biotechnol. Techniq.*, **6**, 353–358.

Galaev, I.Yu. and Mattiasson, B. (1993a) Affinity thermoprecipitation: Contribution of the efficiency of ligand–protein interaction and access of the ligand, *Biotechnol. Bioeng.*, **41**, 1101–1106.

Galaev, I.Yu. and Mattiasson, B. (1993b) Thermoreactive water-soluble polymers, nonionic surfactants, and hydrogels as reagents in biotechnology, *Enzyme Microb. Technol.*, **15**, 354–366.

Galaev, I.Yu. and Mattiasson, B. (1997) New methods for affinity purification of proteins. Affinity precipitation: A review, *Biochemistry (Moscow)*, **62**, 571–577.

Gisser, K.R.C., Geselbracht, M.J., Capellari, A., Hunsberger, L., Ellis, A.B., Perepezko, J., and Lisensky, G.C. (1994) Nickel-titanium memory metal. A 'smart' material exhibiting a solid-state phase change and superelasticity, *J. Chem. Educ.*, **71**, 334–340.

Gold, D.H. and Gregor, H.P. (1960) Metal–polyelectrolyte complexes. VIII. The poly-N-vinylimidazole-copper(II) complex, *J. Phys. Chem.*, **64**, 1464–1467.

Guoqiang, D., Kaul, R., and Mattiasson, B. (1993) Purification of *Lactobacillus bulgaricus* D-lactate dehydrogenase by precipitation with an anionic polymer, *Bioseparation*, **3**, 333–341.

Guoqiang, D., Kaul, R., and Mattiasson, B. (1994a) Integration of aqueous two-phase extraction and affinity precipitation for the purification of lactate dehydrogenase, *J. Chromatogr. A*, **668**, 145–152.

Guoqiang, D., Lali, A., Kaul, R., and Mattiasson, B. (1994b) Affinity thermoprecipitation of lactate dehydrogenase and pyruvate kinase from porcine muscle using Eudragit bound Cibacron blue, *J. Biotechnol.*, **37**, 23–31.

Guoqiang, D., Batru, R., Kaul, R., Gupta, M.N., and Mattiasson, B. (1995a) Alternative modes of precipitation of Eudragit S 100: A potential ligand carrier for affinity precipitation of protein, *Bioseparation*, **5**, 339–350.

Guoqiang, D., Benhura, M.A.N., Kaul, R., and Mattiasson, B. (1995b) Affinity precipitation of yeast alcohol dehydrogenase through metal ion promoted binding with Eudragit bound Cibacron blue 3 GA, *Biotechnol. Progress*, **11**, 187–193.

Gupta, M.N., Guoqiang, D., Kaul, R., and Mattiasson, B. (1994) Purification of xylanase from *Trichoderma viride* by precipitation with anionic polymer Eudragit S 100, *Biotechnol. Techniq.*, **8**, 117–122.

Gupta, M.N., Guoqiang, D., and Mattiasson, B. (1993) Purification of endo-polygalacturonase by affinity precipitation using alginate, *Biotechnol. Appl. Biochem.*, **18**, 321–327.

Gupta, M.N. and Mattiasson, B. (1994) Affinity precipitation. In Street, G., (ed.), *Highly Selective Separations in Biotechnology*, Blackie Academic & Professional, London, pp. 7–33.

Harris, P.A., Karlström, G., and Tjerneld, F. (1991) Enzyme purification using temperature-induced phase formation, *Bioseparation*, **2**, 237–246.

Homma, T., Fujii, M., Mori, J., Kawakami, T., Kuroda, K., and Taniguchi, M. (1993) Production of cellobiose by enzymatic hydrolysis: Removal of β-glycosidase from cellulase by affinity precipitation using chitosan, *Biotechnol. Bioeng.*, **41**, 405–410.

Hoshino, K., Taniguchi, M., Kitao, T., Morohashi, S., and Sasakura, T. (1998) Prepration of a new thermo-responsive adsorbent with maltose as a ligand and its application to affinity precipitation, *Biotechnol. Bioeng.*, **60**, 568–579.

Izumrudov, V.A., Galaev, I.Yu., and Mattiasson, B. (1999) Polycomplexes – potential for bioseparation, *Bioseparation*, 7, 207–220.

Kamihira, M., Kaul, R., and Mattiasson, B. (1992) Purification of recombinant Protein A by aqueous two-phase extraction integrated with affinity precipitation, *Biotechnol. Bioeng.*, 40, 1381–1387.

Kaul, R. and Mattiasson, B. (1992) Secondary purification, *Bioseparation*, 3, 1–26.

Kitano, S., Hisamitsu, I., Koyama, Y., Kataoka, K., Okano, T., Yokoyama, M., and Sakurai, Y. (1993) Preparation of glucose-responsive polymer complex system having phenylboronic acid moiety and its application to insulin-releasing device. In Takagi, T., Takahashi, K., Aizawa, M. and Miyata, S. (eds.), *Proceedings of the First International Conference on Intelligent Materials*, Technomic Publishing Co., Inc., Lancaster, pp. 383–388.

Kokufuta, E. and Matsukawa, S. (1995) Enzymatically induced reversible gel–sol transition of a synthetic polymer system, *Macromolecules*, 28, 3474–3475.

Kondo, A. and Fukuda, H. (1997) Preparation of thermo-sensitive magnetic hydrogel microspheres and application to enzyme immobilization, *J. Fermentation Bioeng.*, 84, 337–341.

Kondo, A., Imura, K., Nakama, K., and Higashitani, K. (1994) Preparation of immobilized papain using thermosensitive latex particles, *J. Fermentation Bioeng.*, 78, 241–245.

Kumar, A., Agarwal, R., Batra, R., and Gupta, M.N. (1994), Effect of polymer concentration on recovery of the target proteins in precipitation methods, *Biotechnol. Techniq.*, 8, 651–654.

Kumar, A., Galaev, I.Yu., and Mattiasson, B. (1998a) Affinity precipitation of alpha-amylase inhibitor from wheat meal by metal chelate affinity binding using Cu(II)-loaded copolymers of 1-vinylimidazole with N-isopropylacrylamide, *Biotechnol. Bioeng.*, 59, 695–704.

Kumar, A., Galaev, I.Yu., and Mattiasson, B. (1998b) Isolation of α-amylase inhibitors I-1 and I-2 from seeds of ragi (Indian finger millet, *Elusine coracana*) by metal chelate affinity precipitation, *Bioseparation*, 7, 129–136.

Kumar, A., Galaev, I.Yu., and Mattiasson, B. (1999) Metal chelate affinity precipitation: A new approach to protein purification, *Bioseparation*, 7, 185–194.

Kumar, A. and Gupta, M.N. (1994) Affinity precipitation of trypsin with soybean trypsin inhibitor linked Eudragit S-100, *J. Biotechnol.*, 37, 185–189.

Lali, A., Balan, S., John, R., and D'Souza, F. (1999) Carboxymethyl cellulose as a new heterobifunctional ligand carrier for affinity precipitation of proteins, *Bioseparation*, 7, 195–205.

Larsson, P.-O., Flygare, S., and Mosbach, K. (1984) Affinity precipitation of dehydrogenases, *Methods Enzymol.*, 104, 364–369.

Larsson, P.-O. and Mosbach, K. (1979) Affinity precipitation of enzymes, *FEBS Letters*, 98, 333–338.

Lee, S.J. and Park K. (1996) Glucose-sensitive phase-reversible hydrogels. In Ottenbrite, R.M., Huang, S.J., and Park, K. (eds.), *Hydrogels and Biodegradable polymers for Bioapplications*, American Chemical Society, Washington, DC, pp. 2–10.

Lee, S.J. and Park, K. (1997) Synthesis and characterization of sol-gel phase reversible hydrogels sensitive to glucose, *J. Molec. Recogn.*, 9, 549–557.

Linné, E., Garg, N., Kaul, R., and Mattiasson, B. (1992) Evaluation of alginate as a carrier in affinity precipitation, *Biotechnol. Appl. Biochem.*, 16, 48–56.

Linné-Larsson, E. and Mattiasson, B. (1994) Isolation of concanavalin A by affinity precipitation, *Biotechnol. Techniq.*, 8, 51–56.

Linné-Larsson, E., Galaev, I.Yu., Lindahl, L., and Mattiasson, B. (1990) Affinity precipitation of concanavalin A with *p* amino-α-D-glycopyranoside modified Eudragit S-100. I. Initial complex formation and build-up of the precipitate, *Bioseparation*, 6, 273–282.

Linné-Larsson, E. and Mattiasson, B. (1996) Evaluation of affinity precipitation and a traditional affinity chromatography for purification of soybean lectin, from extracts of soya flour, *J. Biotechnol.*, 49, 189–199.

Liu, F., Liu, F.H., Zhou, R.X., Peng, Y., Deng, Y.Z., and Zeng, Y. (1995) Development of a polymer-enzyme immunoassay method and its application, *Biotechnol. Appl. Biochem.*, **21**, 257–264.

Liu, K.-J. and Gregor, H.P. (1965) Metal–polyelectrolyte complexes. X. Poly-N-vinylimidazole complexes with zinc (II) and with copper (II) and nitrilotriacetic acid, *J. Phys. Chem.*, **69**, 1252–1259.

Lu, M., Albertsson, P.-Å., Johansson, G., and Tjerneld, F. (1996) Ucon-benzoyl dextran aqueous two-phase systems: Protein purification with phase component recycling, *J. Chromatogr. B*, **680**, 65–70.

Maeda, M., Nishimura, C., Inenaga, A., and Takagi, M. (1993) Modification of DNA with poly(N-isopropylacrylamide) for thermally induced affinity precipitation, *Reactive Functional Polymers*, **21**, 27–35.

Mattiasson, B. and Kaul, R. (1993) Affinity precipitation. In Ngo, T. (ed.), *Molecular Interactions in Bioseparations*, Plenum Press, New York, pp. 469–477.

Mattiasson, B. and Kaul, R. (1994) 'One-pot' protein purification by process integration *BIO/TECHNOLOGY*, **12**, 1087–1089.

Mattiasson, B., Kumar, A., and Galaev, I.Yu. (1998) Affinity precipitation of proteins: Design criteria for an efficient polymer, *J. Molec. Recogn.*, **11**, 211–216.

Mori, S., Nakata, Y., and Endo, H. (1994) Purification of rabbit C-reactive protein by affinity precipitation with thermosensitive polymer, *Protein Expression Purific.*, **5**, 151–156.

Nguyen, A.L. and Luong, J.H.T. (1989) Syntheses and application of water-soluble reactive polymers for purification and immobilization of biomolecules, *Biotechnol. Bioeng.*, **34**, 1186–1190.

Okubo, M. and Ahmad, H. (1998) Enzymatic activity of trypsin adsorbed on temperature-sensitive composite polymer particles, *J. Polym. Sci: Part A: Polym. Chem.*, **36**, 883–888.

Pécs, M., Eggert, M., and Schügerl, K. (1991) Affinity precipitation of extracellular microbial enzymes, *J. Biotechnol.*, **21**, 137–142.

Porath, J. and Olin, B. (1983) Immobilized metal ion affinity adsorption and immobilized metal ion affinity chromatography of biomaterials. Serum protein affinities for gel-immobilized iron and nickel ions, *Biochemistry*, **22**, 1621–1630.

Röhm Pharma GMBH (1993) *Information Materials on Eudragit.*

Scopes, R.K. (1994) *Protein Purification: Principles and Practice*, Springer-Verlag, New York.

Senstad, C. and Mattiasson, B. (1989a) Affinity precipitation using chitosan as a ligand carrier, *Biotechnol. Bioeng.*, **33**, 216–220.

Senstad, C. and Mattiasson, B. (1989b) Purification of wheat germ agglutinin using affinity flocculation with chitosan and a subsequent centrifugation or flotation step, *Biotechnol. Bioeng.*, **34**, 387–393.

Shiroya, T., Tamura, N., Yasui, M., Fujimoto, K., and Kawaguchi, H. (1995a) Enzyme immobilization on thermosensitive hydrogel microspheres, *Colloids Surfaces B: Biointerfaces*, **4**, 267–274.

Shiroya, T., Yasui, M., Fujimoto, K., and Kawaguchi, H. (1995b) Control of enzymatic activity using thermosensitive polymers, *Colloids Surfaces B: Biointerfaces*, **4**, 275–285.

Shu, H.-C., Guoqiang, D., Kaul, R., and Mattiasson, B. (1994) Purification of the D-lactate dehydrogenase from *Leuconostoc mesenteroides ssp. cremoris* using a sequential precipitation procedure, *J. Biotechnol.*, **34**, 1–11.

Sun, Y., Yu, K., Jin, X.H., and Zhou, X.Z. (1995) Polymerized liposome as a ligand carrier for affinity precipitation of proteins, *Biotechnol. Bioeng.*, **47**, 20–25.

Taniguchi, M., Kobayashi, M., Natsui, K., and Fujii, M. (1989) Purification of staphylococcal protein A by affinity precipitation using a reversibly soluble–insoluble polymer with human IgG as a ligand, *J. Fermentation Bioeng.*, **68**, 32–36.

Taniguchi, M., Tanahashi, S., and Fujii, M. (1990) Purification of staphylococcal protein A by affinity precipitation: Dissociation of protein A from the adsorbent with chemical reagents, *J. Fermentation Bioeng.*, **69**, 362–364.

Todd, R.J., Johnson, R.D., and Arnold, F. (1994) Multiple-site binding interactions in metal-affinity chromatography. I. Equilibrium binding of engineered histidine-containing cytochromes c, *J. Chromatogr.*, **662**, 13–26.

Tyagi, R., Kumar, A., Sardar, M., Kumar, S., and Gupta, M.N. (1996) Chitosan as an affinity macroligand for precipitation of N-acetyl glucosamine binding proteins/enzymes, *Isolation & Purification*, **2**, 217–226.

Umeno, D., Mori, T., and Maeda, M. (1998) Single stranded DNA-poly(N-isopropylacrylamide) conjugate for affinity precipitation separation of oligonucleotides, *Chem. Commun.*, 1433–1434.

Wu, K.-Y.A. and Wisecarver, K.D. (1992) Cell immobilization using PVA crosslinked with boric acid, *Biotechnol. Bioeng.*, **39**, 447–449.

4 Aqueous two-phase systems with smart polymers

Hans-Olof Johansson, Mårten Svensson,
Josefine Persson and Folke Tjerneld

4.1. Introduction

The use of smart polymers in separation systems is attracting increasing interest. One reason for this is the recent advances in the physico-chemical characterization of the solution behavior of amphiphilic copolymers. The knowledge of how these polymers behave opens the way to applications. In biotechnology there is a strong interest in use of 'smart polymers' in separation systems. The need here is for polymers which can react on external influence, such as temperature or pH change. With such polymers it is possible from the outside to affect the properties of a separation system. The interest has been directed towards amphiphilic copolymers and polyampholytes. As we show in this chapter, the amphiphilic copolymers show drastic changes in solubility properties, such as self-association and phase separation at, e.g., temperature increase. With polyampholytes the solubility can be affected by changing the solution pH. These characteristic properties of copolymers in solution can be utilized for external control of separations.

An important separation technique in biotechnology is aqueous two-phase extraction. The two-phase systems are based on the phase separation in water solution of two polymers with different structure, and these systems are used for extraction of sensitive biomolecules or particles, e.g. proteins or membranes. In this chapter we show studies where amphiphilic block copolymers have been used as phase forming polymers in two-phase systems. The possibility is hereby created for micelle formation in one of the phases. The micelle formation depends on the temperature and the partitioning of substances between the phases can thus be regulated by, e.g., the temperature.

The random (statistical) copolymers of EO and PO units have been used in aqueous two-phase separations. The main interest in this application has been to utilize temperatures above the copolymer cloud point for separation of the copolymer from the purified biomolecule and to utilize these effects for copolymer recycling. The phase formation by temperature increase offers the possibility to create separation systems with only one polymer in water. We describe studies of these novel systems, where the partitioning of molecules between a water phase and a polymer phase formed above the copolymer cloud point is utilized for separation. Similar principles are applied in systems with thermoprecipitating copolymers; e.g. vinyl or acrylamide based copolymers. For these polymers the temperature or pH is used to trigger the precipitation. If the target biomolecule can be bound to the copolymer, e.g. by affinity interaction, it is then possible to precipitate the biomolecule-copolymer complex by temperature or pH change, which leads to effective separation of biomolecule from contaminants.

The rapid expansion of modern biotechnology creates new demands on effective separations. The amphiphilic copolymer systems are very attractive because of the multitude of possibilities for design of polymers with capability for capture and release of the target biomolecule and which also can be reused in the process. Furthermore these polymers are water-soluble, mild and not denaturing towards biomolecules. They can be derivatized, e.g. with charged groups for introduction of pH sensitivity or with affinity ligands for the creation of specific binding to the copolymer. As we show in this chapter the amphiphilic copolymers have a range of properties which can be utilized in polymer-based separation systems.

4.2. Thermoseparating Polymers in Aqueous Two-phase Systems

Thermoseparating polymers (TSP) separate from water solutions above a certain temperature, the cloud point temperature or the lower critical solution temperature (LCST) (Saeki *et al.*, 1976). These polymers have been used either in thermosep-arated aqueous two-phase systems (see Section 4.2.1.) or in segregating polymer-polymer two-phase systems (Section 4.2.2.). Structurally the TSPs are a diverse group including poly N-isopropylacrylamid (poly-NIPAM) (Schild, 1992), polyvinyl-caprolactam (poly-VCL) (Galaev and Mattiasson, 1993), cellulose ethers such as ethyl(hydroxyethyl)cellulose (EHEC) (Thuresson *et al.*, 1995), ethylene-propylene random copolymers (EOPO copolymers) (Johansson *et al.*, 1993, 1996) and EOPO-block copolymers (Samii *et al.*, 1991; Zhang and Khan, 1995). The number of ther-moseparating polymers can also be extended by copolymerization with other types of monomers or by having aliphatic tails grafted on the backbone or at the polymer ends. Examples are the copolymers of NIPAM-VI (vinylimidazole) and VCL-VI (Persson *et al.*, 2000a) and hydrophobically modified (HM) polymers such as HM-EHEC (Thuresson *et al.*, 1995) and HM-EOPO (Thuresson *et al.*, 1999; Bagger-Jörgensen *et al.*, 1997; Johansson *et al.*, 1999). The alteration of TSPs by copolymerization with other monomer types results in new separation properties such as lower cloud points, pH or ion dependency of separation temperature. An extensive list of thermoseparating polymers is given in a review by Galaev and Mattiasson (1992).

Depending on the type, the thermoseparated polymer may separate from the solution as a solid precipitate, a gel or a liquid phase (Galaev and Mattiasson, 1992). The polymers, which separate as a liquid phase, are used preferentially, since in these cases the phases are better defined and more easily handled. Examples of polymers with this property are ethylene oxide-propylene oxide random copolymers.

Three different models have been proposed for explanation of the phenomenon of thermoseparation. These models explain the thermoseparation for EO-containing polymers by either breakdown of water structures around the EO-segment (Kjellander and Florin, 1981) or hydrogen bonds between water and EO-segment (Goldstein, 1984) or by increased average non-polarity (hydrophobicity) due to conformational changes of the EO segments (Karlström, 1985). This latter model has been applied for explanation of phase behavior of ternary mixtures of thermoseparating polymer and partitioning behavior of biomolecules in these systems (Johansson *et al.*, 1993, 1995 and 1997a and b).

4.2.1. One-polymer Systems

Different factors affect the critical solution temperature for thermoseparating polymer systems. In binary systems these are the ratio of polar/non-polar groups of the polymer molecule, molecular weight and solvent nature. The higher the propylene oxide (PO) content, the lower the cloud point. PEG, which contains 100% ethylene oxide (EO) has a cloud point above 100 °C (Saeki *et al.*, 1976) while EO50PO50, EO30PO70 and EO20PO80 (the numbers indicate weight % of the group in the composition of the polymer) have the cloud points of 50, 40 and 30 °C respectively, see Figure 4.1 (Persson *et al.*, 1999d). Increasing molecular weight decreases the LCST although the LCST seems to reach a limiting value for very large molecular weights of the polymer (Saeki *et al.*, 1976). Thermoseparation is also possible in non-aqueous formamide solutions. However, since the difference in polarity between polymer and formamide is smaller than between polymer and water, thermoseparation occurs at higher temperature in formamide solutions (Samii *et al.*, 1991).

The two-phase formation starts at the cloud point. However, two macroscopic phases are formed quickly (within one hour) only if the temperature of the system is at least a few degrees above the cloud point. For a two-component polymer-water system the bottom phase is usually polymer enriched, where the polymer concentration ranges from 40% to 80% for the random EOPO copolymers (Figure 4.1), 20%–25% for EOPO block copolymers (Samii *et al.*, 1991; Zhang and Khan, 1995) and 4%–12% for HM-EOPO (Johansson *et al.*, 1999). The top phases in these thermoseparated systems usually contain almost 100% water (see Figure 4.1). The phase diagrams of thermoseparating cellulose ethers, Poly-NIPAM and poly-VCL are only partially investigated due to the very high viscosity of these polymers (Persson *et al.*, 2000a).

In ternary or more complex systems additives as given for the case of ethylene oxide based polymeric systems below affect the thermoseparation:

(1) Hydrophilic additives decrease the cloud point of the system and the additive partitions strongly to the water phase, e.g. salt (Ananthapadmanabhan and Goddard, 1987), glycine (Johansson *et al.*, 1997a) and sugars (Sjöberg *et al.*, 1989).

(2) Hydrophobic additives decrease the cloud point and partition more to the polymer-rich phase, e.g., butyric acid, phenol (Johansson *et al.*, 1993, 1997b, respectively) and n-butanol (Louai *et al.*, 1991a and b).

(3) The cloud point is almost unchanged. The additive has an almost even partitioning between the phases. In this case the additive has an intermediate chemical character

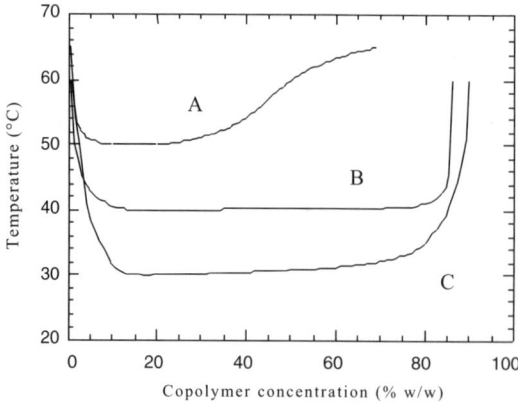

Figure 4.1 Phase diagram of thermoseparating EOPO random copolymers. Binodals: A EO50PO50, B EO30PO70, C EO20PO80. Data from Persson *et al.*, 1999d.

in a hydrophilic-hydrophobic scale, e.g., acetic acid (Johansson *et al.*, 1993), and ethanol (Louai *et al.*, 1991a).

(4) The cloud point is increased or the system may loose the thermoseparating property. This effect can be obtained if the additive is a good solvent for the polymer and is highly concentrated in the system (Johansson *et al.*, 1993) or if the additive is strongly amphiphilic, e.g., sodium valerate (Johansson *et al.*, 1997a) and SDS (Carlsson *et al.*, 1988).

4.2.1.1. *Random EOPO Copolymers*

Random EOPO copolymers have several advantages compared to other thermoseparating polymers namely: they are inexpensive, inert, they have low viscosity and form well defined liquid phases. The low viscosity of the concentrated polymer allows a fast dissolution of the polymer in water. A typical one polymer aqueous two-phase system contains 10–20 weight % polymer. By heating the system above the cloud point (typically ten degrees above), a two-phase system is quickly formed, where the bottom phase is polymer enriched (between 40 and 80% polymer, see Figure 4.1) and the top phase contains almost pure water. The difference in hydrophobicity between the phases is relatively large and hydrophobic cosolutes are easily separated from hydrophilic cosolutes.

One disadvantage of the random copolymers is that they cannot be used for protein extraction in one-polymer two-phase systems. Most studied proteins are totally excluded from the polymer-rich phase (Harris *et al.*, 1991; Alred *et al.*, 1992, 1994; Berggren *et al.*, 1995; Johansson *et al.*, 1996). The non-ideal partial molar mixing entropy of the cosolutes is lower in the polymer-rich phase than in the water rich phase. This is due to the relatively low number of molecules per volume unit in the polymer-rich phase (Johansson *et al.*, 1998). This difference in non-ideal, partial, molar-mixing entropy between the phases corresponds to a driving force, which tends to drive the cosolutes to partition to the water phase. One could say that the polymer-rich phase exerts an 'entropic repulsion' against all cosolutes.

Ucon 50 HB-5100 (denoted as Ucon) which has the composition EO50PO50 and molecular weight of 4000, has been much studied in applications of aqueous two-phase systems (Harris *et al.*, 1991; Alred *et al.*, 1993, 1994). The system has been used to separate amino acids, oligopeptides and polypeptides (Johansson *et al.*, 1995, 1997a). The partition coefficient of amino acids in Ucon/water two-phase systems has been studied, from which a hydrophobicity scale can be constructed, where tryptophan partitions more to the polymeric phase and is thus the most hydrophobic amino acid. Hydrophilic amino acids such as lysine and glycine are strongly excluded from the polymer-rich phase. The Ucon/water two-phase system can therefore be used to purify tryptophan from other amino acids or other hydrophilic cosolutes. Oligopeptides have also been separated according to differences in hydrophobicity (Johansson *et al.*, 1997b). For homopeptides the preference for the water phase or the polymer-rich phase becomes more pronounced with increasing degree of polymerization of the peptide, which can be seen in Figure 4.2.

If the partitioning is performed at a higher temperature the polymer-rich phase becomes more concentrated with polymer which leads to an increased entropic repulsion. This effect can be exemplified by tryptophan, which has a K value of 0.64 at 60 °C, and 3.3 at 100 °C, in a thermoseparated Ucon/water two-phase system

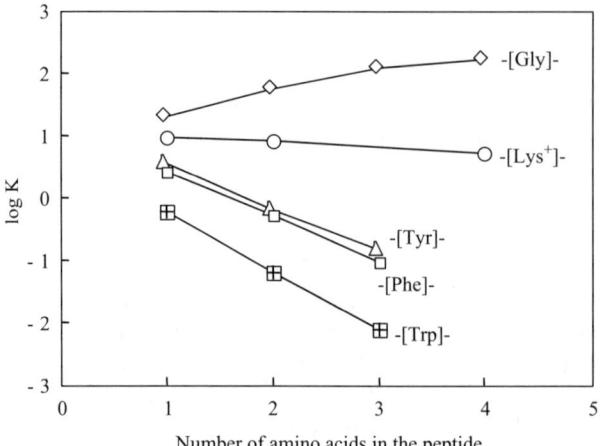

Figure 4.2 Partition coefficients of oligopeptides and amino acids in a Ucon/water two-phase system as function of the number of amino acids in the peptide. System composition: 20% Ucon 50 HB-5100, 1 mg/ml oligopeptide. 100 mM NaClO$_4$, and 10 mM Na-phosphate buffer pH 6. Temperature 60 °C. Data from Johansson *et al.*, 1997b.

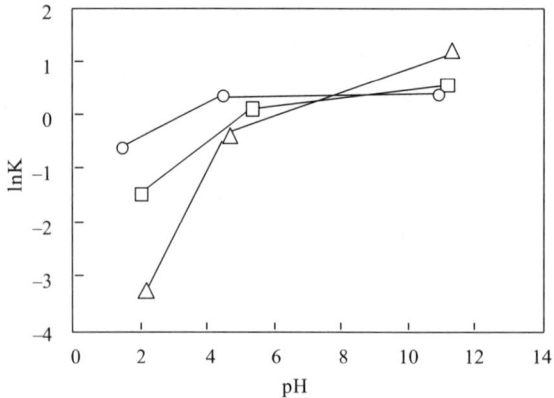

Figure 4.3 Partition coefficients of tryptophan in Ucon/water two-phase system at different pH and addition of salts. System composition: 20% Ucon 50 HB-5100, 2 mM tryptophan, salts: △ 100 mM NaClO$_4$, ☐ 100 mM NaCl, ○ 50 mM Na$_2$SO$_4$. Temperature 60 °C. Data from Johansson *et al.*, 1995.

(Johansson *et al.*, 1997b). However, if the cosolute is sufficiently hydrophobic, the partitioning of the cosolute to the polymer-rich phase may increase upon temperature increase. This effect has been observed for polytryptophan in the Ucon/water two-phase system (Johansson *et al.*, 1997b).

The partitioning of a charged cosolute to the EOPO-rich phase is facilitated by counter ions, which have relatively low hydrophilicity. These effects are exemplified in Figure 4.3 where the partitioning of positively charged tryptophan (pH 2) to the polymer-rich phase was enhanced when its counter ion was exchanged from SO$_4^{2-}$ to ClO$_4^-$ (Johansson *et al.*, 1995). In another experiment the addition of ionic surfactants such as dodecylsulphate (SDS) and cetyltrimethylammonium (CTAB) lead to dramatically enhanced partitioning of positively and negatively charged tryptophan, respectively to the polymer-rich phase (Johansson *et al.*, 1995). Strong partitioning to either the polymer-rich or the water-rich phase can be obtained for amphiphilic polyelectrolytes upon changing the salt composition in thermoseparated one-polymer systems. A positively charged amphiphilic polypeptide composed of randomly distributed lysinyl and tryptophanyl residues was quantitatively transferred to the polymer-rich or the water phase with the addition of $NaClO_4$ and Na_2SO_4, respectively (Johansson *et al.*, 1997b).

4.2.1.2. *Hydrophobically Modified EOPO Copolymer*

Hydrophobically modified (HM) random EOPO copolymers differ structurally from ordinary random EOPO copolymers by having aliphatic tails at both ends of the polymer molecule (Bagger-Jörgensen *et al.*, 1997; Johansson *et al.*, 1999). From this point of view they resemble more block copolymers of the pluronic type. The physicochemical properties of HM-copolymers differ significantly from ordinary random EOPO copolymers. One newly studied HM-EOPO copolymer (Johansson *et al.*, 1999; Persson *et al.*, 1999b, 2000a,b) where the EOPO chain is end-capped with myristyl groups ($C_{14}H_{29}$) and where the EOPO chain consists predominantly of EO groups is shown in Figure 4.4. This polymer is denoted HM-EOPO. The LCST for HM-EOPO is only 14 °C and a phase diagram is shown in Figure 4.5. The most striking

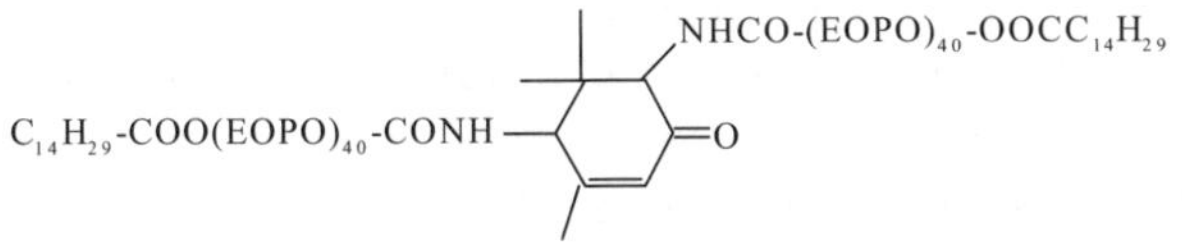

$$C_{14}H_{29}\text{-COO(EOPO)}_{40}\text{-CONH} \quad\quad NHCO\text{-(EOPO)}_{40}\text{-OOCC}_{14}H_{29}$$

Figure 4.4 Structure of hydrophobically modified random ethylene oxide propylene copolymer: HM-EOPO. From Johansson *et al.*, 1999.

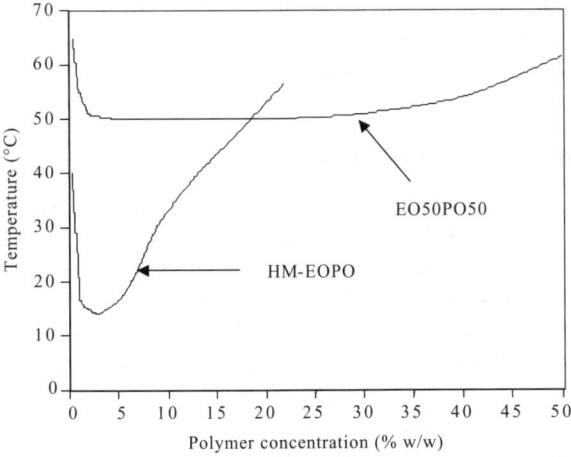

Figure 4.5 Cloud point curves of HM-EOPO and EO50PO50 random copolymer (Breox PAG50A 1000) in water. From Persson *et al.*, 1999b.

difference between HM-EOPO and the random EOPO copolymers is the low LCST of the HM-polymer and the low polymer concentration in the thermoseparated polymer-rich phase (7–10%). It is this property that makes it useful for protein separation and a purification process has been developed, described below. Another unusual property is the strong tendency to segregate from other polymers in solutions (Persson et al., 2000a). This is discussed in Section, 4.2.2.4. The HM-EOPO is considerably more viscous than random EOPO copolymers but is nevertheless easily handled. There are several indications that the HM-EOPO self-associates and forms micellar structures in water solutions, for instance the relative high viscosity for such a small polymer (Mw = 8000) and a CMC at 12 µM (0.01%) (Johansson et al., 1999).

Recombinant human apolipo protein A-1 has been purified from an E. coli extract in a water/HM-EOPO system (Johansson et al., 1999). In this process E. coli extract containing apolipo protein A-1 and HM-EOPO was thermoseparated at 30 °C. Practically all A-1 protein was partitioned to the polymer-rich phase with a good purification factor. The polymer-rich phase was then mixed with a NaClO$_4$ solution and re-thermoseparated at 50 °C. In this back extraction step most of the A-1 was partitioned to the water phase and thus completely separated from the polymer. The extreme partitioning of the apolipo protein A-1 to the HM-EOPO rich phase in the extraction step at 30 °C, can be explained by the affinity of A-1 protein to the aliphatic micellar structures in HM-EOPO solutions.

4.2.2. Two-polymer Systems

Aqueous two-phase-systems-based polymer-polymer segregation have been well studied and used for biomolecule partitioning (Albertsson, 1986). Most of these systems are mixtures of polymers based on EO and PO groups and polysaccharides, notably dextran and starch derivatives. The utility of these systems is due to the mild non-denaturing phase conditions where even living cells can be partitioned (Albertsson, 1986). In these systems however, the entropic and enthalpic differences of the phases are small due to the high concentration of water in both phases (typically 90% in each phase). Because of this, no extreme partitioning can be obtained for small molecules even in the case of relatively high hydrophobicity of the cosolute (Bringmann et al., 1994). Another disadvantage of these systems is the relatively high cost of the polysaccharides, for which no efficient recycling process is known.

4.2.2.1. Random EOPO Copolymer/Polysaccharide Systems

The random EOPO copolymers have been used in aqueous two-phase systems for separation of proteins (Alred et al., 1992, 1994), lactic acid (Planas et al., 1997) and ectysteroids (Alred et al., 1993). The first type of system studied contained often dextran in the two-phase system, e.g. random EOPO copolymer/dextran (Alred et al., 1992, 1994; Berggren et al., 1995). Recently more work has been done to develop low cost starch derivatives for replacement of dextran, e.g. hydroxypropyl starch-containing systems (Persson et al., 1998, 1999a). In these types of systems the EOPO copolymer is enriched in the top phase and the polysaccharide is enriched in the bottom phase. The top phase is more hydrophobic than the bottom phase.

Hydrophobic target proteins can be strongly partitioned to the top phase without special additives (Persson et al., 1998). Less hydrophobic but highly charged proteins

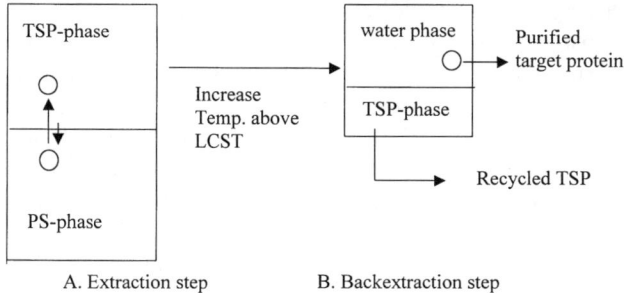

A. Extraction step   B. Backextraction step

Figure 4.6 A purification process with an aqueous two-phase system where one of the polymers is a thermoseparating polymer (TSP). A. Extraction step: a target protein partitions to the TSP (due to hydrophobicity or affinity partitioning). B. The TSP-rich top phase is collected and thermoseparated by heating 10 °C above the Lower Critical Solution Temperature (LCST). A new two-phase system is formed where the target protein is separated from the polymer (back extraction).

can be partitioned to the EOPO-rich phase by addition of hydrophobic counterions such as $(C_2H_5)_3NH^+$, I^-, SCN^- and ClO_4^- (Berggren *et al.*, 1995; Johansson *et al.*, 1996). The separation of the target protein from the EOPO-copolymer is accomplished by heating the top phase above the cloud point. This will induce the formation of a new two-phase system where the new top phase consists mainly of water and almost no polymer and the new bottom phase is strongly enriched with polymer. The target protein is completely excluded from the polymer-rich phase. Thus the thermoseparation property in this system is used to separate polymer from protein and to recover the polymer. This principle is shown in Figure 4.6.

4.2.2.2. *Affinity Ligands Coupled to Thermoseparating Polymers in Two-polymer Systems*

In order to improve the partitioning of a target protein to one of the phases in an aqueous two-phase system, affinity ligands to the target protein have been covalently attached to one of the polymers. Affinity partitioning in aqueous two-phase systems is extensively discussed in the book by Walter and Johansson (1994). Most studies have been done in EO-based polymers and dextran systems. Nguyen and Luong (1989) studied affinity partitioning in copolymer/dextran or copolymer/pullulan systems. They synthesized a copolymer from N-isopropylacrylamide (NIPAM) and glycidyl acrylate. To this polymer a strong trypsin inhibitor was coupled, *p*-benzamidine. Trypsin was strongly partitioned to the phase containing the ligand-polymer. Recovery of the polymer was achieved by dissociating the trypsin ligand-polymer complex at low pH and adding ammonium sulfate, which precipitated the polymer but not the trypsin. A new way to recover the ligand-polymer has been examined by Alred *et al.* (1992), by coupling the ligand to a thermoseparating EOPO random copolymer. The ligand was a triazine dye, procion Yellow HE-3G. The ligand has two reactive chlorine atoms where the end groups of the random copolymer Ucon were added. Before the reaction the hydroxyl end group of Ucon was modified to an amino group in order to improve the reaction. The partition coefficient of glucose-6-phosphate dehydrogenase (G6PD) in a Ucon/dextran system, increased from 0.065 to 12 when ligand polymer was added (0.4% of total system). This system was exploited for purification of the enzyme from yeast homogenate. In a subsequent separation the enzyme was separated from the ligand-polymer. In the first step the enzyme was separated from the bulk proteins in a Ucon/dextran two-phase system (the extraction step in Figure 4.6). In the second step 0.2 M of Na_2SO_4 and 0.2 M NaCl were added in order to decrease the cloud point of the polymer and to dissociate the enzyme from the Ucon-ligand. The temperature was raised to 40 °C. In this step a new two-phase system which contained a Ucon-rich bottom phase and a water-salt enriched top phase was formed (the back extraction step, Figure 4.6). No protein was found in the Ucon rich bottom phase. The recovery of the enzyme in the top phase and Ucon-ligand in the bottom phase was 79% and 85%, respectively (Alred et al., 1992).

A similar type of purification process was studied by Garg et al. (1994). A triazine dye, Cibacron Blue 3GA, was coupled to a thermoseparating nonionic surfactant, Triton X-114. A three-phase system was formed with Triton, Reppal (a hydroxypropyl starch) and PEG. The top phase contained PEG, the middle phase Triton and the bottom phase contained Reppal. In this system lactate dehydrogenase (LDH) was strongly partitioned to the Triton-rich phase, which also contained the Triton-ligand conjugate. Separation of LDH was performed by thermoseparation of the Triton-rich phase at 45 °C. No salt or cofactor such as NADH was needed for separation of enzyme from ligand-Triton.

4.2.2.3. Block Copolymers in Two-phase Systems

By replacing one of the phase-forming components in an aqueous two-phase system with an associating component, e.g. a block copolymer or a non-ionic surfactant, an additional pseudophase is introduced, namely the interior of the aggregate/micelle. Due to its nature, a micelle should have a favorable effect on the partitioning of hydrophobic biomolecules. This has been verified experimentally on partition of membrane proteins in two-phase systems with thermoseparating non-ionic surfactants (Bordier, 1981; Minuth et al., 1995; Liu et al., 1996). The partitioning of biomaterials in an aqueous two-phase system with Pluronic P105/hydroxypropyl cellulose (Klucel L, molecular weight 100 000) was recently investigated by Skuse et al. (1992). Pluronic P105 is a block copolymer with the composition $(EO)_{37}(PO)_{56}(EO)_{37}$ (subscripts denote number of units). Physical characterizations (phase diagrams, interfacial tension, phase viscosities, and phase-separation times) as well as partitioning of cells (*Salmonella enteriditis* 3b and human erythrocytes) and proteins (glucose-6-phosphate dehydrogenase (G6PD)) were performed. The binodal was determined and three systems were chosen for further investigation (see Figure 4.7). The three systems had the same Pluronic P105 concentration (10 wt%) but different Klucel L concentrations (2, 4, and 6 wt%).

The results from interfacial tension (IFT) measurements revealed that systems 2 and 3 were unsuitable for cell partitioning, due to the high values of IFT, 2.23 and 3.04×10^{-2} mNm^{-1}, respectively, whereas the IFT for system 1 (2.9×10^{-3} mNm^{-1} was in the range typical for cell separations. Approximately 57% of the *Salmonella enteriditis* 3b cells were found in the P105-rich phase and the rest in the Klucel L-rich phase. In the case of human erythrocytes, *ca.* 100% of the cells partitioned to the top phase. The partitioning coefficients (K) of G6PD in systems 1–3 were 1.6, 1.52,

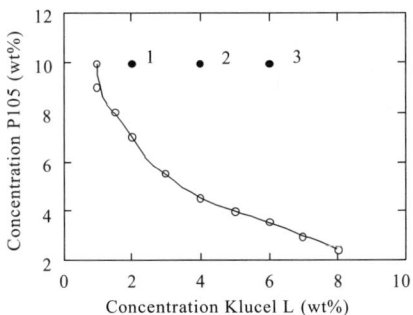

Figure 4.7 Phase diagram for the system Pluronic P105/Klucel L-system at 21 °C. System chosen for partitioning studies are denoted 1, 2 and 3. Data extracted from Skuse *et al.*, 1992.

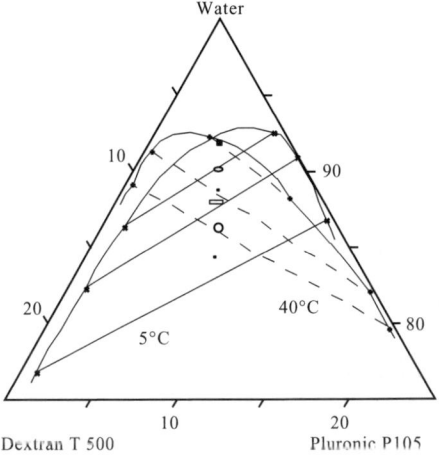

Figure 4.8 Triangular phase diagram for the Pluronic P105/dextran T500/water system at 5 and 40 °C with tie lines. Concentrations in weight percent. The composition of the corners are as follows: top (100% water), bottom left (25% Dextran and 75% water) and bottom right (25% Pluronic P105 and 75% water). From Svensson *et al.*, 1995.

and 1.94, respectively. The high *K* value of system 3 was likely to reflect the higher polymer concentration difference between the phases compared to systems 2 and 3. Similar results have been obtained for partitioning of G6PD in PEO/dextran-system (Johansson *et al.*, 1983).

When Pluronic block copolymers are used as one of the phase forming components the temperature plays an important role in the phase behavior, since the onset of micelle-formation is very temperature dependent. Svensson *et al.* (1995) studied the phase diagrams of three Pluronic copolymers (F68, P105, and L64) together with dextran T500 as the other phase-forming component. They found that the binodals as well as the tielines changed dramatically with temperature (see Figure 4.8). The ternary system is presented in a triangular phase diagram contrary to the rectangular diagrams in Figure 4.7 and Figures 4.10, 4.11 below. The composition of a point in the

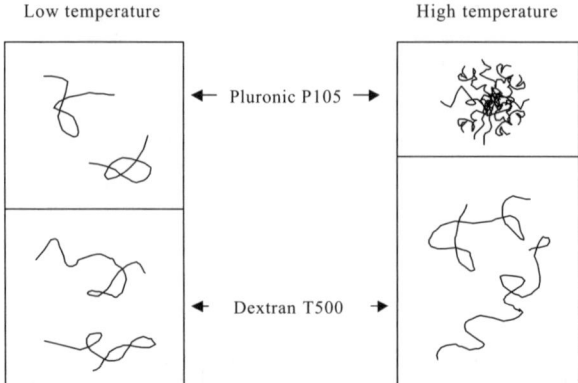

Low temperature   High temperature

Figure 4.9 Illustration of temperature dependent two-phase systems containing micelle-
forming block copolymer.

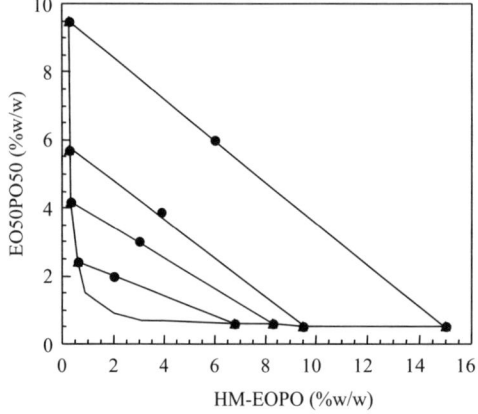

Figure 4.10 Phase diagram for the HM-EOPO/random EOPO system with tie lines.
Temperature 4 °C. Concentrations in weight percent. From Persson et al.,
1999b.

diagram is given by drawing three lines through the point where each line is parallel to
the sides of the triangle. The composition is then read by the intersection of the lines
with the triangular axes. Formally ternary systems should be presented in triangular
diagrams, which can show all possible concentrations of the three components in a
system. However, for systems where the two-phase region of interest is highly water
enriched, it is more convenient to depict the system in rectangular phase diagrams (as
in Figures 4.7, 4.10, 4.11).

When the temperature is elevated, water is transferred from the Pluronic-rich
phase to the dextran-rich phase. This transfer is caused by the temperature-triggered
micellization of the Pluronic molecules. In principal, it is possible at a fix starting-
concentration, to form a two-phase system with block copolymer micelles and
solubilized unimer at high and low temperature, respectively. This situation is illustrated in Figure 4.9. As a consequence the relative difference in hydrophobicity between
the phases increases at higher temperature. Partitioning of hydrophobic amino acids

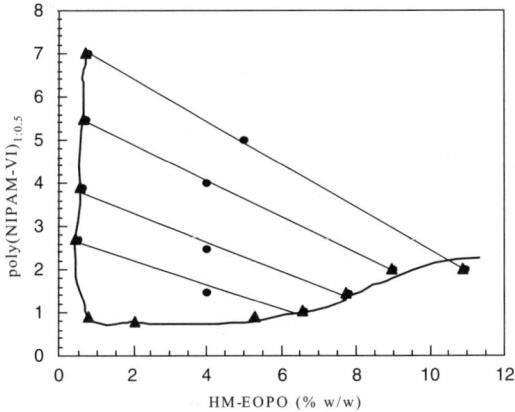

Figure 4.11 Phase diagram for the poly(NIPAM-VI)1 : 0.5/HM-EOPO with tie lines. Temperature 4 °C. Concentrations in weight percent. From Persson *et al.*, 1999c.

and oligopeptides in such a situation has been performed (Svensson *et al.*, 1997). It was found that the partitioning coefficient (K) increased when the temperature was elevated from 5 to 40 °C at a fix point in the P105/dextran T500/water-system. This increase can originate from two effects: (i) the top phase (Pluronic rich) is depleted of water at higher temperatures, causing a more concentrated phase, (ii) micelle-formation in the top phase. Additional work is needed to clarify the potential applications of block copolymers in aqueous two-phase separations. Nevertheless, the use of block copolymers in this technique for bioseparation is promising. The fact that many block copolymers exhibit self-aggregating tendency will introduce more parameters (temperature and composition of the blocks) than ordinary segregating two polymer systems to control the partition behavior.

4.2.2.4. HM-EOPO/Random EOPO Two-polymer System

One of the limiting factors of using large-scale aqueous two-phase systems based on polymer-polymer segregation is the difficulty in efficiently recycling the phase forming polymers. Thermoseparating polymers with low cloud points have been used as replacement for PEG in for instance EO50PO50/polysaccharide systems, where in contrast to the polysaccharide the EOPO polymer can be easily recycled. Recently new systems where all of the phase-forming polymers in the system can be thermoseparated (Persson *et al.*, 1999b, 2000a, b) were developed. An aqueous mixture of HM-EOPO (see Figure 4.4 for structure) and a random EO50PO50 copolymer demix due to effective polymer-polymer repulsion. A phase diagram at 4 °C is shown in Figure 4.10. Interestingly, the pure mixture of these polymers forms a one-phase system. Similar properties have also been observed for aqueous and binary mixtures of PEO and EO50PO50 copolymers (Zhang *et al.*, 1994). However, contrary to the PEO/EO50PO50 system the HM-EOPO/random EOPO segregates at low polymer concentrations. This is probably due to the strong segregating property of the HM-EOPO polymer, which is also observed in other HM-EOPO/polymer mixtures (Persson *et al.*, 2000a).

Recombinant apolipoprotein A-1 was purified from *E. coli* extract in a HM-EOPO/random EOPO two-phase system. Phase inversion was observed with increasing protein concentration. In systems with 20 and 63 mg/ml total protein concentration the top phase was composed of EO50PO50 and HM-EOPO respectively. In the former case the partition coefficients (K-value) of apolipoprotein A-1 and total protein were 0.45 and 2.2 respectively. The separation of apolipoprotein A-1 from a HM-EOPO rich phase was obtained by thermoseparating the polymeric phase at 55 °C, giving a K-value of 2.61. The corresponding K-value for the A-1 protein in the thermoseparated EOPO rich phase was very high (>100). Thus the thermoseparation of protein from polymer is less efficient for a polymer like HM-EOPO as compared to a random EOPO copolymer. However, the recovery of HM-EOPO is more efficient than the recovery of EO50PO50: 97% and 73%, respectively.

4.3. Ionic Smart-polymers in Aqueous Two-phase Systems

Ionic smart polymers are getting more attention as either phase-forming polymers (Hughes and Lowe, 1988; Patrickios *et al.*, 1992; Persson *et al.*, 1999c) or as charged additives (Planas *et al.*, 1998) or as ligand carriers (Guoqiang *et al.*, 1994) in aqueous two-phase systems. Two advantages can be gained by using these polymers: (1) modulation of the coulombic electrostatic attraction and repulsion between polymer and target biomolecule by simple pH changes; (2) polymer recovery by pH induced precipitation of the polymer. It is however not always possible to obtain these two properties in all ionic two-phase systems. The complexity of these systems is large, which explains the difficulty of predicting phase behavior of these systems.

4.3.1. Polyampholytes in Two-phase Systems

A polyampholyte is a polymer, which contains acidic and basic groups. A recent review concerning general solution properties of synthetic polyampholytes is given by Bekturov *et al.* (1990). The polyampholyte contains charged groups at all pH values. However, at the isoelectric point (pI) the net charge of the polyampholyte is zero. At pI the solubility of the polyampholyte is usually very low. This property has been utilized in order to recover the polyampholyte from a water solution by adjusting the pH to the pI of the polyampholyte. The polyampholyte may also contain uncharged groups which makes the polymer less hydrophilic. With a sufficient amount of uncharged groups the polyampholyte may have a low solubility in the whole pH interval. Polyampholytes based on copolymers of polyacrylic derivatives have been studied as phase-forming polymers in aqueous two-phase systems, for separation of proteins (Hughes and Lowe, 1988; Patrickios *et al.*, 1992). Examples of polyacrylic residues are given in Table 4.1.

Hughes and Lowe investigated the partitioning behavior of human serum albumin and trypsin (from pancreatic extract) in two-phase systems, composed of either two different polyampholytes or one polyampholyte and the uncharged polymer polyvinyl alcohol (PVA). The polyampholytes had different pI (see Table 4.2). With two polyampholytes (Amph.1 and Amph.2) with similar net charge a two-phase system could be obtained at pH 6.6, where the more negatively charged Amph. 2 was enriched in the bottom phase. In this system the total protein and positively charged trypsin were partitioned to the bottom phase ($K_{protein} = 0.34$, $K_{trypsin} = 0.05$). Upon addition of

Table 4.1 Monomers of acrylic polymers. Acidic groups are acrylic acid (AA), methacrylic acid (MA) and ethylacrylic acid (EA). Basic groups are aminoethylmethacrylic acid (AEMA) and dimethylaminoethylmethacrylic acid (DMAEMA). Uncharged groups are methylmethacrylic acid (MMA) and hydroxyethylmethacrylic acid (HEMA).

$$\text{Acrylic monomers} \qquad -[CH_2-\underset{\underset{COOR_2}{|}}{\overset{\overset{R_1}{|}}{C}}-]-$$

Name	Abbreviation	Composition	
acrylic acid	AA	$R_1 = H$	$R_2 = H$
methacrylic acid	MA	$R_1 = CH_3$	$R_2 = H$
methylmethacrylic acid	MMA	$R_1 = CH_3$	$R_2 = CH_3$
ethylacrylic acid	EA	$R_1 = C_2H_5$	$R_2 = H$
hydroxyethylmethacrylic acid	HEMA	$R_1 = CH_3$	$R_2 = C_2H_4OH$
aminoethylmethacrylic acid	AEMA	$R_1 = CH_3$	$R_2 = C_2H_4NH_2$
dimethylaminoethylmethacrylic acid	DMAEMA	$R_1 = CH_3$	$R_2 = C_2H_4N(CH_3)_2$

Table 4.2 Acrylic polyampholytes. The chemical structures of the monomers are given in Table 4.1. (1) Data extracted from Hughes and Lowe, 1988. (2) Data extracted from ref. Patrickios *et al.*, 1992)

Polyampholyte	Composition	Monomer ratio	pI	Mw
(1) Amph. 1	AA, EA, HEMA, AEMA, DMAEMA	$2.5:2:0.1:0.1:1$	4.8	44 000
(1) Amph. 2	AA, MA, DMAEMA	$4:1:1$	4.1	64 000
(1) Amph. 3	AA, MA, DMAEMA	$3.5:1:1$	4.4	50 000
(1) Amph. 4	AA, MMA, DMAEMA	$1:1:1$	6.5	not available
(2) Amph. A	AA, MMA, DMAEMA	$0.9:1:1$	6.3	88 500
(2) Amph. B	AA, MMA, DMAEMA	$3.7:1:1$	3.8	43 000

1 M NaCl the protein was partitioned to the top phase ($K_{protein} = 2.4$) and trypsin remained in the bottom phase ($K_{trypsin} = 0.5$).

Two types of PVA/polyampholyte (PVA in the top phase) two-phase systems were also studied at pH 6.0: the PVA/(negatively charged Amph. 2) and the PVA/(positively charged Amph.4) systems. These systems formed one-phase systems for salt concentration lower than 50 mM. In the PVA/Amph.2 system trypsin and protein were partitioned to the bottom phase. The preference for the bottom phase decreased for trypsin and increased for protein with increasing salt concentration. In the PVA/Amph.4 system the proteins were partitioned to the bottom phase while trypsin was almost evenly distributed between the phases. With increasing salt concentration the partitioning of proteins became more even while the trypsin partitioned more to the bottom phase. Effect of pH was also studied. HSA and plasma protein were partitioned in a PVA/Amph.3 system. At pH 5 and 6 HSA and proteins partitioned almost quantitatively to the Amph.3 phase while at higher pH the HSA and the protein partitioned to the PVA rich phase. With the addition of salt (NaCl) at high pH the proteins and HSA were again strongly partitioned to the Amph.3 phase.

In the two types of systems, Amph.1/Amph.2 and PVA/Amph.2, the polyampholytes could be precipitated out of the solution by adjusting the pH to pI. The proteins remained in the solution. However, proteins can be precipitated together with an oppositely charged polyampholyte (Morawetz and Hughes, 1952) This has been explained as an isoelectric complex between protein and polyampholyte (Bekturov et al., 1990; Morawetz and Hughes, 1952).

A more recent study of similar systems has been performed by Patrickios et al. (1992). Data for the synthesized polyampholytes used in their study are given in Table 4.2. They found a very complex phase behavior of the system PVA/Amph.A. At pH < 5.3 they obtained only one-phase systems if the ionic strength was less than 1 M (KCl). At higher pH two-phase systems were obtained if the ionic strength was higher than 0.1 M. At low salinity and pH less than 7.2 the polyampholyte was found in the bottom phase. At pH 7.2 and 0.7 M KCl phase inversion was observed with the polyampholyte in the top phase. Qualitatively similar results to the study of Hughes and Lowe (1988) were observed for the partitioning behavior of proteins. At low salinity (about 0.1 M) positively charged chymotrypsinogen was partitioned to the negatively charged polyampholyte phase, while negatively charged ovalbumin was partitioned to the PVA phase.

Effect of cation on protein partitioning was also studied in PVA/Amph.B system at pH 4.6 where Amph.B is negatively charged. Positively charged chymotrypsinogen and ovalbumin were partitioned to the Amph.B phase in the presence of alkali chloride salts. Changing the type of alkali cation had no significant influence on the partitioning of the proteins. However, changing from monovalent to divalent cation by using $MgCl_2$ or $SrCl_2$ instead of alkali salts raised the K values substantially. In this case ovalbumin was partitioned to the PVA phase.

4.3.2. Ionic Thermoseparating Segregating Polymer Two-phase System

Polymers that have chargeable groups and that exhibit thermoseparation are usually copolymers of thermoseparating and ionic polymers such as poly-NIPAM and poly-(methacrylic acid) (Chen and Hoffman, 1995) and poly-NIPAM and vinylimidazol (denoted as VI) (Persson et al., 2000a). In these examples poly-NIPAM is temperature sensitive and the poly-(methacrylic acid) and vinylimidazole are pH sensitive. Recently, the ionic conjugate of NIPAM-DNA has been studied in an affinity precipitation process (Umeno and Maeda, 1999).

Persson et al. (2000a) studied an aqueous segregating two-polymer-phase system where both polymers exhibit thermoseparation. The phase system was composed of aqueous solution of NIPAM-VI (copolymer) and HM-EOPO (see phase diagram in Figure 4.11). The thermoseparation of NIPAM-VI is dependent on ratio of monomer type, for instance the cloud points for the polymers with NIPAM/VI monomer ratios of 1, 2 and 20 were 80, 60 and 40 °C, respectively at pH 6.5. The pH has also a strong influence on the cloud point. Thus for NIPAM/VI monomer ratios of 2 and 20 no thermoseparation is observed below pH 5. The imidazole group has a pK_a of ca. 6.5. and thus the NIPAM-VI copolymer becomes a cationic polyelectrolyte at low pH.

Lysozyme and bovine serum albumin (BSA) have been partitioned in a NIPAM-VI (molar ratio 2 : 1)/HM-EOPO aqueous two-phase system. At pH 5.4 the partition coefficient of BSA was 5, while at pH 8 it was 0.8. For lysozyme the corresponding partition coefficients at pH 5.4 and 8 were 0.7 and 1.1, respectively. These changes indicate the possibility to direct the partitioning of proteins by charging or neutralizing the polymers. Separation of BSA from NIPAM-VI was achieved by increasing pH of the NIPAM-VI to 8 and then thermoseparating at 50 °C. In this back extraction step a water phase and a polymer-rich bottom phase was formed and all BSA was partitioned in the water-rich top phase.

4.3.3. Ionic Polymers as Molecular Carriers in Aqueous Two-phase System

A usual case of biomolecule separation in aqueous two-phase system is the fact that the target molecule has weak or no preference for either phase. Two strategies to improve the partitioning to a specific phase is to add a molecular carrier in the system, i.e. a molecule that simultaneously has a strong preference for one of the phases and for the target molecule. Affinity ligands coupled to polymers and polyelectrolytes are molecular carriers for target molecules which have an affinity for the ligand or an opposite charge to the polyelectrolyte. One polyelectrolyte used as molecular carrier is polyethyleneimine (PEI). Planas *et al.* (1998) studied the partitioning of lactic acid in EO30PO70/Dextran T500 systems. They found that partitioning of lactic acid to a dextran phase at pH 6.0 could be improved by adding 7.2% PEI. The PEI being positively charged at this pH partitioned to the dextran-rich phase. The K-values of lactic acid without PEI and with PEI were 0.8 and 0.09, respectively. At pH 2.8, however, the K-value of lactic acid was 1.04 and the PEI partitioned also almost evenly at this pH. The more even partitioning of PEI at low pH was probably a consequence of low salt concentration in the system which caused an even polyelectrolyte distribution, which in turn is due to the big entropic penalty of phase compartmentalization of the small counter ions of the polyelectrolyte. With the addition of salt a more extreme partitioning of PEI was obtained.

One advantage of using affinity ligands coupled to ionic polymers is the possibility for an almost complete recovery of the affinity-polymer by pH induced precipitation. Affinity precipitation is discussed elsewhere in this book. Guoqiang *et al.* (1994) developed an ionic molecular carrier by coupling Cibacron blue to Eudragit S 100. The latter is a copolymer of methacrylic acid and methylmethacrylate (Mw $\sim$ 135 000, ratio of carboxyl to ester groups; 1:2). This affinity polymer was added to a PEG8000/Dextran T250 two-phase system (6% PEG and 8% dextran) used for purification of lactate dehydrogenase (LDH) from porcine muscle extract. Without the affinity-polymer the log K-value of LDH was -1. Upon addition of the affinity polymer (0.05%) the log K-value was raised to 1.45. The partitioning of the total protein was practically unaffected by the affinity-polymer and remained in the dextran rich phase (log $K_{protein} = -0.5$). The LDH was subsequently separated from the affinity-polymer by first coprecipitating the affinity-polymer and the LDH by lowering the pH to 5.1, at which Eudragit S 100 becomes insoluble. In the next step the LDH was eluted out by incubation in 0.5 M NaCl, pH 7 at 30 min and then the affinity polymer was reprecipitated by lowering the pH to 5.1 This process gave purification factors from 5–12 and a yield of 46%.

4.4. Conclusion

The use of smart polymers in aqueous two-phase systems has opened the way to more applications of two-phase extractions of biomolecules. One driving force for this development is the possibility to separate target molecule from polymer and thus recover the polymer in the process. Another driving force is the numerous possibilities to modulate the interaction between polymer and target molecule by changing environmental variables such as temperature, pH and ionic strength and specific affinity. Furthermore, composition of the smart polymer can be varied for optimizing the purification process. However, one of the limiting aspects in using smart polymers is the lack of a deeper knowledge of how these polymers interact with biomolecules. This knowledge is essential for selecting or developing a suitable polymer for a purification system.

4.5. Abbreviations

A-1	Apolipoprotein A-1
Amph.	Ampholytic polymer. See compositions in Tables 4.1 and 4.2
EHEC	Ethyl(hydroxyethyl)cellulose
EOPO	random copolymer of ethylene oxide and propylene oxide
EO20POS0	random copolymer of ethylene oxide and propylene oxide composed of 20 wt% ethylene oxide and 80 wt% of propylene oxide. Analogous for EO50PO50 and EO30PO70, respectively
G6PD	glucose-6-phosphate dehydrogenase
HM	Hydrophobically modified
HM-EOPO	Hydrophobically modified random copolymer of ethylene oxide and propylene oxide.
LCST	Lower critical solution temperature
LDH	Lactate dehydrogenase
NIPAM	N-isopropylacryamide
PEI	Poly(ethyleneimine)
PEO	Poly(ethyleneoxide)
PVA	Poly(vinylalcohol)
TSP	Thermoseparating polymer
Ucon	Ucon 50 HB-5100 (an EO50PO50 polymer, Mw 4000)
VCl	N-vinyl caprolactam
VI	1-vinylimidazole

4.6. References

Albertsson, P.-Å. (1986) *Partitioning of Cell Particles and Macromolecules*, 3rd ed., Wiley, New York.

Alred, P.A., Tjerneld, F., Kozlowski, A., and Harris, J.M. (1992) Synthesis of dye conjugates of ethylene oxide-propylene oxide copolymers and application in temperature-induced phase partitioning, *Bioseparation*, 2, 363–373.

Alred, P.A., Modlin, R.F., and Tjerneld, F. (1993) Partitioning of ecdysteroids using temperature-induced phase separation, *J. Chromatogr.*, 628, 205–214.

Alred, P.A., Kozlowski, A., Harris, J.M., and Tjerneld, F. (1994) Application of temperature-induced phase partitioning at ambient temperature for enzyme purification, *J. Chromatogr.*, 659, 289–298.

Ananthapadmanabhan, K.P. and Goddard, E.D. (1987) The relationship between clouding and aqueous biphase formation in polymer solution, *Colloid. Surf.*, 25, 393–396.

Bagger-Jörgensen, H., Coppola, L., Thuresson, K., Olsson, U., and Mortensen, K. (1997) Phase behaviour, microstructure, and dynamics in nonionic microemulsion on addition of hydrophobically end-capped poly(ethylene oxide), *Langmuir*, 13, 4204–4218.

Bekturov, E.A., Kudaibergenov, S.E., and Rafikov, S.R. (1990) Synthetic polymeric ampholytes in solution, *J. Macromol. Sci.*, C30, 233–303.

Berggren, K., Johansson, H.-O., and Tjerneld, F. (1995) Effects of salts and the surface hydrophobicity of proteins on partitioning in aqueous two-phase systems containing thermoseparating ethylene oxide-propylene oxide copolymers, *J. Chromatogr. A*, 718, 67–79.

Bordier, C. (1981) Phase separation of integral membrane proteins in Triton X-114 solution, *J. Biol. Chem.*, 256, 1604–1607.

Bringmann, J., Keil, B., and Pfennig, A. (1994) Partition of dipeptides in aqueous polymer two-phase systems as a function of pH in the presence of salts, *Fluid Phase Equilibria*, 101, 211–225.

Carlsson, A., Karlström, G., Lindman, B., and Stenberg, O. (1988) Interaction between ethyl(hydroxyethyl)cellulose and sodium dodecylsulphate in aqueous solution, *Colloid. Polym. Sci.*, 266, 1031–1036.

Chen, G. and Hoffman, A.S. (1995) Graft copolymers that exhibit temperature-induced phase transitions over a wide range of pH, *Nature*, 373, 49–52.

Galaev, I.Yu. and Mattiasson, B. (1993) Thermoreactive water-soluble polymers, nonionic surfactants, and hydrogels as reagents in biotechnology, *Enzyme Microb. Technol.*, 15, 354–366.

Garg, N., Galaev, I.Yu., and Mattiasson, B. (1994) Use of a temperature-induced phase forming detergent (Triton X-114) as ligand carrier for affinity partitioning in an aqueous three-phase system, *Biotechnol. Appl. Biochem.*, 20, 199–215.

Guoqiang, D., Kaul, R., and Mattiasson, B. (1994) Integration of aqueous two-phase extraction and affinity precipitation for the purification of lactate dehydrogenase, *J. Chromatogr. A*, 668, 145–152.

Goldstein, R.E.J. (1984) On the theory of lower critical solution points in hydrogen-bonded mixtures, *J. Chem. Phys.*, 80, 5340–5341.

Harris, P.A., Karlström, G., and Tjerneld, F. (1991) Enzyme purification using temperature-induced phase formation, *Bioseparation*, 2, 237–246.

Hughes, P. and Lowe, C.R. (1988) Purification of proteins by aqueous two-phase partition in novel acrylic co-polymer systems, *Enzyme Microb. Technol.*, 10, 115–122.

Johansson, G., Kopperschläger, G., and Albertsson, P.-Å. (1983) Affinity partitioning of phosphofructokinase from baker's yeast using polymer-bound Cibacron blue F3G-A, *Eur. J. Biochem.*, 131, 589–594.

Johansson, H.-O., Karlström, G., and Tjerneld, F. (1993) Experimental and theoretical study of phase separation in aqueous solutions of clouding polymers and carboxylic acids, *Macromolecules*, 26, 4478–4483.

Johansson, H.-O., Karlström, G., Mattiasson, B., and Tjerneld, F. (1995) Effects of hydrophobicity and counter ions on the partitioning of amino acids in thermoseparating Ucon-water two-phase systems, *Bioseparation*, 5, 269–279.

Johansson, H.-O., Lundh, G., Karlström, G., and Tjerneld, F. (1996) Effects of ions on partitioning of serum albumin and lysozyme in aqueous two-phase systems containing ethylene oxide/propylene oxide co-polymers, *Biochim. Biophys. Acta*, 1290, 290–298.

Johansson, H.-O., Karlström, G., and Tjerneld, F. (1997a) Effect of solute hydrophobicity on phase behaviour in solutions of thermoseparating polymers, *Colloid Polym. Sci.*, 275, 458–466.

Johansson, H.-O., Karlström, G., and Tjerneld, F. (1997b) Temperature-induced phase partitioning of peptides in water solutions of ethylene oxide and propylene oxide random copolymer, *Biochim. Biophys. Acta*, 1335, 315–325.

Johansson, H.-O., Karlström, G., Tjerneld, F., and Haynes, C.A. (1998) Driving forces for phase separation and partitioning in aqueous two-phase systems, *J. Chromatog.*, 711, 3–17.

Johansson, H.-O., Persson, J., and Tjerneld, F. (1999) Thermoseparating water/polymer system: a novel one polymer aqueous two-phase system for protein purification, *Biotech. Bioeng.*, **66**, 247–257.

Karlström, G. (1985) A new model for upper and lower critical solution temperatures in poly(ethylene oxide) solutions, *J. Phys. Chem.*, **89**, 4962–4964.

Kjellander, R. and Florin, E. (1981) Water structure and changes in thermal stability of the system polyethylene oxide-water, *J. Chem. Soc. Faraday Trans. 1*, **77**, 2053–2077.

Liu, C-L., Nikas, J., and Blankschtein, D. (1996) Novel bioseparations using two-phase aqueous micellar systems, *Biotechnol. Bioeng.*, **52**, 185–192.

Louai, A., Sarazin, D., Pollet, G., François, J., and Moreaux, F. (1991a) Properties of ethylene oxide-propylene oxide statistical copolymers in aqueous solution, *Polymer*, **32**, 703–712.

Louai, A., Sarazin, D., Pollet, G., François, J., and Moreaux, F. (1991b) Effect of additives on solution properties of ethylene oxide-propylene oxide statistical copolymer, *Polymer*, **32**, 713–720.

Minuth, T., Thömmes, J., and Kula, M.-R. (1995) Extraction of cholesterol oxidase from *Nocardia rhodochrous* using a nonionic surfactant-based aqueous two-phase system, *J. Biotechn.*, **38**, 151–164.

Morawetz, H. and Hughes, W.L. (1952) The interaction of proteins with synthetic polyelectrolytes. I. Complexing of bovine serum albumin, *J. Phys. Chem.*, **56**, 64–69.

Nguyen, A.L. and Luong, J.H.T. (1989) Syntheses and applications of water-soluble reactive polymers for purification and immobilization of biomolecules, *Biotech. Bioeng.*, **34**, 1186–1190.

Patrickios, C.S., Abbot, N.L., Foss, R.P., and Hatton, T.A. (1992) Synthetic polyampholytes for protein partitioning in two-phase aqueous polymer systems, *AIChE Symp. Ser.*, **290**, 80–88.

Persson, J., Nyström, L., Ageland, H., Tjerneld, F. (1998) Purification of recombinant apolipoprotein A-1$_{Milano}$ expressed in *E. coli* using aqueous two-phase extraction followed by temperature induced phase separation, *J. Chromatogr.*, **711**, 97–109.

Persson, J., Nyström, L., Ageland, H. and Tjerneld, F. (1999a) Purification of recombinant apolipoprotein A-1$_{Milano}$ using surfactant micelles in aqueous two-phase systems; recycling of thermoseparating polymer and surfactant with temperature induced phase separation, *Biotech. Bioeng.*, **65**, 371–381.

Persson, J., Johansson, H.-O., and Tjerneld, F. (1999b) Purification of protein and recycling of polymers in a new aqueous two-phase system using two thermoseparating polymers, *J. Chromatogr. A*, **864**, 31–48.

Persson, J., Johansson, H.-O., Galaev, I.Yu., Mattiasson, B., and Tjerneld, F. (1999c) Aqueous polymer two-phase systems formed by new thermoseparating polymers, *Bioseparation*, **9**, 105–116.

Persson, J., Kaul, A., and Tjerneld, F. (1999d) Polymer recycling in aquoeous two-phase extractions using thermoseparating ethylene oxide–propylene oxide copolymers, *J. Chromatogr. B*, **743**, 115–126.

Planas, J., Lefebvre, D., Tjerneld, F., and Hahn-Hägerdal, B. (1997) Analysis of phase composition in aqueous two-phase systems using a two-column chromatographic method: Application to lactic acid production by extractive fermentation, *Biotech. Bioeng.*, **54**, 303–311.

Planas, J., Varelas, V., Tjerneld, F., and Hahn-Hägerdal, B. (1998) Amine-based aqueous polymers for the simultaneous titration and extraction of lactic acid in aqueous two-phase systems, *J. Chromatogr. B*, **711**, 265–275.

Saeki, S., Kuwahara, N., Nakata, M., and Kaneko, M. (1976) Upper and lower critical solution temperature in poly (ethylene glycol) solutions, *Polymer*, **17**, 685–689.

Samii, A., Karlström, G., and Lindman, B. (1991) Phase behavior of poly(ethylene oxide)-poly(propylene oxide) block copolymers in nonaqueous solution, *Langmuir*, **7**, 1067–1071.

Schild, H.G. (1992) Poly(N-isopropylacrylamide): Experiment, Theory and Application. *Prog. Polym. Sci.*, **17**, 163–249.

Sjöberg, Å., Karlström, G., and Tjerneld, F. (1989) Effects on the cloud point of aqueous poly(ethylene glycol) solutions upon addition of low molecular weight saccharides, *Macromolecules*, **22**, 4512–4516.

Skuse, D.R., Norris-Jones, R., Yalpani, M., and Brooks, D.E. (1992) Hydroxypropyl cellulose/poly(ethylene glycol)-copol(propylene glycol) aqueous two-phase systems: System characterization and partition of cells and proteins, *Enzyme Microb. Technol.*, **14**, 785–790.

Svensson, M., Linse, P., and Tjerneld, F. (1995) Phase behavior in aqueous two-phase systems containing micelle-forming block copolymers, *Macromolecules*, **28**, 3597–3603.

Svensson, M., Joabsson, F., Linse, P., and Tjerneld, F. (1997) Partitioning of hydrophobic amino acids and oligopeptides in aqueous two-phase system containing self-aggregating block copolymer, *J. Chromatogr. A*, **761**, 91–101.

Thuresson, K., Karlström, G., and Lindman, B. (1995) Phase diagrams of mixtures of a nonionic polymer, hexanol and water. An experimental and theoretical study of the effect of hydrophobic modification, *J. Phys. Chem.*, **99**, 3823–3829.

Thuresson, K., Nilsson, S., Kjøniksen, A.-L., Walderhaug, H., Lindman, B., and Nyström, B. (1999) Dynamics and rheology in aqueous solutions of associating diblock and triblock copolymers of the same type, *J. Phys. Chem. B*, **103**, 1425–1436.

Umeno, D. and Maeda, M. (1999) Temperature-Induced Precipitation of Specific DNA Fragments Using DNA-Poly-(N-isopropylacrylamide) Conjugate, *Chemistry Letters (The Chemical Society of Japan)*, 381–382.

Zhang, K., Carlsson, M., Linse, P., and Lindman, B. (1994) Phase behavior of copolymer-homopolymer mixtures in aqueous solution, *J. Phys. Chem.*, **98**, 2452–2458.

Zhang, K. and Khan, A. (1995) Phase behavior of poly(ethylene oxide)-poly(propylene oxide)-poly(ethylene oxide) triblock copolymers in water, *Macromolecules*, **28**, 3807–3812.

Walter, H. and Johansson, G. (eds.) (1994) Aqueous two-phase systems, *Methods in Enzymology*, **228**, Academic Press, London.

5 Polycomplexes for bioseparation and bioprocessing

Vladimir A. Izumrudov

5.1. Introduction

This chapter discusses the cooperative reactions between oppositely charged partners in dilute aqueous solutions. One of the partners is the most highly charged polyelectrolyte with ionic groups in nearly all monomer units. The counterpart could be either an oppositely charged polyion or a globular protein. The products of these reactions are known as polyelectrolyte complexes (PEC) and protein-polyelectrolyte complexes (PPC), respectively. There is every reason to regard PEC as macromolecular compounds produced as a result of equilibrium reactions with inherent permanent exchange of polyions in water–salt solutions. They combine two properties that might appear at first sight to be mutually exclusive, i.e. rather high stability and lability. The latter manifests itself in the ability to take part in competitive interpolyelectrolyte reactions that could be rather selective. It provides self-assembly of the complex particles in solutions and ensures their behavior as self-adjustment systems. This view on PEC and their properties is shown to be fundamental and can be extended to different complexes, e.g. DNA-containing PEC, PPC, complexes of polyelectrolyte–protein (enzyme, antibody) conjugates and so on. Introduction of bioaffinity ligands endows PEC with the recognition capacity sufficient for the purposes of bioseparation, bioanalysis and bioprocessing. Antibody-PEC conjugates were successfully used in the immunoassay combining the advantages of both homogeneous and heterogeneous assays and for simulating of chaperone action. The unique properties of polyelectrolyte complexes in combination with bioaffinity ligands makes them promising for the development of highly efficient means for protein isolation, new immunoassay procedures and creation of reversibly soluble biocatalysts.

The ability of synthetic and natural polyelectrolytes to interact with globular proteins forming protein-polyelectrolyte complexes (PPC) is well known. Products of these interactions might be water–soluble PPC, coacervates or insoluble PPC. Rather high stability of PPC in water-salt solutions and their pH-sensitivity are intriguing for application of PPC in bioseparation. In particular, if PPC is specifically formed with one of the proteins in the crude extract followed by a phase separation, this process could be used as a rather convenient and straightforward way for isolation and purification of the target protein. Besides, such approach has potential for use in immobilization and stabilization of enzymes.

Most of the work done so far in this field (for reviews see Xia and Dubin, 1994; Shieh and Glatz, 1994; Kokufuta, 1994) deals with the screening of polymers with varying charge density and chain length or with the choice of conditions (pH, ionic strength, concentration of polyelectrolyte, composition of the mixtures and so on) where the most complete protein separation is achieved without denaturation of the protein.

To make PPC an attractive means for protein separation, two additional problems should be solved. The first problem is protein recovery from the PPC and regeneration of the polymer. One of the potential solutions to this problem is the use of reversibly soluble systems based on the better understanding of factors affecting phase state and stability of PPC formed. Some of these factors are discussed in this chapter.

The second problem is the selectivity of the polymer–protein interaction. The higher the discriminating power between the target protein and protein impurities during the complex formation the greater is the potential of PPC for bioseparation. This problem could be addressed by tailor-made polyelectrolytes with some specific interactions introduced by affinity ligands coupled to the polymer, so called macroligands, so that a reversible biospecific water-soluble complex is formed with the desired target protein. Discrimination of the complex from the rest of contamination supposed to be done using distinctive features of the parent polymer, i.e. high molecular weight (biospecific ultrafiltration and biospecific gel filtration), high density of charges (affinophoresis), surface tension properties (affinity partitioning) and reversible solubility (affinity precipitation) (for review see Hubert and Dellacherie, 1994).

A promising approach in this direction is the use of cooperative reactions between oppositely charged polymers giving rise to the formation of polyelectrolyte complexes (PEC). Unique properties of PEC, in particular reversible solubility in water solution controlled by pH and/or ionic strength, can serve as the basis for the separation process (for reviews see Kabanov, 1994a and b).

This chapter describes the main results of PEC studies that allow one to formulate general principles of interpolyelectrolyte reactions and their potential applications in bioseparation and bioprocessing. The reactions between globular proteins and synthetic polyelectrolytes are shown to obey the same principles. These principles could be developed into highly efficient means of protein isolation and purification, new immunoassay procedures and reversibly soluble biocatalysts.

5.2. Polyelectrolyte Complexes and Interpolyelectrolyte Reactions

Polyelectrolytes are polymers that contain many charged groups. The most highly charged polymers with ionic groups in each monomer unit are considered in this chapter. A family of these polyions is presented by numerous synthetic and natural polyanions and polycations that, in turn, are subdivided into those of the 'integral' type (i.e. with the charge sites integrated into the backbone chain) and those of the 'pendant' type (i.e. with the charge sites attached as side-groups or at side chains). Both types of polyions are described in a review of Philipp *et al.* (1989).

PEC are formed as a result of cooperative coupling reactions between two oppositely charged polyions. The degree of conversion in the reaction, θ, is determined as the ratio of equilibrium number of interpolymer salt bonds to the ultimate one. In the discussion that follows the systems with θ close to 1 will be dealt with. If one of the interacting polyelectrolytes is a weak polyacid or a weak polybase, the degree of conversion can be easily controlled by variation of pH. The comparison of experimental dependencies of θ on pH and pH dependencies of ionization degree of the weak polyelectrolyte indicated that a highly charged partner of the PEC could induce charges on a weak acidic or basic partner (for reviews see Kabanov, 1994a and b).

It suggests a pronounced widening of pH region corresponding to the existence of PEC as a product of complete interpolyelectrolyte reaction, $\theta \approx 1$, and hence an increase in the number of polyelectrolyte pairs suitable for application in bioseparation processes as a component of PEC.

PEC combine two main properties that might appear at first sight to be mutually exclusive, i.e. being both rather highly stabile and a labile. The latter property is manifested by the ability of PEC to participate readily in interpolyelectrolyte reactions accompanied by a transfer of polyions from one complex particle to another one.

5.2.1. Stability of PEC

A cooperative character of multisite interpolyion binding makes PEC extremely stable with respect to dissociation. The theoretical treatment of the cooperative interaction in polymer-oligomer systems resulting from hydrogen bonding was described in detail in a review of Papisov and Litmanovich (1989). The following equilibrium constant relationship was proposed:

$$K_n = K_1^n = \exp\left(-n\Delta F_1^0 / RT\right) \tag{1}$$

where K_n is the overall equilibrium constant for the formation of interpolymer complexes composed of a high polymer and an oligomer with n repeating units. K_1 and $\Delta F_1{}^0$ are the equilibrium constant and the free energy change, respectively for the reaction of one repeating unit.

The applicability of such theoretical treatment of cooperative interactions to PEC was demonstrated experimentally and discussed in a review by Tsuchida and Abe (1982). For instance, in the system of poly(methacrylate) polyanion (PMA) and quaternized oligo(ethyleneimine) the relationship between K_n or $-\Delta F^0$ and n below n = 4 was described by the following equations:

$$K_n = A \exp(Bn) \tag{2}$$

$$-\Delta F^0 = \alpha n + \beta \tag{3}$$

where α is comparable with a cooperative coefficient, β is the basic binding constant and A and B are constants corresponding to α and β, respectively.

Thus, the dissociation constant of PEC sharply decreases with increasing oligomer length and reaches practically zero even for relatively short charged oligomers. This 'critical' chain length required to form a stable PEC in aqueous solution slightly varies for different pairs of polyions being n = 4 − 6 (Tsuchida and Abe, 1982; Kabanov, 1994a and b). Therefore, one can assume that after this small 'critical' length of polyions is exceeded, the reactions of complementary polyelectrolytes become apparently irreversible. The oppositely charged polyions with a more broad molecular mass distribution also could be used for bioseparation if PEC are formed by the mixing of salt-free water solutions of them.

Addition of simple salt destroys interpolymer salt bonds in PEC increasing the 'critical' chain length. It was demonstrated for PEC formed by relatively long PMA polyanion and poly(N-ethyl-4-vinylpyridinium) polycations (PEVP) of different degree of polymerization (D.P.) in NaCl solutions (Pergushov et al., 1995). Critical concentration of the salt, [NaCl]*, at which the expulsion of PEVP chains from PEC is observed, increases drastically with the increase of D.P. of the polycation and then changes only slightly (Figure 5.1). Thus, in the presence of low-molecular-weight electrolyte, bioseparation should be accomplished by relatively long charged chains, in particular PEVP with D.P. > 100.

Stability of PEC is dependent on the nature of the salt added. It was shown (Pergushov *et al.*, 1993) that with respect to their ability to induce PEC(PMA-PEVP) dissociation, the ions exhibit the following order: $Br^- > Cl^- > F^-$ and $Li^+ > Na^+ > K^+ \gg (CH_3)_4N^+$. Both the affinity of the halide anions for PEVP polycations and the affinity of these cations for PMA polyanions appear to follow the same orders. Substitution of one halide anion with another resulted in a more

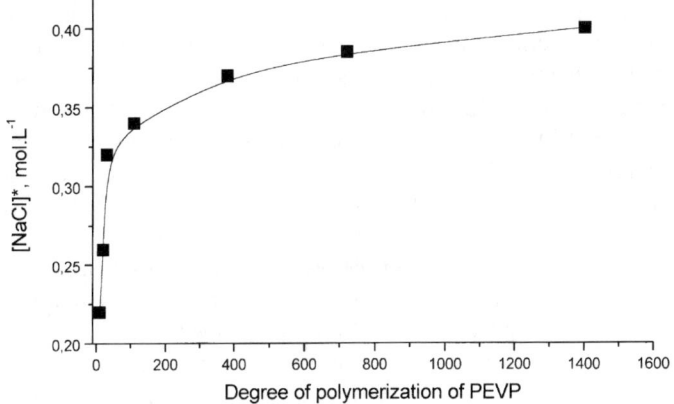

Figure 5.1 Dependence of the critical salt concentration, [NaCl]* corresponding to the expulsion of PEVP from the complex with PMA on the degree of polymerization of the polycation (redrawn from Pergushov *et al.*, 1995).

Table 5.1 Critical salt concentration, corresponding to dissociation of PEC(PMA-PEVP) and of PEC(DNA · EB-PEVP) in solutions of different salts

Salt	Critical concentration, M	
	Dissociation of PEC(PMA-PEVP)*	Dissociation of PEC(DNA · EB-PEVP)**
LiCl	0.41	0.22
NaCl	0.47	0.23
KCl	0.56	0.25
KBr		0.14
KJ		0.07
KF		1.0
$(CH_3)NBr$	0.72	0.18
$(CH_3)NCl$	1.6	0.43
$(CH_3)NF$	>6.0	>2.0
$MgCl_2$		0.18
$CaCl_2$		0.15

* Calculated from the data presented by Pergushov *et al.* (1993); D.P.(PMA) = 2150, D.P.(PEVP) = 390, 20 °C.
** Calculated from the data presented by Izumrudov and Zhiryakova (1999); D.P.(PEVP) = 100, 20 °C.

pronounced change of PEC(PMA-PEVP) stability than the substitution of alkaline metal cations (Table 5.1). Tetramethylammonium salts were used in the experiments because bulky $(CH_3)_4N^+$ cation scarcely affects polyanions and the contribution of $(CH_3)_4N^+$ to the PEC destruction is negligible. So, concentration and nature of added low-molecular-weight electrolyte could be used as a tool to fine tune bioseparation in PEC solutions.

5.2.2. Lability of PEC

In spite of relatively high stability with respect to dissociation, PEC are able to partic-ipate in interpolyelectrolyte reactions. It follows from the experiments with insoluble stoichiometric PEC (SPEC) formed by aromatic ionenes and poly(styrene sulfonate) polyanion (PSS) (Tsuchida *et al.*, 1972). Addition of excess of the polyanion to sus-pension of the SPEC resulted in dissolution of the suspension. It was interpreted as the transfer of ionene chains from SPEC to the added polyanion and the formation of new water-soluble nonstoichiometric polyelectrolyte complexes (NPEC) as a result of uniform distribution of ionenes among all PSS chains.

The discovery of soluble polyelectrolyte complexes marked a serious breakthrough in studies of interpolyelectrolyte reactions. It brought the studies to a qualitatively new level, making it possible to apply modern techniques developed for the investi-gation of macromolecules in solution. The phenomenon, on the other hand, should be taken into account when bioseparation is achieved by PEC, because the formation of water-soluble PEC could impair the bioseparation process. So, the identification of the conditions of NPEC formation is worth consideration.

The complete dissolution of insoluble SPEC(PSS-ionene) was observed in the pres-ence of three-fold excess of PSS charged units relative to charged units of ionene. On the contrary, the addition of excess of the ionene to the solution of sodium PSS did not lead to the formation of water-soluble NPEC. Under these conditions, the insoluble SPEC coexisted with the excess of added ionene remaining in solution. These observations led Tsuchida *et al.* (1972) to the conclusion that polyelectrolytes of 'integral' type, like the ones formed by ionenes, have a special capacity for the formation of soluble NPEC, which is associated with the location of the charged groups in the backbone of the chains. This conclusion was seemingly confirmed by the failure to obtain soluble NPEC from salt-free aqueous solutions of sodium PSS and poly(trimethylbenzylvinylammonium chloride), i.e. from a pair of oppositely charged polyions of 'pendant' type. It was shown for the pair sodium poly(acrylate)-5,6-ionene bromide that aliphatic ionenes are also capable of forming soluble NPEC (Gulyaeva *et al.*, 1976). This finding favored the hypothesis of the particular properties of ionenes as polyions of 'integral' type.

However, it was subsequently reported that soluble NPEC could be obtained, whilst maintaining certain conditions, from pairs of oppositely charged polyions of any type and having a wide variety of chemical structures (Izumrudov and Zezin, 1976; Izumrudov *et al.*, 1978; Kharenko *et al.*, 1979). These data and conditions of NPEC formation are summarized in a review of Kabanov and Zezin (1982).

A simple and universal method of NPEC preparation proved to be a direct mixing of aqueous solutions of oppositely charged polyelectrolytes taken in non-equivalent proportions in the pH range where they are both fully charged and in the presence of small amounts of a low-molecular weight electrolyte.

It is apparent that this way of NPEC preparation suggests formation of PEC of different composition, including insoluble SPEC, as a consequence of inevitable local supersaturation. However, the final product – soluble NPEC – is homogeneous with regards to composition. The study of aqueous salt solutions of various NPEC by light scattering (Kharenko *et al.*, 1979) showed that their characteristics were independent on the way of preparation, i.e. such complexes should be considered as macromolecular compounds. These findings strongly suggest the proceeding of interpolyelectrolyte reactions accompanied by a transfer of polyions from one complex to another one.

Evidently, the exchange of polyions that in the case of PSS-ionene pair takes place even in salt-free solutions, was not due to peculiarities of ionene structure, as it was stated by Tsuchida *et al.* (1972) but was caused by a relatively low charge density of ionene molecules. Poor electrostatic complementary of pair PSS-ionene weakens the polyanion–polycation interaction and accelerates the interpolyelectrolyte exchange followed by formation of soluble NPEC(PSS-ionene). Analogously, if in salt-free aqueous solutions of electrostatic complementary PMA polyanion (D.P. = 3000) and PEVP polycation (D.P. = 100) the transfer of polyions is kinetically restricted and virtually forbidden, then in the system with poor complementary, i.e. PMA (D.P. = 3000) and half-alkylated poly(N-ethyl-4-vinylpyridinium) polycation (D.P. = 100, contains only a half of positively charged pyridinium units in the chain) the equilibrium state corresponding to formation of water-soluble NPEC is reached in minutes (Izumrudov *et al.*, 1986a). Substitution of the above polycations by longer ones, D.P. = 600 led to the same acceleration of interpolyelectrolyte exchange reaction in the case of the half-alkylated polycation followed by a formation of soluble NPEC. The decisive factor of the kinetics is suggested to be the difference in arrangement of the charges (or charge density) of the chains rather than a number of interpolymer salt bonds linking the polyions. Breakdown of the complementary salt bonds could be also achieved by the addition of a rather small amount of low-molecular-weight electrolyte. Thus, the addition of NaCl in the aqueous solution of kinetically 'frozen' mixture of PMA and PEVP accelerates drastically the exchange and 0.05 M NaCl proved to be quite enough for the formation of soluble NPEC(PMA-PEVP) in a time of the mixing (Izumrudov *et al.*, 1984a).

According to modern terminology (Kabanov, 1994a and b), a soluble NPEC particle contains solubilizing host polyelectrolyte (HPE) and oppositely charged guest polyelectrolyte (GPE). The charged units of the HPE are incorporated in the NPEC particle in excess as compared with charged units of GPE. Molar ratio, φ, GPE and HPE units should not exceed a definite critical value, $\varphi = [GPE]/[HPE] < \varphi_{cr}$. The value φ_{cr} for most of the investigated pairs of polyions varies over the range 0.2–0.7 and depends mainly on the chemical structure of the polyelectrolytes constituents. The fragments of HPE and GPE chains, linked by interpolymer salt bonds, form rather hydrophobic double-stranded sequences, which tend to from in the hydrophobic core of complex particles. The hydrophilic periphery of NPEC consists of charged HPE segments, ensuring the solubility of the complex particle as a whole. These segments stick out in solution as loops and/or tails (Figure 5.2).

Negatively charged NPEC of 'A' type with relatively long chains of solubilizing host polyanion (Figure 5.2A) is one of the most extensively studied NPEC (for a reviews see Kabanov, 1994a and b). NPEC with relatively short host polyanions (Figure 5.2B)

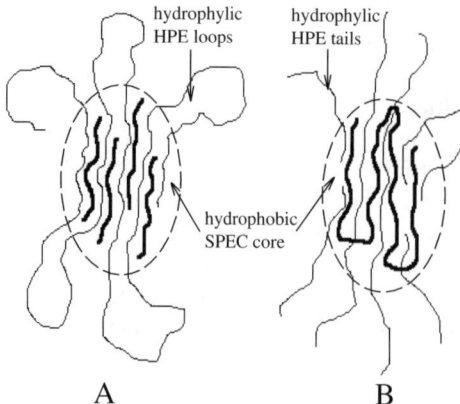

Figure 5.2 Structure of water-soluble NPEC particle with long host polyelectrolyte (HPE) chains and short oppositely charged guest polyelectrolyte (GPE), D.P.(HPE) > D.P.(GPE) (A) and with short HPE and long GPE chains, D.P.(HPE) < D.P.(GPE) (B). D.P.(HPE) and D.P.(GPE) are the degree of polymerization of HPE and GPE, respectively.

were obtained either from the pair PMA-PEVP (Izumrudov *et al.*, 1987a) or the pair poly(phosphate) polyanion (PPh)-PEVP (Izumrudov *et al.*, 1988). Formation of NPEC of 'B' type is achieved with an unfavorable change of entropy of the system due to the immobilization of a large amount of short HPE in NPEC particle. The shorter the HPE chain, the more these chains should be immobilized to reach the critical molar ratio, φ_{cr} of the polyions to provide NPEC solubility and hence, the more is the loss of entropy due to the decrease of the total particle number in solution. The result is a rather poor stability of such NPEC in water–salt solutions. Addition of salt leads to binding of counterions by NPEC, decrease of HPE solubilizing ability and transformation of NPEC into insoluble SPEC by releasing excessive HPE chains from NPEC in solution. The shorter the HPE chains are, the less the critical salt concentration corresponding to the transformation is. The phenomenon caused problems in preparation of such NPEC by direct mixing of water-salt solutions of polyions and even led to the mistaken conclusion (Kasaikin *et al.*, 1979; Zezin *et al.*, 1984) about the impossibility to obtain water-soluble NPEC with D.P.(HPE) < D.P.(GPE).

Positively charged NPEC with host polycation were obtained from PEVP and PMA (Nefedov *et al.*, 1985). The authors managed to prepare the NPEC only after substitution of bromide counterions of PEVP polycation for fluoride ones. Halide anions have a high affinity to PEVP, decreasing in the order $I^- > Br^- > Cl^- > F^-$. The high affinity impairs solubilizing ability of PEVP fluoride and reduces it to a minimum in the case of PEVP bromide. I^- anions bind with the polycation to an extent that long PEVP chains are salted out and precipitated in even relatively dilute KI solution. Most likely, the failure to obtain soluble NPEC(PSS-ionene) in excess amount of ionene (Tsuchida *et al.*, 1972) was due to either relatively low D.P. of ionene chains and/or the use of Br^- (or Cl^-) as counterions.

To provide a complete bioseparation, the above features of the PEC should be taken into account. In principle, in order to avoid the formation of soluble NPEC the mixing of water–salt solutions of polyions should be done in a strictly equivalent ratio of positive and negative charges of the polyion pair. However, in practice, relatively short polycations or polycations with Br^- counterions are available and hence an overdose of the polycation would not lead to the formation of soluble NPEC. On the other hand, by using a relatively short polyanion one can quite – safely overdose the polyanion without hindering the precipitation in water–salt solutions. The solubility of NPEC in water facilitated the investigation of competitive interpolyelectrolyte reactions in homogeneous aqueous solutions avoiding difficulties arising when studying colloidal dispersions of insoluble polyelectrolyte complexes. Interpolyelectrolyte reactions imply the competition between different HPE molecules for binding with GPE chains, as it is represented below in a general schematic form.

$$\text{NPEC(HPE.nGPE)} + \text{HPE}^* \longleftrightarrow \text{NPEC(HPE.(n} - \text{x)GPE)} \tag{4}$$
$$+ \text{NPEC(HPE}^*\text{.xGPE)}$$

The transfer of GPE chains from initial NPEC to HPE* of the same chemical nature as HPE in the NPEC is referred to as the interpolyelectrolyte exchange reaction. Otherwise, if HPE and HPE* have different chemical structures, then the transfer (4) is denoted as interpolyelectrolyte substitution reaction.

Studies of kinetics and mechanism of interpolyelectrolyte reactions summarized by Bakeev *et al.* (1992)) revealed that in salt-free solutions the reactions are kinetically restricted. Addition of low-molecular-weight electrolyte triggers the polyions transfer. Relatively small amounts of simple salt proved to be enough for systems to reach rather quickly an equilibrium state characterized by permanent exchange of polyions between NPEC particles. This mutual exchange is nothing more than a form of thermal motion in polyelectrolyte complexes that are the products of reversible ionic reactions between polyelectrolytes.

It is highly improbable that the exchange of polyions goes on via dissociation of the very stable NPEC and the transfer of released chain to another complex particle. This mechanism should not be excluded in a special case of short oligomers, but generally the transfer proceeds by a so called 'contact' mechanism. According to Bakeev *et al.* (1992), the transfer is carried out via the collision of two likely charged NPEC coils, their interpenetration into each other to form a joint coil, the transfer of GPE to the neighboring NPEC, and breaking of the joint coil down to two NPEC coils of new composition. The lifetime of the joint coils is very short and the segmental mobility of polyions is relatively low. Therefore, the efficiency of the collisions is low, and the amounts of joint coils in the solutions are so insignificant that they slip away from experimental identification.

The mechanism of interpolyelectrolyte substitution suggests the collision and mutual penetration of the NPEC coils, but the lifetime of the joint coils increases significantly in this case. Thus, the addition of poly(vinylsulfate) polyanion (PVS) to the solution of NPEC(PMA-PEVP) is accompanied by increase of specific viscosity caused by the appearance of associated joint coils. These associates are observed as rapidly sedimenting particles (Izumrudov *et al.*, 1987b). A gradual decrease of the viscosity occurs with time due to the break down of the joint coils proving that interpolyelectrolyte reactions proceed according to the 'contact' mechanism.

The substitution reaction can be fully terminated at the stage of formation of the ternary polycomplex (joint coils) by the introduction of a small amount (1–2 mole%)

of covalent cross-links between the components in the original NPEC species (Kabanov *et al.*, 1990). Addition of PVS to such modified NPEC is followed by a much sharper increase in the viscosity that does not change with time. The higher the concentration of polyions, the higher is the specific viscosity. When polyelectrolyte concentration exceeds some critical value, *ca.* 1 g/l, the overall reaction mixture becomes a gel. Both the gelation phenomenon and properties of the obtained gel appear to be of an obvious biochemical interest. Indeed, the gelation takes place in a matter of seconds by the mere mixing of relatively diluted solutions of modified NPEC and PVS. The presence of proteins (antibodies, enzymes) in the above solutions assures protein immobilization in the gel. The properties of such systems should be determined by the properties of the gels. In turn, the gels collapse under acidification and recover under the addition of alkali. Introduction of either low-molecular-weight electrolyte or an additional amount of polycation also results in the contraction of the gel. The gel is able to undergo a reversible transition to the viscous state under applied stresses. These features seem to be rather promising for the development of reversible self-adjusting systems for bioseparation and bioprocessing.

The position of the equilibrium (4) might be quite different, from the uniform distribution of GPE chains between HPE and HPE* chains of the same length ($x = n/2$) to a highly selective binding of GPE with HPE* chains ($x \approx n$). The latter case is of primary concern for the design of target drug delivery and self-adjusting systems and merits detailed consideration here.

5.2.3. Molecular 'Recognition'

The phenomenon of molecular 'recognition' in NPEC solutions and factors leading to the selective binding of polyelectrolytes are described in a review of Izumrudov *et al.* (1991). Polycation recognizes and selects that polyanion which forms more stable NPEC even if the difference between the free energy change counted per pair of oppositely charged units is very small.

Thus, in solutions containing relatively long and relatively short PMA polyanions, PEVP polycation prefers to bind with the long chains. If the difference in the chains length is significant, this competitive interaction leads to a complete displacement of short PMA chains (D.P. = 50) from the NPEC by long ones (D.P. = 4100) (Izumrudov *et al.*, 1987a). Similar regularity is observed in a substitution reaction (5). Relatively long poly(phosphate) chains (D.P. = 230) compete strongly with PMA for binding with PEVP, whereas short PPh chains (D.P. = 20) are less efficient (Figure 5.3a) (Izumrudov *et al.*, 1988).

Modification of PMA polyanion by hydrophobic pyrenyl groups capable of also stabilizing NPEC(PMA-PEVP) can assure preferential binding of PEVP polycation with the modified PMA chains. A minute quantity of these groups in PMA chain (*ca.* 1 pyrenyl group per 400 monomer units) proved to be enough for PEVP to unmistakably choose the modified polyanion among the unmodified ones and to bind with it (Bakeev *et al.*, 1987). It suggests that a small number of molecular 'anchors' with energies of the order of several kT is apparently sufficient to ensure a crucial advantage in the binding for the modified polyanion, whose length is greater by a factor of several hundreds than the length of the 'anchor'. This result appears to be rather promising for the development of bioseparation and assay procedures using conjugates of polyions with a small number of antibodies, antigens or antigen determinants.

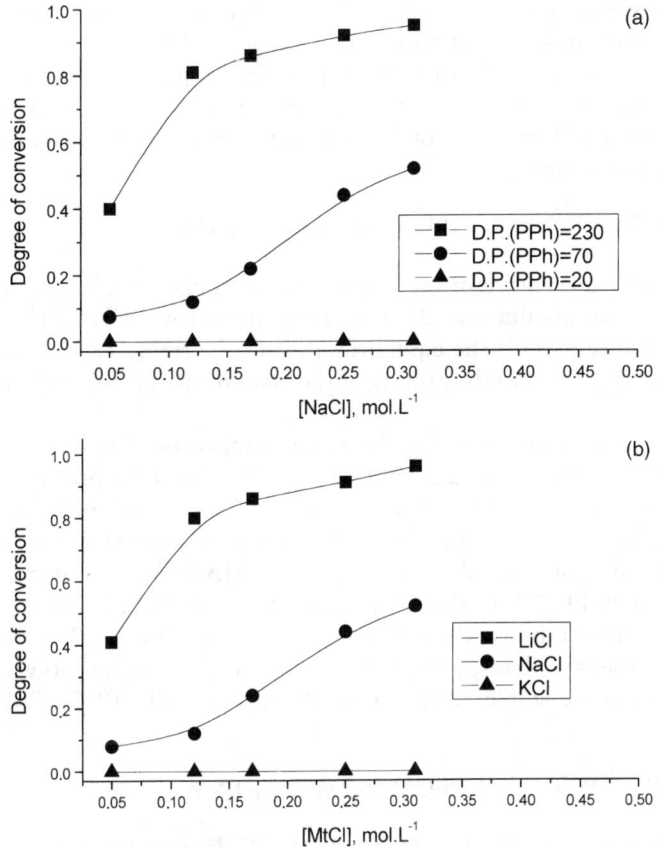

Figure 5.3 Equilibrium conversion in substitution reaction (equation 5) as a function of
NaCl concentration (a) and concentration of different alkaline metals chlorides,
MtCl (b). D.P. (PPh) is the degree of polymerization of PPh (redrawn from
Izumrudov *et al.*, 1988).

Sulfonate (or sulfate) groups could play the role of these 'anchors'. Introduction of
these groups in the molecule endows the polyanions with the ability to be particularly
attractive to the polycations. Thus, PEVP polycation recognizes negatively charged
modified polysaccharides carrying such groups among other modified polysaccharides,
even though the latter ones are the chains with a much higher density of COO⁻ groups
(Izumrudov *et al.*, 1999). Thus, PEVP presents a promising probe to elucidate the
segments of modified polysaccharides bearing sulfonate (or sulfate) groups and to
block these sites. In turn, it might be a key point in the study of structure–function
relation of the heparin-like polyanions.

Polyanions with a high density of sulfonate or sulfate groups in the chains (e.g.,
potassium PVS, sodium poly(anetholesulfonate), sodium PSS, heparin) form extremely
stable polyelectrolyte complexes with polycations and irreversibly substitute carboxylate polyanions in their NPEC with the polycations (Izumrudov *et al.*, 1985; Kabanov
et al., 1985). This property of heparin was used as a basis for the application of soluble NPEC as a depot for polycations playing the role of antiheparin agents (Kabanov,

1994a and b). The same features of the polyanions could be exploited in bioseparation for the regeneration of polyanions containing carboxylate (or phosphate) groups (or conjugates of proteins with these polyanions) from precipitated SPEC.

If the affinity of both competitive polyanions to the polycation is quite similar, the direction of the interpolyelectrolyte substitution depends drastically on the environmental conditions. So, the equilibrium (5) of the competitive binding of PMA and PPh polyanions with PEVP polycation

$$NPEC(PMA - PEVP) + PPh \longleftrightarrow NPEC(PPh\text{-}PEVP) + PMA \qquad (5)$$

is effectively controlled by the nature of the added low-molecular-weight electrolyte (Izumrudov *et al.*, 1988). The alkaline metal cations range in the row $Li^+ > Na^+ > K^+$ (Figure 5.3b) in their ability to shift the equilibrium (5) to the right. These data are in accordance with the differences in the affinity of the alkaline metal cations to PMA and PPh polyanions.

The equilibrium of reaction (5) proved to be rather temperature sensitive in the studied range 5–60 °C (Izumrudov *et al.*, 1996a). PEVP polycation preferentially binds with polyphosphate anion at lower temperatures and with PMA polyanion at higher temperatures. This change in the direction of the reaction (5) is consistent with a change in the stability of either the starting NPEC(PMA-PEVP) or the product of the reaction, NPEC(PPh-PEVP), in the same temperature interval. In the case of significant differences in the stability of the initial and resulting NPEC, the temperature sensitivity is not observed. Thus, poly(N-methyl-4-vinylpyridinium) polycation (PMVP) forms distinctly more stable NPEC(PPh-PMVP) than NPEC(PMA-PMVP), and equilibrium (6)

$$NPEC(PMA\text{-}PMVP) + PPh \longleftrightarrow NPEC(PPh\text{-}PMVP) + PMA \qquad (6)$$

is completely shifted to the right in the studied range 5–60 °C (Izumrudov *et al.*, 1998).

The revealed possibility to control the equilibrium of interpolyelectrolyte reactions by changing the media suggests one might be able to develop 'intelligent' systems capable of solving bioseparation and bioassay problems.

The study of the equilibrium, kinetics and mechanism of interpolyelectrolyte reactions resulted in an understanding of interpolyelectrolyte complexes as macromolecular compounds, permanently exchanging polyions in water-salt solutions. The complexes have unique features combining high stability with the ability to take part in interpolyelectrolyte reactions. High selectivity of cooperative interpolyelectrolyte reactions, sensitivity to the change of environmental conditions and a rather high reaction rate define a self-adjusting property of the systems. Both, formation of the complex particles and their transformation following a change of the media are accomplished by the method of trial and error via interpolyelectrolyte reactions until the equilibrium state is achieved.

This view of interpolyelectrolyte complexes and their properties is fundamental and quite widespread. The principles of the competitive binding and chain transfer discovered in solutions of synthetic polyelectrolytes can be extended to the reactions with biopolymers. Highly cooperative supramolecular NPEC particles appear to be similar to self-assembled supramolecular complexes of biopolymers. The interactions of both types of complexes in aqueous solutions follow the same regularities and are controlled by similar specific factors as was shown by the studies of complexes formed by synthetic polyelectrolyte and biopolymers, i.e. DNA, globular proteins (enzymes, antibodies) and conjugates of the proteins with polyions. The main results of these studies are discussed below.

5.3. Interpolyelectrolyte Reactions in Solutions of DNA-containing PEC

Being of immense biological importance, DNA is a polyacid that should be attributed to the family of highly charged polyions. The high charge density of the DNA double helix is provided by negatively charged phosphate groups. Recently, the complexes of DNA with PEVP or other polycations were successfully used for increasing the efficiency of transformation of the cells by the plasmides and for protection of the DNA from splitting by cell nucleases, for a review see (Kabanov and Kabanov, 1994). The prospects for the delivery of DNA packed in PEC species to the target cell has motivated extensive study of DNA-containing PEC in order to give a precise control over their stability.

Monitoring of DNA-containing PEC dissociation was carried out by the fluorescence quenching technique (Izumrudov et al., 1995b). It is based on the ability of cationic dye ethidium bromide (EB) to intercalate with DNA double helix. The intercalation is accompanied by the increase of ethidium fluorescence. The capacity of added cations and anions to dissociate PEC(DNA-PEVP) decreases in the orders $Ca^{++} > Mg^{++} \gg Li^+ > Na^+ > K^+ \gg (CH_3)_4N^+$ (Izumrudov et al., 1995a and b; Izumrudov et al., 1996b) and $I^- > Br^- > Cl^- \gg F^-$ (Izumrudov and Zhiryakova, 1999) (Table 5.1). These series coincide with a decrease of affinity of the counter ions to DNA and to the polycation, respectively. The same interrelation between the ability of cations and anions to destroy PEC and their affinity as a counter ions to the components of PEC was established for PEC(PPh-PEVP) (Izumrudov et al., 1988) and PEC(PMA-PEVP) (Pergushov et al., 1993). Subsequently, the similar row $Ca^{++} \geq Sr^+ \gg Li^+ > Na^+ \geq K^+ > Cs^+$ was reported by Schindler and Nordmeier (1997) for PEC formed by DNA and polycations containing quaternized amino groups, i.e. PMVP bromide, poly(N,N'-dimethyldiallylammonium chloride) and ionene chloride.

The dependence of the stability of PEC(DNA-PEVP) on PEVP chain length is a perfect analogy to that of PEC(PPh-PEVP). In water–salt solutions of both PEC, the stability increases sharply up to D.P.(PEVP) $\approx$ 100 and then goes up only slightly (Figure 5.4). The stability of PEC is virtually independent on the length of the nucleic acid in the studied region 500–10000 base pairs. The PEC stability was independent of D.P. of a relatively long PMA polyanion in the case of PEC(PMA-PEVP) and the dissociation of PEC was controlled by the chain length of the shorter partner. (Pergushov et al., 1995). In general, stability of DNA-containing PEC could be effectively controlled by changing the length, charge density and the structure of the polycation chain as well as by addition of different low-molecular-weight electrolytes. The combination of above factors could be used to achieve a more pronounced effect (Izumrudov and Zhiryakova, 1999).

Both the stability of DNA-containing PEC and the changes in response to changing conditions are quite close to that of PEC formed by the polycations and flexible synthetic polyanions like PMA or PPh. This fact does not favor the view of Schindler and Nordmeier (1997) that the dissociation of DNA-containing PEC in water-salt solutions is determined by the special properties of DNA, in particular, by the high

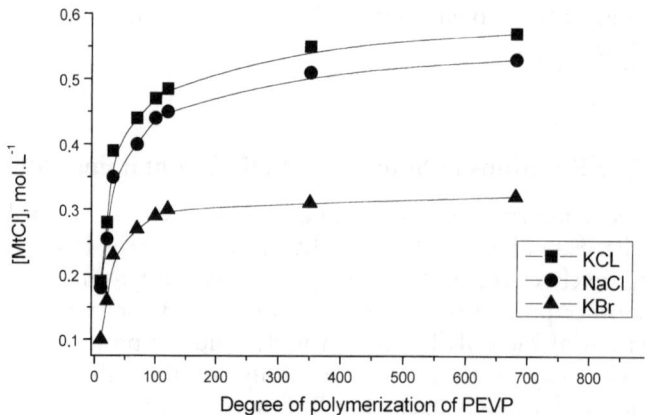

Figure 5.4 Dependence of the critical salt concentration, [MtCl]* corresponding to the expulsion of PEVP from the complex with DNA on the degree of polymerization of the polycation (redrawn from Izumrudov and Zhiryakova, 1999).

stiffness of a double helix, which was supposed to be responsible for the destruction of PEC at the ionic strength predicted by theoretical models.

As might be expected from the presence of phosphate groups in both PPh and DNA, competitive interpolyelectrolyte reactions in solutions of DNA-containing PEC have much in common with the PPh reactions. Just as the equilibrium of reaction (5) is determined by concentration and the nature of the low-molecular-weight electrolyte added, the interpolyelectrolyte substitution reaction

$$\text{NPEC(PMA-PEVP)} + \text{DNA} \longleftrightarrow \text{NPEC(DNA-PEVP)} + \text{PMA} \tag{7}$$

is sensitive to the same factors. The cations are arranged in the series $Ca^{++} > Mg^{++} \gg Na^+ > K^+ > Li^+$ according to their ability to shift the equilibrium (7) to the right (Izumrudov *et al.*, 1996b). This series differs from the row $Li^+ > Na^+ > K^+$ that was obtained for the reaction (5) (Figure 5.3b). The antipodal position of Li^+ in the above series of alkali metal cations is probably caused by its abnormally high affinity to native DNA. Denatured DNA interacts with Li^+ without any abnormalities and in solutions of denatured DNA equilibrium (7) obeys the expected order $Li^+ \approx Na^+ > K^+$.

Direction of the substitution (7) is determined by the ratio $\psi = \text{D.P.(PMA)}/\text{D.P.(PEVP)}$ (Izumrudov *et al.*, 1995a). If NPEC(PMA-PEVP) with $\psi > 1$, i.e. NPEC of 'A' type (Figure 5.2A) is used, the equilibrium (7) is shifted to the left. In the case of NPEC(PMA-PEVP) of 'B' type (Figure 5.2B), the decrease of ψ leads to a shift of the equilibrium to the right, which can be explained by the gain of entropy due to the increase in the total number of polymeric particles in the solution.

So, both reactions (5) and (7) are controlled by the concentration and the nature of the low-molecular-weight counter ions added. The competition between interacting polyions in these reactions depends substantially on the ratio of D.P. of the polyions. Binding of DNA with positively charged macromolecules or the release of DNA from the complexes can be controlled by the splitting or elongation of the chains of inter-acting macromolecular partners. In addition, our recent findings show that similar to the model reaction (5), the equilibrium of the substitution (7) in solutions of DNA-containing PEC is rather temperature sensitive (Izumrudov, 1998). In conclusion, DNA-based PEC has a great potential as a polyanion-partner in bioseparation and bioprocessing.

5.4. Protein–polyelectrolyte Complexes and their Properties

Complexes of proteins with nucleic acids, i.e. chromatin, ribosomes and other cell components as well as viruses, are widely distributed in nature and fulfill important functions. It is assumed that the main contribution to the stabilization of the complexes is provided by cooperative electrostatic interactions involving ionic groups of proteins exposed on the surface of the globules. To reveal kinetic and thermodynamic factors controlling the reconstruction and rearrangement of these rather complicated macromolecular structures the model protein–polyelectrolyte complexes have been investigated. These results are summarized in the review of Izumrudov (1996).

Since the complex of bovine serum albumin (BSA) with PEVP polycation has received the most study, let us consider this model system to demonstrate main features of protein-polyelectrolyte complexes.

5.4.1. Water-soluble PPC

Interaction of PEVP polycation with negatively charged BSA globules might lead to the formation of either water-soluble or insoluble polyelectrolyte–protein complexes. Positively charged water-soluble PPC are formed at more than 3-fold excess of PEVP charged units relative to the negatively charged groups of BSA located on a surface of the protein globules. Segments of PEVP chains carrying excess charge act as solubilizing fragments of PPC particle (Figure 5.5A) (Kabanov *et al.*, 1980a). These soluble PPC consist of one PEVP chain and the corresponding number of BSA globules (Izumrudov *et al.*, 1981). Negatively charged water-soluble PPC with protein as solubilizing component, are formed at more than 2.5-fold excess of the negatively charged BSA groups (Zaitsev *et al.*, 1992a). These soluble PPC consisted of one chain of the polycation and the minimum amount of BSA globules to provide solubility of PPC particle as a whole (Figure 5.5B). The system is heterogeneous at intermediate compositions of PEVP and BSA mixture.

Formation of water-soluble PPC suggests the transfer of BSA globules from one PPC particle to another. Direct experimental evidence of the transfer was obtained for PPC(PEVP-BSA) of type 'A' (Izumrudov *et al.*, 1984b). Kinetics of the protein

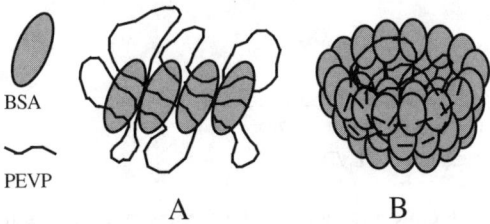

Figure 5.5 Structure of positively charged (A) and negatively charged (B) particles of water-soluble protein-polyelectrolyte (BSA-PEVP) complex.

transfer was studied in solutions of negatively charged PPC of type 'B' with fluorescence labeled protein, soybean trypsin inhibitor (Izumrudov et al., 1986b). The transfer of the protein proceeds relatively quickly even in salt-free media and is accelerated by the addition of simple salts. The reaction of competitive binding of soybean trypsin inhibitor with two different polycations was shown to be reversible and effectively controlled by the concentration of the polycation chains.

The phenomenon of 'molecular recognition' in PPC solutions was observed in the mixture of PEVP polycation with different proteins of serum (Kabanov et al., 1980b). The equilibrium state of the system corresponded to a coexistence of soluble PPC particles of type 'A' each consisting of one PEVP chain and protein globules of only one sort. Competition between PEVP polycations (D.P. = 650) and PEVP oligocations (D.P. = 30) for binding with BSA was in favor of the polymer. The equilibrium of the reaction (8)

$$\text{Complex (oligocation-BSA)} + \text{polycation} \longleftrightarrow \text{Complex (polycation-BSA)}$$
$$+ \text{oligocation} \qquad (8)$$

is almost entirely shifted to the right (Zaitsev et al., 1992b). The driving force of the shift appears to be the entropy-favorable increase of the particle number in solutions. The entropy factor determines a decrease of PPC(BSA-PEVP) stability upon a shortening of PEVP chains. The shorter the PEVP chains, the lower the critical salt concentration, $[NaCl]^*$ sufficient to liberate them from the complex (Figure 5.6). The entropy factor determines a decrease of solubilizing ability of short PEVP chains and prevents formation of soluble PPC(BSA-PEVP) of type 'A' in the case of oligocations. The potential uses of this thermodynamic limitation of the formation of soluble oligocation–protein complexes of type 'A' are intriguing. Oligocations could be used as overdose proof precipitate agents for proteins separation and the recovery of the components could be achieved by addition of a relatively small amount of simple salts.

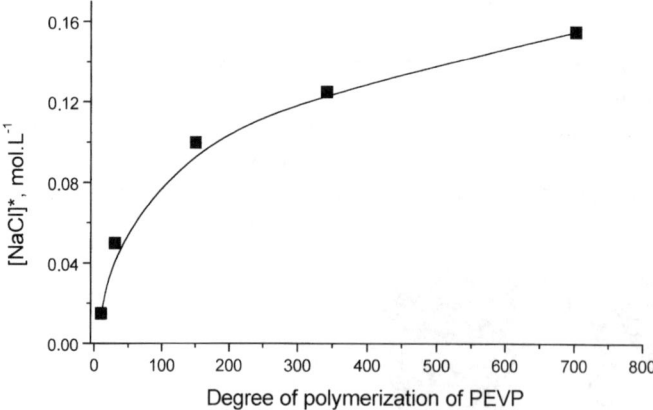

Figure 5.6 Dependence of the critical salt concentration, $[NaCl]^*$ corresponding to the salt concentration, when the expulsion of PEVP from the complex with BSA started, on the degree of polymerization of the polycation (redrawn from Zaitsev et al., 1992b).

The data obtained show that properties of soluble NPEC and PPC as well as self-assembly and self-adjustment phenomena in their solutions are quite similar. Formation of complexes of both types is determined by the hydrophilic–hydrophobic balance of their particles. The complexes are composed of a few molecules of the components providing their solubility. The similarity is also observed in the course of the competitive reactions. Direction and rate of the macromolecular transfer proved to be controlled by the same factors reflecting the similarity of driving forces of these reactions.

The revealed distinctions between PPC and PEC appear to be determined by the differences in structure and charge density of the molecules of globular proteins and flexible chains of linear synthetic polyelectrolytes. Thus, negatively charged groups of BSA are unevenly distributed and rigidly fixed on the surface of the globule. They form on the average 55 salt bonds with PEVP chain (Izumrudov *et al.*, 1981), whereas in NPEC a segment of PMA chain of the same molecular mass is bound with PEVP polycation by *ca.* 800 salt bonds. These differences might be the reason for the formation of soluble PPC when mixing aqueous solutions of the partners. In this particular case, counter ions of interacting polyion and protein play the role of 'triggering' salt. Mixing of rather dilute aqueous solutions of BSA and PEVP containing minute concentrations of the above counter ions lead to the formation of insoluble PPC. The relatively low stability of PPC in water–salt solutions (compare the values of [NaCl]* in Figures 5.1 and 5.6) is another point in favor of either a less extended or a less ordered system of salt bonds in a pair polyion-globule as compared to a pair polyion-polyion.

The studies of competitive reactions in the mixtures of proteins, polyanions and polycations strongly support this conclusion. Under certain conditions, synthetic polyelectrolytes with a high charge density displace globular proteins from their complexes with oppositely charged polyelectrolytes. This effect was observed for PPC, formed by either α-chymotrypsin or penicillin amidase with PMA polyanion (Margolin *et al.*, 1985a). The addition of equivalent amount of PEVP to the solution of these soluble PPC resulted in the precipitation of formed SPEC(PMA-PEVP) and almost entirely squeezed out the proteins from PPC in the solution. Moreover, this process did not depend on ionic strength within the range 0.01–0.2 M NaCl. Protein globules are not able to compete successfully with linear partners characterized by both higher charge density and uniform distribution of the charge along the chains. This result proved to be a key point in the development of systems meeting the important requirement of any biochemical method of assay, i.e. the absence of nonspecific interaction. This requirement can be met by using competitive reactions between proteins and synthetic polyelectrolytes (Figure 5.7).

5.5. Competitive Reactions in Solutions of Polyelectrolyte–protein Conjugates

To be used in bioseparation, polycomplexes should express a sufficient degree of biorecognition to discriminate between the protein of interest and impurities. As it was mentioned above, interpolyelectrolyte reactions have inherent selectivity in regards to differently charged species. Introduction of affinity interactions in the system by coupling an affinity ligand to one of the components could dramatically improve the selectivity of polycomplexes.

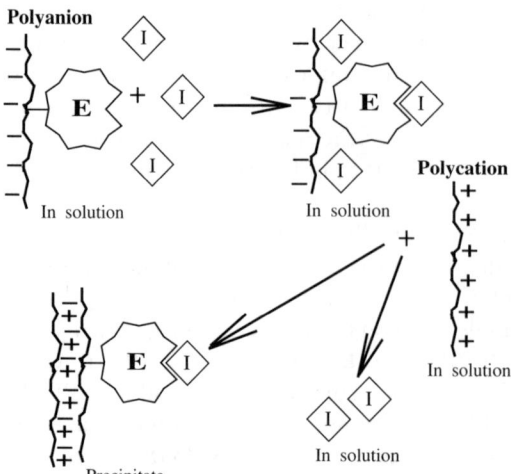

Figure 5.7 Scheme of protein–protein, protein–polyelectrolyte and polyelectrolyte–polyelectrolyte interactions in solutions of the protein–polyelectrolyte conjugate.

Coupling of the ligand, in particular the protein to the polyion significantly affects the interaction with its counterpart. Covalent attachment of α-chymotrypsin to PMA was shown (Margolin *et al.*, 1985a) to improve the interaction of the enzyme with the basic pancreatic trypsin inhibitor. The bimolecular rate constant of the association increased, whereas the equilibrium constant of dissociation of the immobilized enzyme-inhibitor complex decreased. This change of constants suggests that positively charged globules of the inhibitor are entirely absorbed on the negatively charged polymer immediately after mixing, and their interaction with the enzyme occurs on the polymer matrix. It is worth noting that the diffusion of globules of protein inhibitor along the charged matrix proceeds rather quickly even in the absence of a low-molecular-weight electrolyte (Savitskii *et al.*, 1987). Thus, a highly charged polymer serves two important functions promoting protein–protein interaction, i.e. a concentration of globules of free protein in the local volume occupied by the matrix and neutralization of their surface charge. The large effect could be observed in solutions of a wide range of different proteins, and not only for the proteins with considerable charge of the sign opposite to the sign of the matrix. Highly charged segments of PMA-enzyme conjugate induce charges in the amphoteric protein globule in the same way as a strong polyion induce charges in a weak polyelectrolyte (for reviews see Kabanov, 1994a and b). A relatively stable PPC (conjugate–protein) could be formed despite the partners are similarly charged at the given pH. It is self-evident that in the latter case the pH should not be significantly different from the isoelectric point of the protein.

If the conjugate α-chymotrypsin-PMA is incubated with the inhibitor and then the equivalent amount of PEVP is added, a complete transfer of enzyme-bound inhibitor to SPEC precipitate is observed (Figure 5.7). In other words, the specific interaction enzyme-inhibitor proved to be sufficient to prevent the squeezing out of the bound inhibitor into solution. Figure 5.7 illustrates a way of improving biochemical assays, in particular, immunochemical methods. Obviously, in such systems the assay can be done in solution, followed by the isolation of the components in a precipitate, i.e. the advantages of both the homogeneous and heterogeneous assays could be combined.

The study of the interaction of labeled antigens (human IgG, α-amylase) with antibodies covalently attached to PMA polyanion (Dzantiev *et al.*, 1988) reinforced the mechanism of competitive binding. The data obtained for proteins with different isoelectric points, structure and molecular mass indicated the absence of any detectable nonspecific binding in the system. At the same time, virtually all antigens specifically bound with PMA-antibody conjugates were precipitated together with the polycation, ensuring the high selectivity of the separation. It was found that water-soluble polymer carriers exerted little influence on the rate of antigen–antibody binding. The presence of components of the blood serum had no significant influence, either on the concentration ratio of the polyions at which phase separation was observed, or on the completeness and time of the separation process.

5.5.1. *Immunoassay*

Dzantiev *et al.* (1995) developed a new enzyme immunoassay based on interpolyelectrolyte reactions. Since the immunoassay is discussed in detail in a separate chapter of this book, we restrict ourselves to outlining the main features and advantages of the assay. The approach is efficient enough to decrease considerably the time of determination without any loss of assay sensitivity. The lowest detectable concentrations were about 5×10^{-10} M for different antigens (human testosterone, insulin, α-amylase, and IgG). This is comparable in sensitivity to the known competitive solid-phase ELISA developed for these compounds. The total duration of the assay did not exceed 15 min which is an order less than the duration of the majority of the well-known competitive immunoassays.

The selective separation of components of the analytical mixtures is determined to a great extent by the absence of nonspecific entrapment of labeled proteins into insoluble SPEC. The entrapment proved to be dependent on their molecular mass (Blintsov *et al.*, 1995) and did not exceed 1% in the case of 'monomer' enzyme–antibody conjugates (240 kDa). It correlates well with the conclusion on squeezing out protein globules from the SPEC (Figure 5.7). However, the nonspecific entrapment increased with the increase in molecular mass and in the case of bulky enzyme–antibody conjugates (1000 kDa) was as high as 10% of the initial concentration, precluding their use in the assay proposed. This entrapment of large protein species, being a serious handicap to the above mentioned immunoassay, proved to be a base for the development of immunoassay of viruses in solution (Dzantiev *et al.*, 1990).Under certain conditions, a soluble complex PEVP-virus interacts efficiently with PMA and forms insoluble SPEC that quantitatively entrapped huge virus particles. This was demonstrated for three phytoviruses with different morphology and structure. The specifity and sensitivity of this new express enzyme immunoassay of phytoviruses (Blintsov *et al.*, 1995) are not affected by the components of the plant sap or its extract. The method allows the detection of quite a wide range of different phytoviruses at concentrations of 5–10 ng/ml, which is comparable with the sensitivity of solid-phase enzyme immunoassay for phytovirus detection. The analysis is much more rapid (total time 25–30 min) than other enzyme immunoassays, which require at least 4 hours.

5.5.2. Enzymes Immobilized in PEC

Covalent attachment of proteins to polyelectrolytes endows conjugates with an attractive property. They form, under certain conditions, either insoluble SPEC or soluble NPEC. Direct mixing of the corresponding amounts of water–salt solutions of enzyme-polyion conjugate and oppositely charged polyion results in self-assembly of the NPEC particles in solution. The presence of the enzyme was not accompanied with noticeable changes in the properties of NPEC (Margolin *et al.*, 1981) whereas the immobilized enzyme acquired the unique abilities of NPEC, i.e. to undergo reversible and quantitative phase transitions under slight changes in ionic strength (Izumrudov and Lim, 1998), pH or composition of the mixture of polyions (Margolin *et al.*, 1983). The features of these reversibly soluble immobilized enzymes are described in detail by Margolin *et al.* (1985b) and reviewed by Zezin *et al.* (1989). The application of reversibly soluble immobilized enzymes allows: (a) the combination of all advantages of the homogeneous catalysis with an easy separation of the enzyme molecules from the reaction mixture at the end of the process; (b) the significant increase of the thermostability of immobilized enzyme via its precipitation by pH decrease; (c) the protection of the enzyme from high-molecular-weight inhibitors either in precipitate or in a homogeneous solution; (d) the creation of self-regulating enzyme systems, in which an attainment of the certain conversion degree results in the termination of the process; (e) the design of reversibly soluble polyenzymatic systems.

5.5.3. Artificial Chaperones

In the last decade, a lot of papers devoted to investigation of chaperones and chaperonines, in particular the elucidation of their role in protein folding, were published. For reviews see (Ellis and Van der Vies, 1991; Guise *et al.*, 1996; Ellis, 1998). Chaperones are able to distinguish a misfolded polypeptide from a native one, to bind this non-native form, to prevent aggregation of misfolded species, and then to release these forms in solution.

A number of works describe the attempts to create artificial chaperones. In the simplest cases, any compounds stimulating a process of correct protein folding are called 'artificial chaperones', for instance, detergent with cyclodextrin (Rozema and Gellman, 1996), glycerol (Rariy and Klibanov, 1997) etc. The ultimate objective of this line of inquiry is to design more sophisticated systems capable of simulating various steps of chaperone action, i.e. (i) the recognition and binding of non-native protein forms, (ii) the deposition of the non-native forms in a complex with chaperones and (iii) the dosed liberation of the non-native forms into solution after folding to the native conformation.

A system based on conjugates of antigens and antibodies with synthetic polyelectrolytes appears to be a rather promising model of the chaperone action. The required features of the system might be conditioned by the properties of the conjugates, in particular their ability to interact with an oppositely charged partner to form PEC which, in turn, could be readily and reversibly transformed from the soluble to the insoluble state by a changing of pH, ionic strength and temperature.

Recently we demonstrated the advantage of this approach. NPEC with coupled antibodies was used as a model of chaperon action in the living cell (Dainiak *et al.*, 1998). Monoclonal antibodies specific for misfolded dimeric forms of glyceraldehyde-3-phosphate dehydrogenase but not binding the native tetrameric form of the enzyme were covalently coupled to PMA polyanion. Partially inactivated enzyme was treated with the immobilized antibodies to bind inactivated dimers and then the complex of PMA-antibodies with the misfolded protein was precipitated by the addition of an equimolar amount of PEVP and removed by centrifugation. The sequential removal of insoluble enzyme aggregates by centrifugation resulted in complete restoration of the specific activity proving that all inactivated species (enzyme aggregates and misfolded dimers) have been removed. Immobilized antibodies recognize misfolded dimers like the chaperone recognizes the misfolded pholypeptides. The misfolded protein is removed from the reaction medium by precipitation of PEC. So, the first stage of chaperone functioning was imitated.

The next stage suggests the design of a system as a depot of denatured protein forms. On the one hand, this system should have sufficient capacity to bind excess of the misfolded species. On the other hand, the species bound specifically as antigens should release as the concentration of free denatured forms in solution decreases to some 'critical' level. The functioning of such a self-adjusting system could be based on competition between free antigens and the conjugated antigens for binding with the conjugated antibodies. This is an active area of our research and preliminary experimental results are extremely encouraging.

5.6. Conclusion

Polyelectrolyte complexes formed by oppositely charged polyelectrolytes of high charge density present an interesting group of 'smart' or 'intelligent' polymers. They undergo reversible changes in the structure triggered by small changes in the reaction conditions and are capable of self-assembly and self-adjustment. Interpolylectrolyte reactions possess inherent recognition properties and are able to discriminate counterparts in the complex on the basis of their structure, molecular mass and charge. Introduction of bioaffinity ligands endows the conjugate with the recognition capacity sufficient for the purposes of bioseparation and bioprocessing. The unique properties of polyelectrolyte complexes in combination with bioaffinity ligands make them promising for the development of highly efficient means for protein isolation, new immunoassay procedures, creation of reversibly soluble biocatalysts and biomimetic.

5.7. Symbols and Abbreviations

BSA	bovine serum albumin
D.P.	degree of polymerization
GPE	guest polyelectrolyte
HPE	host polyelectrolyte
NPEC	soluble nonstoihiometric polyelectrolyte complex(es)
[Salt]*	critical concentration of the salt, corresponding to dissociation of the complex
PEC	polyelectrolyte complex(es)
PEVP	poly(N-ethyl-4-vinylpyridinium) polycation
PMVP	poly(N-methyl-4-vinylpyridinium) polycation
PMA	poly(methacrylate) polyanion
PPC	protein-polyelectrolyte complex(es)
PPh	poly(phosphate) polyanion

PSS	poly(styrene sulfonate) polyanion
PVS	poly(vinylsulfate) polyanion
SPEC	insoluble stoichiometric polyelectrolyte complex(es)
θ	degree of conversion in interpolyelectrolyte reaction (ratio of equilibrium number of interpolymer salt bonds to the ultimate one)
φ	molar ratio of charged units of the components, [GPE]/[HPE] in NPEC particle
φ_{cr}	critical molar ratio of charged units of the components, [GPE]/[HPE] providing solubility of the NPEC particle

5.8. References

Bakeev, K.N., Izumrudov, V.A., Zezin, A.B., and Kabanov, V.A. (1987) Phenomenon of molecular recognition in interpolyelectrolyte reactions, *Vysokomolek. Soed.* (in Russian), **29B**, 483–484.

Bakeev, K.N., Izumrudov, V.A., Kuchanov, S.I., Zezin, A.B., and Kabanov, V.A. (1992) Kinetics and mechanism of interpolyelectrolyte exchange and addition reactions, *Macromolecules*, **25**, 4249–4254.

Blintsov, A.N., Dzantiev, B.B., Bobkova A.F., Izumrudov, V.A., Zezin, A.B., and Atabekov, I.G. (1995) A new method for enzyme immunoassay of phytoviruses, based on interpolyelectrolyte reactions, *Doklady Biochemistry*, **345**, 175–178.

Dainiak, M.B., Izumrudov, V.A., Muronetz, V.I., Galaev, I.Yu., and Mattiasson, B. (1998) Conjugates of monoclonal antibodies with polyelectrolyte complexes – an attempt to make an artificial chaperone, *Biochim. Biophys. Acta*, **1381**, 279–285.

Dzantiev, B.B., Blintsov, A.N., Tsivileva, L.S., Berezin, I.V., Egorov, A.M., Izumrudov, V.A., Zezin, A.B., and Kabanov, V.A. (1988) Interaction of antigens with antibodies immobilized on a synthetic water-soluble polyelectrolyte, *Doklady Akademii Nauk* (in Russian), **302**, 222–226.

Dzantiev, B.B., Blintsov, A.N., Izumrudov, V.A., Zezin, A.B., Bobkova, A.F., Egorov, A.M., Atabekov, I.G., and Kabanov, V.A. (1990) Complexes of viruses with synthetic polyelectrolytes and their interaction with antibodies, *Doklady Akademii Nauk* (in Russian), **311**, 1482–1486.

Dzantiev, B.B., Blintsov, A.N., Bobkova, A.F., Izumrudov, V.A., and Zezin, A.B. (1995) New enzyme immunoassays based on interpolyelectrolyte reactions, *Doklady Biochemistry*, **342**, 77–80.

Ellis, R.J. and Van der Vies, S.M. (1991) Molecular chaperones, *Ann. Rev. Biochem.*, **60**, 321–347.

Ellis, R.J. (1998) Steric chaperones, *Trends in Biochemical Sciences*, **23**, 43–45.

Guise, A.D., West, S.M., and Chaudhuri, J.B. (1996) Protein folding *in vivo* and renaturation of recombinant proteins from inclusion bodies, *Mol. Biotechnol.*, **6**, 53–64.

Gulyaeva, G.G., Poletayeva, O.A., Kalachov, A.A., Kasaikin, V.A., and Zezin, A.B. (1976) Study of water-soluble polyelectrolyte complexes formed by sodium poly(acrylate) and 5,6-ionene bromide, *Vysokomolek. Soed.* (in Russian), **18A**, 2800–2805.

Hubert, P. and Dellacherie, E. (1994) Water-soluble biospecific polymers for new affinity purification techniques. In Dubin, P.L. (ed.) *Macromolecular Complexes in Chemistry and Biology*, Springer-Verlag, Berlin-Heidelberg, pp. 229–246.

Izumrudov, V.A. and Zezin, A.B. (1976) Conformation of polyelectrolytes and formation of polyelectrolyte complexes, *Vysokomolek. Soed.* (in Russian), **18A**, 2488–2494.

Izumrudov, V.A., Kasaikin, V.A., Ermakova, L.N., and Zezin, A.B. (1978) Study of soluble nonstoichiometric polyelectrolyte complexes, *Vysokomolek. Soed.* (in Russian), **20A**, 400–406.

Izumrudov, V.A., Kasaikin, V.A., Ermakova, L.N., Mustafaev, M.I., Zezin, A.B., and Kabanov, V.A. (1981) Study of structure of water-soluble complexes of bovine serum albumin with poly(N-ethyl-4-vyniepyridinium) bromide by light-scattering, *Vysokomolek. Soed.* (in Russian), **23A**, 1365–1373.

Izumrudov, V.A., Savitskii, A.P., Bakeev, K.N., Zezin, A.B., and Kabanov, V.A. (1984a) A fluorescence quenching study of interpolyelectrolyte reactions, *Macromol. Chem., Rapid Commun.*, **5**, 709–714.

Izumrudov, V.A., Zezin, A.B., and Kabanov, V.A. (1984b) Macromolecular exchange in solutions of complexes of globular proteins with synthetic polyelectrolytes, *Doklady Akademii Nauk* (in Russian), **275**, 1120–1123.

Izumrudov, V.A., Bronich, T.K., Zezin, A.B., and Kabanov, V.A. (1985) The kinetics and mechanism of intermacromolecular reactions in polyelectrolyte solutions, *J. Polym. Sci., Polym. Lett. Ed.*, **23**, 439–444.

Izumrudov, V.A., Bakeev, K.N., Zezin, A.B., and Kabanov, V.A. (1986a) Influence of charge density of polyion on the rate of interpolyelectrolyte reactions, *Doklady Akademii Nauk* (in Russian), **286**, 1442–1445.

Izumrudov, V.A., Zezin, A.B., and Kabanov, V.A. (1986b) Kinetics of macromolecular exchange in solutions of protein-polyelectrolyte complexes, *Doklady Akademii Nauk* (in Russian), **291**, 1150–1153.

Izumrudov, V.A., Nyrkova, T.Yu., Zezin, A.B., and Kabanov, V.A. (1987a) Influence of chain length of lyophilizing polyion on direction and kinetics of interpolyelectrolyte exchange reaction, *Vysokomolek. Soed.* (in Russian), **29B**, 474–478.

Izumrudov, V.A., Bronich, T.K., Zezin, A.B., and Kabanov, V.A. (1987b) Features of the interpolyelectrolyte substitution reaction, *Vysokomolek. Soed.* (in Russian), **29A**, 1224–1230.

Izumrudov, V.A., Bronich, T.K., Saburova, O.S., Zezin, A.B., and Kabanov, V.A. (1988) The influence of chain length of a competitive polyanion and nature of monovalent counterions on the direction of the substitution reaction of polyelectrolyte complexes, *Macromol. Chem., Rapid Commun.*, **9**, 7–12.

Izumrudov, V.A., Zezin, A.B., and Kabanov, V.A. (1991) Equilibrium of interpolyelectrolyte reactions and the phenomenon of molecular "recognition" in solutions of interpolyelectrolyte complexes, *Russian Chemical Reviews*, **60**, 792–806. Translated from *Uspekhi Khimii* (1991), **60**, 1570–1595.

Izumrudov, V.A., Kargov, S.I., Zhiryakova, M.V., Zezin, A.B., and Kabanov, V.A. (1995a) Competitive reactions in solutions of DNA and water-soluble interpolyelectrolyte complexes, *Biopolymers*, **35**, 523–531.

Izumrudov, V.A., Zezin, A.B., Kargov, S.I., Zhiryakova, M.V., and Kabanov, V.A. (1995b) Competitive displacement of ethidium cations intercalated in DNA by polycations, *Doklady Physical Chemistry*, **342**, 150–153. Translated from *Doklady Akademii Nauk.*, **342**, 626–629.

Izumrudov, V.A., Ortega Ortiz H., Zezin, A.B., and Kabanov, V.A. (1996a) Temperature-controlled reversible subtitution of polyions in an interpolyelectrolyte complex, *Doklady Physical Chemistry*, **349**, 190–192. Translated from *Doklady Akademii Nauk*, **349**, 630–633.

Izumrudov, V.A., Zhiryakova, M.V., Kargov, S.I., Zezin, A.B., and Kabanov, V.A. (1996b) Competitive reactions in solutions of DNA-containing polyelectrolyte complexes, *Makromol. Chem., Macromol. Symp.*, **106**, 179–192.

Izumrudov, V.A. (1996) Competitive reactions in solutions of protein-polyelectrolyte complexes, *Ber. Bunsenges. Phys. Chem.*, **100**, 1017–1023.

Izumrudov, V.A. (1998) Temperature effects in interpolyelectrolyte reactions. In *Book of Abstracts of German-Russian Workshop on Interpolymer Reactions in Homogeneous Systems and at Interfaces; Structure and Properties of Interpolymer Complexes*, p. 7.

Izumrudov, V.A. and Lim, S. H. (1998) The effect of charge and length of a blocking polycation on phase separation in aqueous salt-containing solutions of nonstoichiometric polyelectrolyte complexes, *Polymer Science*, **40A**, 276–282. Translated from *Vysokomolek. Soed.*, **40A**, 459–465.

Izumrudov, V.A., Ortega Ortiz, H., Zezin, A.B., and Kabanov, V.A. (1998) Temperature controllable interpolyelectrolyte substitution reactions, *Macromol. Chem. Phys.*, **199**, 1057–1062.

Izumrudov, V.A., Chaubet, F., Clairbois, A.-S., and Jozefonvicz, J. (1999) Interpolyelectrolyte reactions in solutions of functionalized dextrans with negatively charged groups along the chains, *Macromol. Chem. Phys.*, **200**, 1753–1760.

Izumrudov, V.A. and Zhiryakova, M.V. (1999) Stability of DNA-containing interpolyelectrolyte complexes in water-salt solutions, *Macromol. Chem. Phys.*, **200**, 2533–2540.

Kabanov, A. V. and Kabanov, V.A. (1994) Interpolyelectrolyte complexes of nucleic acids as a means for target delivery of genetic material to the cell, *Polymer Science*, **36**, 157–168. Translated from *Vysokomolek. Soed.*, **36**, 198–211.

Kabanov, V.A., Zezin, A.B., Mustafaev, M.I., and Kasaikin, V.A. (1980a) Soluble interpolymer complexes of polyamines and polyammonium salts. In Goethals, E.J. (ed.) *Polymer Amines and Ammonium Salts*, Pergamon Press Oxford and New York, pp. 173–192.

Kabanov, V.A., Mustafaev, M.I., Blokhina, V.D., and Agafieva, V.S. (1980b) About competitive relationships of reactions of serum proteins with polycation, *Mol. Biology* (in Russian), **14**, 64–75.

Kabanov, V.A. and Zezin, A.B. (1982) Water-soluble nonstoichiometric polyelectrolyte complexes: a new class of synthetic polyelectrolytes, *Soviet Sci. Rev.*, **4B**, 207–282.

Kabanov, V.A., Zezin, A.B., Izumrudov, V.A., Bronich, T.K., and Bakeev, K.N. (1985) Cooperative interpolyelectrolyte reactions, *Macromol. Chem. Symp.*, **13**, 137–155.

Kabanov, V.A., Zezin, A.B., Izumrudov, V.A., Bronich, T.K., Kabanov, N.M., and Listova, O.V. (1990) Structure formation and gelation phenomena in solutions of ternary interpolyelectrolyte complexes, *Macromol. Chem., Macromol. Symp.*, **39**, 155–169.

Kabanov, V.A. (1994a) Basic properties of soluble interpolyelectrolyte complexes applied to bioengineering and cell transformations. In Dubin, P.L. (ed.) *Macromolecular Complexes in Chemistry and Biology*, Springer-Verlag, Berlin-Heidelberg, pp. 151–174.

Kabanov, V.A. (1994b) Physicochemical basis and the prospects of using soluble interpolyelectrolyte complexes, *Polymer Sci.*, **36**, 143–156. Translated from *Vysokomolek. Soed.*, **36**, 183–197.

Kasaikin, V.A., Kharenko, O.A., Kharenko, A.V., Zezin, A.B., and Kabanov, V.A. (1979) The principles of formation of water-soluble polyelectrolyte complexes, *Vysokomolek. Soed.* (in Russian), **21B**, 84–85.

Kharenko, O.A., Kharenko, A.V., Kalyugnaya, R.I., Izumrudov, V.A., Kasaikin, V.A., Zezin, A.B., and Kabanov, V.A. (1979) Nonstoichiometric polyelectrolyte complexes as a new water-soluble macromolecular compounds, *Vysokomolek. Soed.* (in Russian), **21A**, 2719–2725.

Kokufuta, E. (1994) Complexation of proteins with polyelectrolytes in a salt-free system. In Dubin, P.L. (ed.) *Macromolecular Complexes in Chemistry and Biology*, Springer-Verlag, Berlin-Heidelberg, pp. 301–325.

Margolin, A.L., Izumrudov, V.A., Shvyadas, V.K., Zezin, A.B., Kabanov, V.A., and Berezin, I.V. (1981) Preparation and properties of penicillin amidase immobilized in polyelectrolyte complexes, *Biochim. Biophys. Acta*, **660**, 359–365.

Margolin, A.L., Izumrudov, V.A., Sherstyuk, S.F., Zezin, A.B., and Shvyadas, V.K. (1983) Enzymes immobilized in polyelectrolyte complexes. Influence of conformational changes of the matrix and phase transitions in solutions on catalytic properties, *Mol. Biology*, **17**, 1001–1008.

Margolin, A.L., Sherstyuk, S.F., Izumrudov, V.A., Shvyadas, V.K., Zezin, A.B., and Kabanov, V.A. (1985a) Protein–protein interaction in systems containing synthetic polyelectrolytes, *Doklady Akademii Nauk* (in Russian), **284**, 997–1101.

Margolin, A.L., Sherstyuk, S.F., Izumrudov, V.A., Zezin, A.B., and Kabanov, V.A. (1985b) Enzymes in polyelectrolyte complexes. The effect of phase transition on thermal stability, *Eur. J. Biochem.*, **146**, 625–632.

Nefedov, N.K., Ermakova, T.G., Kasaikin, V.A., Zezin, A.B., and Lopiryov, V.A. (1985) Influence of a nature of counterion on formation and properties of nonstoichiometric polyelectrolyte complexes, *Vysokomolek. Soed.* (in Russian), **27A**, 1496–1499.

Papisov, I.M. and Litmanovich, A.D. (1989) Molecular recognition in interpolymer interactions and matrix polyreactions, *Adv. Polym. Sci.*, **90**, 139–179.

Pergushov, D.V., Izumrudov, V.A., Zezin, A.B., and Kabanov, V.A. (1993) Effect of low-molecular mass salts on the behaviour of water-soluble nonstoichiometric polyelectrolyte complexes, *Polymer Science*, **35**, 940–944. Translated from *Vysokomolek. Soed.* (1993) **35**, 844–849.

Pergushov, D.V., Izumrudov, V.A., Zezin, A.B., and Kabanov, V.A. (1995) Stability of inter-polyelectrolyte complexes in aqueous saline solutions, *Polymer Science*, **37A**, 1081–1087. Translated from *Vysokomolek. Soed.* (1995) **37A**, 1739–1746.

Philipp, B., Daurzenberg, H., Linow, K.J., Kötz, J., and Dawydoff, W. (1989) Polyelectrolyte complexes – recent developments and open problems, *Prog. Polym. Sci.*, **14**, 91–172.

Rariy, R.V. and Klibanov, A.M. (1997) Correct protein folding in glycerol, *Proc. Natl. Acad. Sci. USA*, **94**, 13520–13523.

Rozema, D. and Gellman, S.H. (1996) Artificial chaperone-assisted refolding of carbonic anhydrase B, *J. Biol. Chem.*, **271**, 3478–3487.

Savitskii, A.P., Izumrudov, V.A., Sivozhelezov, V.S., Papkovskii, D.B., Berezin, I.V., Zezin, A.B., and Kabanov, V.A. (1987) Acceleration of protein–protein interactions in the presence of a polycation. Approximation to one-dimensional diffusion, *Doklady Akademii Nauk* (in Russian), **294**, 1501–1505.

Schindler, T. and Nordmeier, E. (1997) The stability of polyelectrolyte complexes of Calf-Thymus DNA and synthetic polycations: theoretical and experimental investigations, *Macromol. Chem. Phys.*, **198**, 1943–1972.

Shieh, J.Y. and Glatz, C.E. (1994) Precipitation of proteins with polyelectrolytes: role of polymer molecular weight. In Dubin, P.L. (ed.) *Macromolecular Complexes in Chemistry and Biology*, Springer-Verlag, Berlin-Heidelberg, pp. 272–284.

Tsuchida, E. and Abe, K. (1982) Interactions between macromolecules in solution and intermacromolecular complexes, *Adv. Polymer Sci.*, **45**, 1–130.

Tsuchida, E., Osada, Y., and Sanada, K. (1972) Interaction of poly(styrene sulfonate) with polycations carrying charges in the chain backbone, *J. Polym. Sci.*, **10**, 3397–3403.

Xia, J. and Dubin, P.L. (1994) Protein-polyelectrolyte complexes. In Dubin, P.L. (ed.) *Macromolecular Complexes in Chemistry and Biology*, Springer-Verlag, Berlin-Heidelberg pp. 247–271.

Zaitsev, V.S., Izumrudov, V.A., Zezin, A.B., and Kabanov, V.A. (1992a). Water-soluble protein-polyelectrolyte complexes containing excess protein in the role of the lyophilizing component, *Doklady Akademii Nauk* (in Russian), **322**, 318–323.

Zaitsev, V.S., Izumrudov, V.A., Zezin, A.B., and Kabanov, V.A. (1992b) Decisive influence of the degree of polymerization of poly-N-ethyl-4-vinylpyridinium cations on the stability and equilibrium conversions of their complexes with bovine serum albumin, *Doklady Akademii Nauk* (in Russian), **323**, 890–894.

Zezin, A.B., Kasaikin, V.A., Kabanov, N.M., Kharenko, O.A., and Kabanov, V.A. (1984) Influence of the ratio of degrees of polymerization of components on formation of nonstoichiometric polycomplexes, *Vysokomolek. Soed.* (in Russian), **26A**, 1519–1524.

Zezin, A.B., Izumrudov, V.A., and Kabanov, V.A. (1989) Interpolyelectrolyte complexes as a new family of enzyme carriers, *Makromol. Chem., Macromol. Symp.*, **26**, 249–264.

6 Controlled permeation through membranes modified with smart polymers: smart polyelectrolytes that undergo configurational transition on hydrophobic membrane surfaces

Etsuo Kokufuta

6.1. Introduction

Polyelectrolyte ions in aqueous solutions undergo conformational changes in response to alterations in solution conditions, such as pH and concentration of small ions. Thus, we may expect that the configuration of polyions, which had been adsorbed onto solid surfaces from aqueous polyelectrolyte solutions, would change when external factors such as pH and ionic strength are altered. In contrast to conformational changes of polyions in aqueous media,which can be examined by various experimental methods, configurational changes occurring on solid surfaces are very difficult to monitor. A computerized literature survey (Science Citation Index as data base) showed that since 1980, only 21 publications have dealt experimentally or theoretically with the configuration of synthetic polyelectrolytes (but not biopolymers such as proteins) at liquid–solid interfaces (see third section).

The author of this chapter intends initially to provide several bases of the conformational and the configurational changes of polyelectrolytes. The idea for looking at configurational change is based on the examination of the inward permeation of small water-soluble molecules through a semipermeable capsule membrane with an adsorbed polyelectrolyte layer (Kokufuta, 1993, 1999). As illustrated in Figure 6.1, a decrease in permeation is expected when the configuration of adsorbed polyions varies from looped (a) to flat (b) or *vice versa*. This should be realistic in such cases where polyions had adsorbed only onto the outer membrane surface from their aqueous solutions. However, we may observe the opposite situations with the capsule membrane whose outer surface as well as 'pore' has the polyions adsorbed (see the bottom of Figure 6.1). In this case, permeability would be reduced when the adsorbed polyion adopts an extended configuration. In contrast, a flat configuration of collapsed polyions seems to facilitate the permeation.

After discussing in detail the configurational changes of adsorbed polyions occurring on the surface of capsule membranes, one would notice that some of the polyelectrolytes fall naturally under the category of 'smart' polymers. This becomes more clear when looking at an application of polyelectrolyte-coated microcapsules to biotechnology (Kokufuta et al., 1986, 1988). Then we will learn that the regulation of the configuration of the polyelectrolyte coat by externally applied stimuli (such as altering pH) provides a useful tool for controlling the inward or outward permeation of solutes through the capsule membrane. Thus, microcapsules which are loaded with enzymes as well as whose surfaces are coated with a suitable polyelectrolyte will permit the control of enzymatic processes by adjusting the pH of the external liquid.

6.2. Conformational Changes of Polyelectrolytes in Aqueous Solutions

It would be helpful to elucidate pH-dependent conformational changes of polyelectrolytes in their aqueous solutions, prior to studies of pH changes in the adsorbed polyion configuration. This may provide a way to analyze the pH dependence of the permeability of capsule membranes onto which polyions had adsorbed, in connection with their configurational changes on the membrane surfaces (see Figure 6.1)

Now let us introduce a simple 'concept' that the conformation of polyions in aqueous solutions varies depending on the balance between the repulsive and attractive forces along the polymer chains. Repulsive forces are usually electrostatic in nature and are controlled by pH and ionic strength. When repulsive forces overcome attractive forces such as hydrogen bonding or hydrophobic interactions, the polyion chain should extend discontinuously in some cases, or continuously in others, as observed in the volume changes of polyelectrolyte gels (Ilmain *et al.*, 1991). Thus polyelectrolytes whose conformational changes can be explained using this concept should provide good samples for the present purpose.

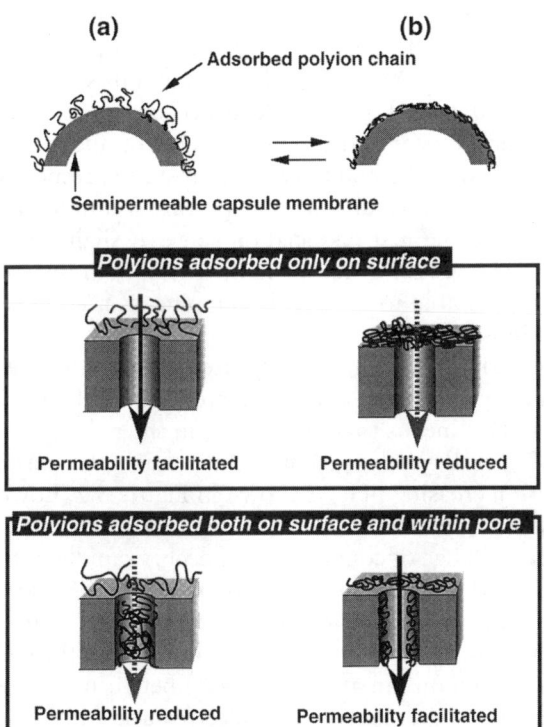

Figure 6.1 Schematic representation of configurational change of polyions adsorbed onto outer membrane of microcapsules: (a) looped form of extended polyions; (b) flat form of collapsed polyions.

A combination of viscometric and electrophoretic measurements is an excellent experimental approach frequently employed for studying the conformational changes of polyelectrolytes in aqueous solutions (e.g., see Kokufuta, 1980). A typical procedure is as follows: A polyelectrolyte is dissolved in a buffer solution (acetate, pH < 6; phosphate, pH 6–8; carbonate, pH > 8) adjusted to ionic strength 0.1, and then dialyzed against the same buffer solution in a cellophane tube until Donnan equilibrium is reached. Sample solutions are prepared at different concentrations by diluting with the buffer against which the polymer solution has been dialyzed. Electrophoretic mobility can be measured by, for example, a Tiselius-type Electrophoresis apparatus and a microcell with a small cross-sectional area ($0.19\,cm^2$ in the present case). Viscosity can also be determined by an Ubbelohde viscometer. In addition, currently developed laser light scattering techniques are also available (for example, Xia *et al.*, 1996); electrophoretic light scattering is used for measurements of the mobility, and dynamic and static light scattering – for polyion configuration.

Figure 6.2 shows the pH dependence of the limiting mobility ($U_{c \to 0}$) and the intrinsic viscosity ($[\eta]$) for several poly(carboxylic acid)s, the results of which have been reported by Kokufuta *et al.* (1974a, 1980). The mobility curves of three poly(dicarboxylic acid)s display an increase over pH 3–5, a plateau near pH 5–6, and a decrease over the alkaline pH range. These results are distinct from that of poly(acrylic acid)(PAA) as the typical poly(monocarboxylic acid). As can be seen from Figure 6.3, there is a pronounced half-neutralization point in each of the potentiometric titration curves for these poly(dicarboxylic acid)s, especially for copolymers of maleic acid (MA) with methyl vinyl ether (MVE), copoly(MA-MVE), and with styrene(St), copoly(MA-St). This indicates that two COOH groups in an MA or itaconic acid unit dissociate independently in two stages below and above pH 5–6, as illustrated in Figure 6.4. The variations in the mobility curves which appear in the pH regions below and above the plateau can thus be assigned to the first and second dissociation stages, respectively, of the polymer-bound COOH groups. As a result, the polyions are found to carry a large portion of the negative charge in the first dissociation stage. A slight decrease in the polyion charge (pH > 7–8), as revealed by lowering the mobility in the second dissociation stage, seems to be attributable to a possible trapping of a counter-ion between the two adjacent COO^- ions.

These charge alterations in poly(carboxylic acid)s directly affect their conformations through strong electrostatic interactions, since the viscosity curves display changes corresponding to the mobility curves. This means that an increase in the polyion charge as a result of dissociation of COOH leads to expansion of the polyion chain as indicated by a rise in the viscosity with increasing pH. As shown in Figure 6.2, however, a detailed comparison of the viscosity curves of copoly(MA-St) with those of other polymers indicates that the viscosity change of copoly(MA-St) occurs rapidly within a very narrow range near pH 4.5. Several previous studies of copoly(MA-St) (Kokufuta, 1980; Ohno *et al.*, 1973; Ferry *et al.*, 1951) have demonstrated that the conformation of this polymer is affected not only by an electrical repulsion between the COO^- ions, but also by a hydrophobic interaction (an attractive force) between the phenyl groups. When the hydrophobic force is overcome by the electrostatic force, a conformational transition from a 'tightly coiled chain' to an extended chain occurs as illustrated in Figure 6.5. This interpretation is supported by the fact that, in contrast to the other poly(dicarboxylic acid)s, the titration curve for copoly(MA-St) in Figure 6.3 exhibits a plateau near $\alpha = 0.2 - 0.4$. Therefore, pH-induced changes

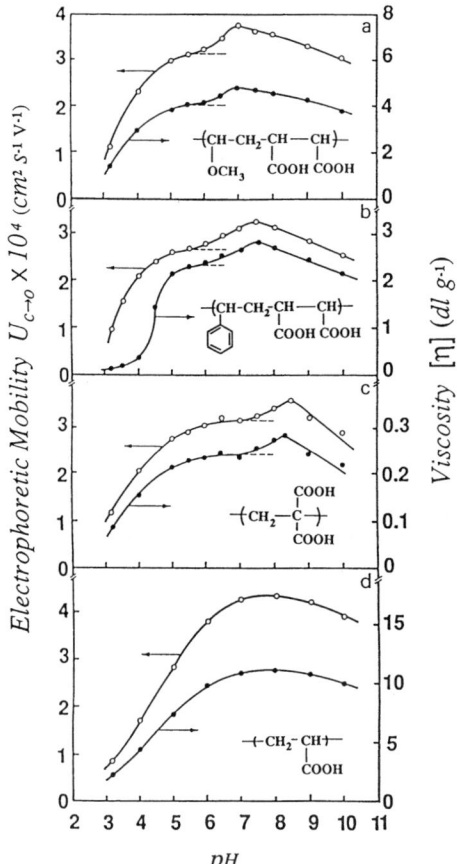

Figure 6.2 Electrophoretic mobility ($U_{c \to 0}$) and viscosity ([η]) against pH for poly(carboxylic acid)s. Both viscometric and electrophoretic measurements were carried out at 25 °C over a polymer concentration range of 0.05 to 0.2 g/dl, and the linear plots of the mobility or viscosity against the concentration were extrapolated to zero concentration. (a) alternating copolymer of maleic acid and methyl vinyl ether ($\overline{M_n} = 4.16 \times 10^5$); (b) alternating copolymer of maleic acid and styrene ($\overline{M_n} = 2.6 \times 10^5$); (c) poly(itaconic acid) [η] = 0.0786 dl/g in 1 M NaCl at 25 °C); (d) poly(acrylic acid) ($\overline{M_n} = 2.81 \times 10^5$). Figure is adapted from Kokufuta *et al.* (1974a, 1980).

in the conformation of copoly(MA-St) are discontinuous. This property is remarkably different from those of other poly(carboxylic acid)s that undergo a continuous conformational change with pH.

Conformational transition was also demonstrated by the mobility and viscosity curves for poly(ethyleneimine) (PEI) as a polybase (Figure 6.6). A rapid increase in the viscosity with decreasing pH may indicate the conformational transition of the polyion from a tightly coiled chain (pH > 6) to an extended one (pH < 6), which takes place *via* a fast change of polyion charge as characterized by the mobility curve. Hydrogen bonding between H_2N- and/or $-NH-$ groups attached to the branched chains seems

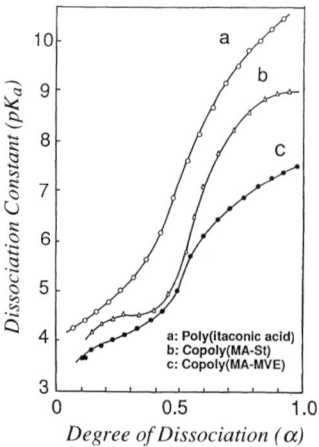

Figure 6.3 Apparent dissociation constant (pK$_a$) against degree of dissociation (α) for the poly(dicarboxylic acid)s as shown in Figure 6.2. The curves were obtained from the results of potentiometric titrations (polymer concentration, 0.01 M COOH group; solvent, 0.1 M NaCl; titrant, 2N NaOH; 25 °C). Figure is adapted from Kokufuta *et al.* (1980).

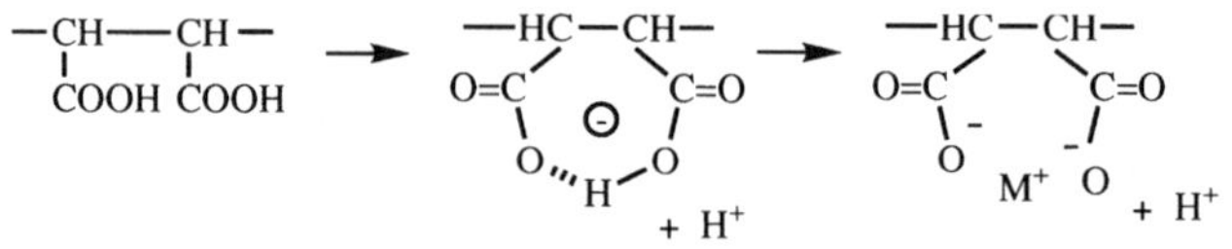

Figure 6.4 Two stage dissociations of poly(dicarboxylic acid)s such as copoly(MA-MVE), copoly(MA-St) and poly(itaconic acid), the structures of which were given in Figure 6.2.

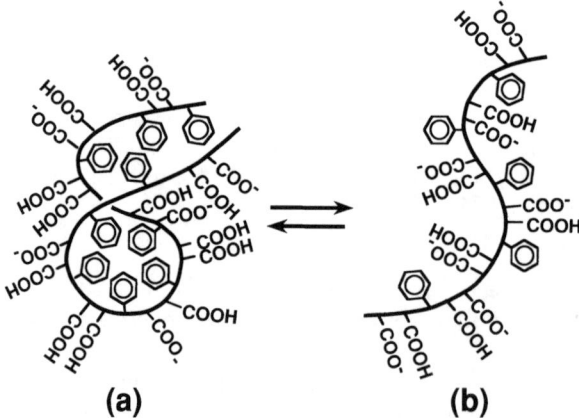

Figure 6.5 Schematic representations of conformational transition of copoly(MA-St): (a) a tightly coiled chain due to hydrophobic interaction between the phenyl groups; (b) an extended chain due to electrostatic repulsion between the charged carboxylic groups.

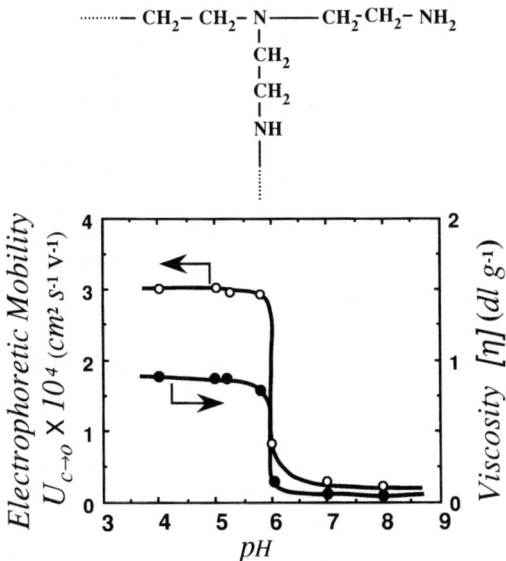

Figure 6.6 Electrophoretic mobility ($U_{c \to 0}$) and viscosity ($[\eta]$) against pH for PEI ($\bar{M}_w = 2.6 \times 10^5$) at 25 °C. Figure is adapted from Kokufuta *et al.* (1974).

to act as an attractive force (Kokufuta *et al.*, 1974b). In the case of PEI, however, both the mobility and viscosity increase rapidly at the same pH. This is different from the results for copoly(MA-St), which showed that the viscosity increases rapidly at pH 4.5 while the mobility increases slowly. Our previous studies (Kokufuta *et al.*, 1975; Kokufuta, 1980) on electrophoresis and viscosity for different polyelectrolytes have shown that when ionic strength is >0.1, the relationship between $[\eta]$ and $U_{c \to 0}$ can be expressed as in Equation 1:

$$[\eta] = AU_{c \to 0}M^a \tag{1}$$

Here, M represents the molecular weight of the polyelectrolyte, and a and A are empirical constants (A is independent on both M and ionic strength). When Equation 1 is combined with the Flory-Fox equation for viscosity (Flory and Fox, 1951), $U_{c \to 0}$ can be related to the expansion factor (α_η) by Equation 2:

$$\alpha_\eta^3 = (AK_0)(U_{c \to 0})M^c$$
$$K_0 = \Phi(\bar{r}_0^2/M)^{3/2} \tag{2}$$
$$c = a - 1/2$$

Here, Φ and $\bar{r}_0$ denote a universal constant and the end-to-end distance at the theta state, respectively. Plots of $[\eta]$ *vs* $U_{c \to 0}$ for three poly(carboxylic acid)s other than copoly(MA-St) in Figure 6.2 and also for PEI in Figure 6.5 were expressed by straight lines passing through the origin (data not shown), as predicted by Equation 1. At pH > 5, such a linearity was also observed in the case of copoly(MA-St). Therefore, the pH-induced conformational changes of these polyelectrolytes can be understood as an alteration in the expansion factor due to the net charge density of the polyions, which can be shown by changes in $U_{c \to 0}$. However, the plots of $[\eta]$ *vs.* $U_{c \to 0}$ for copoly(MA-St) at pH < 5 deviated from a straight line. It is thus likely that the 'tightly coiled chains' of copoly(MA-St) and PEI are different from one another. This can probably be explained by assuming that PEI is tightly coiled at pH > 6 due to hydrogen bonding between the amino and/or imino groups attached to branched chains, while copoly(MA-St) forms a coil at pH < 4 due to the hydrophobic interaction between the phenyl groups along the linear chains.

With respect to the role of hydrogen bonding in tighter coiling (or chain collapse) of PEI, a detailed study has recently been performed by Kokufuta *et al.* (1998). A gel was then synthesized using a linear PEI (but not a branched PEI as in Figure 6.6), the preparation of which was performed *via* acid hydrolysis of poly(ethyloxazoline) with HCl. When equilibrium gel swelling was measured as a function of pH, a discontinuous volume change was observed near pH 5.9 (swelling process) and pH 10.7 (deswelling process); thus, a large hysteresis appeared in the swelling curves. This result has been understood in terms of hydrogen bonding, which plays a key part in the precipitation of the polymer from its alkaline solution due to the formation of crystalline hydrates.

In conclusion, copoly(MA-St) and PEI have been found to be suitable for the study of the configurational changes of polyelectrolytes who had been adsorbed onto outer surfaces of microcapsules, because these two polymers undergo discontinuous conformational changes with pH, as compared to other polyelectrolytes that undergo continuous conformational changes with pH.

6.3. Configurational Changes of Polyelectrolytes on Microcapsule Surfaces

A polyion layer adsorbed onto a capsule membrane, as illustrated in Figure 6.1, would result in a resistance to mass transfer. Thus, a study of the inward permeation of low molecular weight solutes through such a capsule membrane would be useful for obtaining information on pH-induced changes in the configuration of the adsorbed polyion. In this section, the effects of pH on the permeability of polyelectrolyte-coated microcapsules are discussed and compared with the results of viscometric and electrophoretic studies described in the preceding section.

6.3.1. Analytical Methods

The configuration of polymers adsorbed on solid surfaces has attracted interest in the field of colloid and interface science. Among the reports dealing experimentally or theoretically with the configuration (or conformation) of synthetic polyelectrolytes at liquid–solid interfaces, the following methods have been employed and tested: IR spectroscopy (Day and Robb, 1980), ESR and IR spectroscopy (Sakai and Imamura, 1980), quasi-elastic light scattering (Kato *et al.*, 1981), surface resolution analysis (Farin *et al.*, 1985), microcalorimetry (Denoyel *et al.*, 1990), a combination of electron-tunneling spectroscopy, ^{13}C-NMR, and infrared reflection–absorption spectroscopy (Konstadinidis *et al.*, 1992), evanescent wave spectroscopy (Trau *et al.*, 1992), and a combination of diffuse reflectance IR spectroscopy, polarization reflection IR spectroscopy, refractive index analysis, and flow microcalorimetry (Yanagisawa, 1993). In addition, phase sensitive AC voltammetry (Temerk *et al.*, 1994), fluorescence spectroscopy (Kowalczyk *et al.*, 1994), and UV-CD (circular dichroism) spectroscopy (Harvey *et al.*, 1995) were very recently employed to study changes in the three-dimensional conformation of proteins synthetic polypeptides, and oligonucleotides at water–solid interfaces. In these studies, unfortunately, it has not yet been demonstrated that the polymers adsorbed at the liquid–solid interface undergo configurational changes in response to experimental conditions. In other works, the studies have focused a 'frozen' configuration of polyions under given experimental conditions; for example, a constant pH as well as a constant ionic strength. Nevertheless, our trial that is reported here is to examine configurational changes of the polyions at the liquid–solid interface through studies of permeability of capsule membranes with the polyions adsorbed. The knowledge of polyelectrolyte conformation in the solution is then helpful for this purpose.

6.3.2. *Preparation of Microcapsules and Polyelectrolyte Adsorption*

Stable microcapsules (mean diameter, 8–10 μm) with a semipermeable polystyrene (PSt) membrane can be prepared by depositing the polymer around emulsified aqueous droplets by the following three-step procedure (Figure 6.7). (1) dispersion of an aqueous solution of sodium dodecylbenzenesulfonate or Triton X-100 as an emulsifier in dichloromethane solution of PSt. (2) dispersion of the resulting water/oil dispersion in an aqueous solution containing either of the above emulsifiers under vigorous agitation. (3) complete removal of dichloromethane from the resulting (W/O)/W

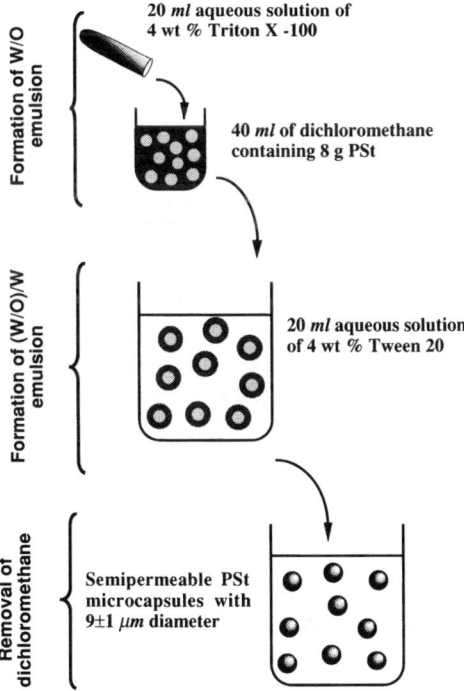

Figure 6.7 Schematic representations for the preparation of stable microcapsules with a semipermeable polystyrene (PSt) membrane.

complex dispersion (Kitajima and Kondo, 1971). The microcapsules are collected by centrifugation, thoroughly washed with distilled water, and then subjected to the polyelectrolyte adsorption procedure.

Polyelectrolyte adsorption is performed by stirring the microcapsules in appropriate buffers containing the desired polyelectrolyte at room temperature. Usually, the volume of microcapsule suspensions is adjusted to 100–120 ml, and the stirring allowed to continue for 10 h. After adsorption, the microcapsules were recovered by centrifugation and purified by repeated washing with the same buffer until no polyelectrolyte was detected in the washing extract, as determined by colloid titration for copoly(MA-MVE) and copoly(MA-St) (Kokufuta and Iwai, 1977), and for PEI (Kokufuta, 1979).

6.3.3. Inward Permeation of n-Propyl Alcohol through Capsule Membrane

Permeability was estimated by measuring the concentration of n-propyl alcohol (PA) as a permeate, after quick mixing of the desired quantity of aqueous PA solution and microcapsule suspension, both of which were previously kept at 25 °C and adjusted to the same pH (3–10) with 0.1 M acetate, phosphate, or carbonate buffer. Samples (usually, 0.1 ml) were filtered from the suspension through a 0.1 μm filter at suitable time intervals and analyzed with a liquid chromatography. The permeability constant (P) was then calculated according to the Equation 3, derived from Fick's first law of diffusion (Kitajima and Kondo, 1971).

$$P = -\frac{C_t V_m}{C_i A t} \ln \frac{C_t - C_f}{C_i - C_f} \tag{3}$$

Here, C_i, C_t and C_f are the initial, intermediary (at time t) and final concentrations of PA, respectively; V_m total volume of microcapsules; and A total surface area of microcapsules. The plot of $\log[(C_t - C_f)/(C_i - C_f)]$ against time showed a straight line for all measurements. Typical examples of the linear plots obtained are represented in Figure 6.8.

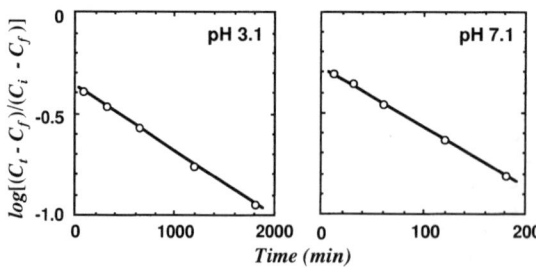

Figure 6.8 Plots of $\log[(C_t - C_f)/(C_i - C_f)]$ against time for polystyrene microcapsules coated with copoly(MA-St) at pH 3.1 and at pH 7.1. Figure is adapted from Kokufuta et al. (1988).

6.3.4. *Effects of pH and Amount of Polyelectrolytes Adsorbed*

Plots of P *vs.* pH for microcapsules coated with copoly(MA-MVE), copoly(MA-St), and PEI are shown in Figure 6.9, together with that for the original PSt micro-capsules. P for the uncoated microcapsules was independent of pH and remained constant (1.59×10^{-5} cm/s), but for the coated microcapsules varied with pH. For copoly(MA-MVE) and copoly(MA-St), an increase in pH brought about an increase in permeability, whereas the permeability of the microcapsules with PEI increased with decreasing pH. Such permeability changes are analogous to the changes in the viscosity curves shown in Figures 6.2 and 6.6. In particular, the microcapsules with copoly(MA-St) and PEI exhibited a rapid change in permeability at the pH at which the conformational transitions of both polyelectrolytes occur in aqueous media. As expected in Figure 6.1, these results are the following indications: (a) the adsorption of the polyions takes place only onto the outer surface (but not within the pores) of the PSt microcapsule membrane; and (b) the polyelectrolyte on the membrane sur-face of microcapsules undergoes configurational changes as a result of pH-induced electrostatic interactions.

Table 6.1 shows the effect of the amount of adsorbed polymer on the permeability for polyelectrolyte-coated microcapsules. P_{max} and P_{min} denote maximum and mini-mum P values, respectively, which were determined by studying the pH dependence of permeability for each polyelectrolyte-coated microcapsule. It is found that the per-meability is affected by the amount of the polyelectrolyte adsorbed onto the outer surface of the microcapsules. In particular, P_{min} for copoly(MA-St)-coated microcap-sules was much smaller than that for PEI-coated capsules, even though the amount of the adsorbed PEI was more than that of copoly(MA-St). As was discussed in the previous section, there is a difference in the tightness of the coiled chains of PEI and copoly(MA-St); that is, copoly(MA-St) seems to be more highly coiled than PEI. Thus, densely packed copoly(MA-St) coils could cover the surface of the microcapsules,

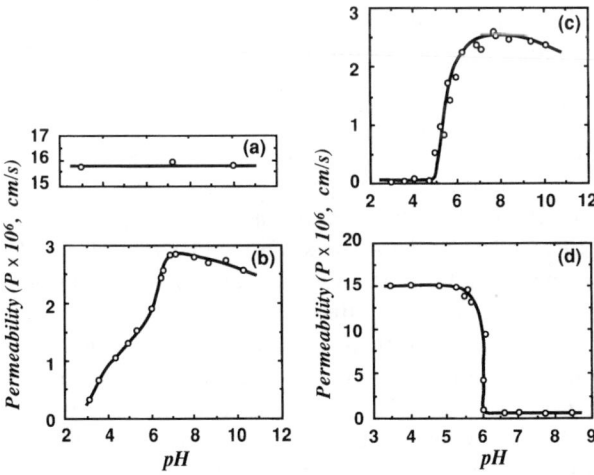

Figure 6.9 Permeability constant (P) against pH for PSt microcapsules: without absorbed polyelectrolyte layer (a); with absorbed layers of copoly(MA-MVE) (b); copoly(MA-St) (c); and PEI (d). Figure is adapted from Kokufuta *et al.* (1986, 1988).

Table 6.1 Effect of adsorbed polyelectrolytes on the premeability of PSt microcapsules

Polyelectrolyte	Adsorption conditions		Adsorbed amount ($\mu g/cm^2$)	P ($\times 10^{-6}$ cm/sec)	
	Initial polymer concn. (w/v %)	pH		P_{max}	P_{min}
PEI	0.1	4.0	4.5	16.2	6.6
PEI	1.0	4.0	24.3	14.0	2.7
PEI	1.0	8.0	23.3	14.6	2.0
PEI	10.0	4.0	235.8	1.1	0.6
Copoly(MA-MVE)	0.005	3.0	0.16	2.8	0.35
Copoly(MA-St)	0.05	3.0	1.7	2.55	0.062

leading to a reduced permeability despite the fact that the amount of polymer adsorbed is smaller than that of PEI. As a result, a study of the permeability of the capsule membrane with adsorbed polyelectrolytes permits an examination of the fine alterations in their configurations on the membrane surface.

6.4. Polyelectrolyte-coated Microcapsules for On/Off Control of Enzyme Reactions

Continued interest in research fields of immobilized biocatalysts has led to several new smart polymer-based methods for constructing 'functional' immobilized biocatalysts that surpass the definition and advantages of conventional immobilized biocatalysts (Kokufuta, 1992). For example, immobilized enzyme systems in which an enzymic process can be controlled by externally applied stimuli such as light, electric fields, pH, temperature, or mechanical force. In such systems, what is crucial in system design is not to rely on a possible improved property of the biocatalyst (e.g., enhanced thermal stability which may result from immobilization), but to add a new capability to the biocatalyst by rational design. Stimuli-sensitive microcapsules as a supporting matrix provide a possible tool for working towards this goal.

6.4.1. Stimuli-sensitive Microcapsules Modified with Smart Polymers

There have been several studies dealing with microcapsules whose outer surfaces were modified by grafting with stimuli-sensitive polymers, such as polyelectrolytes (pH-sensitive) (Okahata *et al.*, 1984), viologen-containing polymers (photo-sensitive) (Okahata *et al.*, 1986a) and thermosensitive polymers (Okahata *et al.*, 1986b). For the purpose of on/off control of enzyme reactions, the most useful preparation among these would be pH-sensitive capsules with surface-grafted poly(4-vinylpyridine) (PVP) and poly(methacrylic acid) (PMA), the permeability of which can thus be varied according to the pH of the external medium. However, as can be seen from Figure 6.10, the permeability gradually changes over a wide pH range (3–8) at which enzyme activities are also changed. This is the case even when the capsule surface was grafted with PMA whose potentiometric titration had indicated a conformational transition near pH 5.5 (e.g., see Morawetz, 1965). Another important observation from Figure 6.10 is that the PAA-grafted capsule does not act as a pH-sensitive capsule, perhaps due to an undesirable configuration of the surface-grafted polymer chains. As a result, we may not simply say that all of the pH-sensitive and polyelectrolyte-modified microcapsules

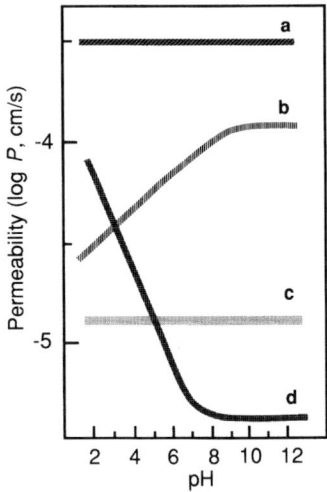

Figure 6.10 Permeability constant (P) against pH for nylon capsules with and without surface-grafted polyelectrolytes: a, ungrafted capsule; b, PMA-grafted capsule; c, PAA-grafted capsule; d, PVP-grafted capsule. The outward permeation of NaCl through the capsule membrane was examined by measuring the electric conductance of the aqueous phase at 25 °C, after dropping one capsule (diameter, 2.5 mm) into an aqueous solution with a desired pH. The curves are drawn using the data reported by Okahata *et al.* (1984).

are adaptable for on/off control of enzyme reactions. In addition, it should be noted that some of grafting procedures require severe chemical conditions. For example, the graft polymerization in an aqueous solution of 0.1 M nitric acid and 0.01 M cerium (IV) ammonium nitrate may lead to deactivation of the enzyme.

Tsai and Moshe (1984) have prepared polyelectrolyte-walled microcapsules under mild conditions by the interfacial crosslinking of an aqueous salt solution of PEI and a toluene solution of brominated poly(2,6-dimethylphenylene oxide). They obtained a stable suspension containing the microcapsules, which could not be isolated but could be cast into a membrane. Using the membrane into which the microcapsules containing 3 M NaCl had been embedded, they examined the release of NaCl from the membrane into distilled water or a 3M NH_4NO_3 solution (see Figure 6.11). The release of Na^+ and Cl^- ions is almost depressed by NH_4NO_3 in the external solution, but initiated by placing the membrane in distilled water because osmotic pressure due to encapsulated NaCl extends the capsule wall to enlarge the pores and thereby allow both Na^+ and Cl^- ions to diffuse through the membrane. Thus, the rates of NaCl release could be controlled by osmotic pressures. Although this technique may be adaptable for on/off control of enzyme reactions, any application has not yet been attempted, presumably because of the difficulty in isolation of the microcapsules from the suspension.

From the above examples, we may learn what requirements should be satisfied in the design and preparation of smart polyelectrolyte-modified capsules, which are truly useful for on/off control of enzyme reactions. Then, it naturally follows that (1) biocatalyst should be immobilized without loss of catalytic activity; and (2) very rapid alterations in the rate of substrate permeability should occur in response to

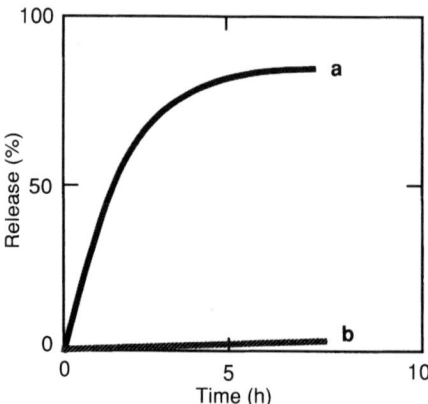

Figure 6.11 Release of Na^+ or Cl^- from membrane with embedded microcapsules containing 3 M NaCl into distilled water (a) and 3 M NH_4NO_3 solution (b). The curves are drawn using the data reported by Tsai and Moshe (1984).

the external pH stimuli with no influence on the activity of the biocatalyst. As was described in the previous section, PSt microcapsules coated with copoly(MA-St) or PEI adequately meet these requirements. This is because the preparation of microcapsules as well as the adsorption of polyelectrolytes onto the capsules can be performed under conditions which have been generally adopted in the microencapsulation of enzymes (e.g., see Chang, 1977). Moreover, the permeability of the capsules rapidly altered over a very narrow pH range as a result of changes in the configuration. Therefore, the pH-sensitive on/off control of enzyme reactions was successfully performed as outlined below.

6.4.2. Stimuli-sensitive Microcapsules with Loaded Enzymes

The procedure used for encapsulating enzymes is the same as that described in Figure 6.7, except for use of an aqueous enzyme solution in the primary dispersion process. For example, in order to entrap invertase (E.C. 3.2.1.26) in PSt microcapsules, the primary dispersion was carried out using 20 ml of 0.1 M acetate buffer containing 600 mg of the enzyme and 800 mg of Triton X-100 (Kokufuta *et al.*, 1998). Separation of the enzyme-loaded microcapsules and polyelectrolyte adsorption can be performed in the same manner as shown in Figure 6.7.

Enzymatic hydrolysis of sucrose was studied at 25 °C by monitoring the concentration of an equimolar mixture of glucose and fructose in an aqueous suspension of the enzyme-loaded microcapsules (see Figure 6.12). The suspension (50 ml) initially contained 100 mM of the substrate and 13.5% (v/v) of the microcapsules with a total of 200 mg of the encapsulated enzyme. On/off control was investigated by batch reaction kinetics at pH 5.5 (at which the reaction occurs) and at pH 4.5 (at which the reaction stops). The adjustment of pH was made by quick addition of a small amount of 2 M HCl or 2 M NaOH.

A typical example of the on/off control of the enzyme reaction is also shown in Figure 6.12. Using the original enzyme-loaded PSt microcapsules without the absorbed

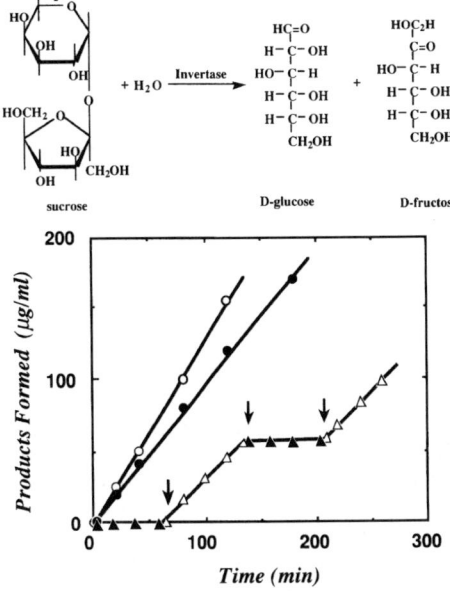

Figure 6.12 Enzymatic hydrolysis of sucrose in aqueous suspension of invertase-loaded PSt microcapsules without coating (●, ○) or with an adsorbed layer of copoly(MA-St) (▲, △). The reaction was carried out at pH 5.5 (○, △) or pH 4.5 (●, ▲). The arrows show pH adjustment by quick addition of a small amount of 2 M HCl or 2 M NaOH to the microcapsule suspension. Figure is adapted from Kokufuta *et al.* (1988).

polymer layer, enzymatic hydrolysis occurred both at pH 5.5 and 4.5, resulting in the formation of both glucose and fructose. In contrast, when the microcapsules with the adsorbed copoly(MA-St) were employed at pH 4.5, the action of the entrapped enzyme was almost completely suppressed (the concentration of the two sugars produced was less than 0.1 μg/ml), but the reaction could be initiated by adjusting the pH of the outer medium to 5.5.

Such on/off control could be repeated reversibly throughout a single run of measurements. Repeated measurements over a period of eight days also gave excellent reproducible results without damage to the microcapsules (see Figure 6.13).

Similar on/off control of the hydrolysis of maltotriose can be performed using β-amylase-loaded PSt microcapsules with a surface layer of adsorbed PEI (Kokufuta *et al.*, 1986). Therefore, pH-sensitive microcapsules covered with an adsorbed polyion layer enable the control of enzyme processes by means of small pH changes in the external medium.

6.5. Conclusions and Future Prospects

The examination of the permeability of polyelectrolyte-coated microcapsules is a useful method for obtaining information on the configurational changes of the polyions on the membrane surface. A dramatic change in the configuration of the polyions such as copoly(MA-St) and PEI on the surface of PSt microcapsules occurs over a narrow pH

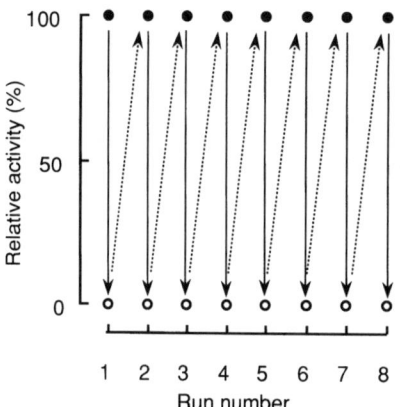

Figure 6.13 Repeated on/off control of hydrolysis of sucrose using invertase-loaded
copoly(MA-St)-coated microcapsules. Relative activity refers to the activity of
the freshly prepared microcapsules (run 1) at pH 5.5 as 100%. The repeated
on/off control was carried out as follows: (1) incubation at pH 5.5 for 2 h;
(2) further incubation at pH 4.5 for 2 h after quick adjustment of pH with
2 M HCl; (3) standing overnight (2 °C) of the capsules which were taken out
of the reaction mixture by centrifugation; and (4) repetition of the procedures
1–3. Figure is adapted from Kokufuta *et al.* (1988).

range. Thus, microcapsules with loaded enzymes and a suitable polyelectrolyte coat
permit the control of enzyme processes by adjusting the pH of the external liquid.

A wide range of polymers are currently available for the preparation of microcapsules, and hence configurational changes of the polyelectrolytes on the capsule
membranes with different hydrophobicities or hydrophilicites present an interesting
area of research. Another related area is the use of polypeptides and proteins as the
polyelectrolyte coat, by which a great deal of information is available on the conformational changes of proteins and polypeptides adsorbed onto hydrophilic or hydrophobic
solid surfaces. This knowledge may provide a better understanding of conformational
changes of proteins in biological membranes in living systems.

Microcapsules coated with a layer of polyelectrolyte whose configuration is dramatically altered in response to small changes in the pH of the outer medium permit the
on/off control of enzyme reactions. Such microencapsulated enzymes are of potential
interest as biochemical sensors and display devices, and constitute 'functional' immobilized enzymes that surpass the conventional functions and advantages of immobilized
enzymes. Thus, the development of pH-sensitive microcapsules containing enzymes
is an interesting interdisciplinary area encompassing enzymology, polymer chemistry
and biomedical engineering.

6.6. Acknowledgments

This work was supported by Grants-in-Aid for Scientific Research from the Japanese
Ministry of Education and New Energy Development Organization (NEDO),
Japan.

6.7. Symbols and Abbreviations

A	empirical constant (Equation 1); total surface area of micro-capsules (Equation 3)
a	empirical constant (Equation 1)
C	concentration of solute
pK_a	apparent dissociation constant
M	molecular weight
$\overline{M}_n$	number-average molecular weight
$\overline{M}_w$	weight-average molecular weight
P	permeability constant
P_{max}	maximum permeability constant
P_{min}	minimum permeability constant
$\bar{r}_o$	the end-to-end distance at the theta state
t	time
$U_{c \to 0}$	limiting electrophoretic mobility
V_m	total volume of microcapsules
α	degree of dissociation
α_η	expansion factor
$[\eta]$	intrinsic viscosity
Φ	universal constant
copoly(MA-MVE)	copolymer of maleic acid with methyl vinyl ether
copoly(MA-St)	copolymer of maleic acid with styrene
MA	maleic acid
MVE	methyl vinyl ether
P	permeability constant
PA	n-propyl alcohol
PAA	poly(acrylic acid)
PEI	poly(ethyleneimine)
PMA	poly(methacrylic acid)
PSt	polystyrene
PVP	poly(4-vinylpyridine)
St	styrene

6.8. References

Chang, T.M.S. (1977) *Biomedical Application of Immobilized Enzymes and Proteins*, Plenum Press, New York, Vol. 1.

Day, J.C. and Robb, I.D. (1980) Conformation of adsorbed poly(vinyl pyrrolidinone) studied by infrared spectrometry, *Polymer*, **21**, 408–412.

Denoyel, R., Durand, D., Lafuma, F., and Audebert, R. (1990) Adsorption of cationic polyelectrolytes onto montmorillonite and silica: Microcalorimetric study of their conformation, *J. Colloid Interface Sci.*, **139**, 281–290.

Farin, D., Volpert, A., and Avnir, D. (1985) Determination of adsorption conformation from surface resolution analysis, *J. Amer. Chem. Soc.*, **107**, 3368–3370.

Ferry, J.D., Udy, D.C., Wu, F.C, Heckler, G.E., and Fordynce, D.B. (1951) Titration and viscosity studies of two copolymers of maleic acid, *J. Colloid Sci.*, **6**, 429–442.

Flory, P.J. and Fox, T.G. (1951) Treatment of intrinsic viscosities, *J. Amer. Chem. Soc.*, **73**, 1904–1098.

Harvey, L.J., Bloomberg, G., and Clark, D.C. (1995) The influence of surface hydrophobicity on the adsorbed conformation of a β-sheet-forming synthetic peptide, *J. Colloid Interface Sci.*, **170**, 161–168.

Ilmain, F., Tanaka, T., and Kokufuta, E. (1991) Volume transition in a gel driven by hydrogen bonding, *Nature*, **349**, 400–401.

Kato, T., Nakamura, K., Kawaguchi, M., and Takahashi, A. (1981) Quasi-elastic light-scattering measurements of polystyrene lattices and conformation of poly(oxyethylene) adsorbed on the lattices, *Polym. J.*, **13**, 1037–1043.

Kitajima, M. and Kondo, A. (1971) Fermentation without multiplication of cells using micro-capsules that contain zymase complex and muscle enzyme extract, *Bull. Chem. Soc. Jpn.*, **44**, 3201–3205.

Kokufuta, E. (1979) Colloid titration behavior of poly(ethyleneimine), *Macromolecules*, **12**, 350–351.

Kokufuta, E. (1980) Electrophoretic and viscometric properties of poly(dicarboxylic acids), *Polymer*, **21**, 177–182.

Kokufuta, E. (1992) Functional immobilized biocatalysts, *Prog. Polym. Sci.*, **17**, 647–697.

Kokufuta, E. (1993) Configuration of polyelectrolytes adsorbed onto the surface of microcapsules. In Dubin, P.L., Tong, T. (eds.), *Colloid-polymer Interactions: Particulate, Amphiphilic, and Biological Surfaces*, ACS Symp. Ser. 532, ACS, Washington, DC, pp. 85–95.

Kokufuta, E. (1999) Polyelectrolyte-coated microcapsules and their applications to biotechnology, *Bioseparation*, **7**, 241–252.

Kokufuta, E., Ito, M., Hirata, M., and Iwai, S. (1974a) Effects of pH and ionic strength on the electrophoretic mobility and viscosity of maleic acid-methyl vinyl ether copolymer, *Kobunshi Ronbunshu* (Jpn. Edn.), **31**, 688–692; *Kobunshi Ronbunshu* (Engl. Edn.), **3**, 1957–1965.

Kokufuta, E., Hirata, M., and Iwai, S. (1974b) Effects of ionic strength and pH on the electrophoretic mobility and viscosity of poly(ethyleneimine), *Kobunshi Ronbunshu* (Jpn. Edn.), **31**, 234–238; *Kobunshi Ronbunshu* (Engl. Edn.), **3**, 1383–1389.

Kokufuta, E., Hirata, M., and Iwai, S. (1975) The relation between electrophoretic mobility and viscosity of sodium polyacrylate, *Nippon Kagaku Kaishi*, 369–373 (in Japanese).

Kokufuta, E. and Iwai, S. (1977) Colloid titration behavior of the maleic acid-methyl vinyl ether copolymer, *Bull. Chem. Soc. Jpn.*, **50**, 3043–3044.

Kokufuta, E., Sodeyama, T., and Katano, T. (1986) Initiation-cessation control of an enzyme reaction using pH-sensitive poly(styrene) microcapsules with a surface-coating of poly(iminoethylene), *J. Chem. Soc., Chem. Commun.*, 641–642.

Kokufuta, E., Shimizu, N., and Nakamura, I. (1988) Preparation of polyelectrolyte-coated pH-sensitive poly(styrene) microcapsules and their application to initiation-cessation control of an enzyme reaction, *Biotechnol. Bioeng.*, **32**, 289–294.

Kokufuta, E., Suzuki, H., Yoshida, R., Yamada, K., Hirata, M., and Kaneko, F. (1998) Role of hydrogen bonding and hydrophobic interaction in the volume collapse of a poly(ethyleneimine) gel, *Langmuir*, **14**, 788–795.

Konstadinidis, K., Thakkar, B., Chakraborty, A., Potts, L.W., Tabbenbaum, R., Tirrell, M., and Evans, J.F. (1992) Segment level chemistry and chain conformation in the reactive adsorption of poly(methyl methacrylate) on aluminum-oxide surfaces, *Langmuir*, **8**, 1307–1317.

Kowalczyk, D., Slomkowski, S., and Wang, F.W. (1994) Changes in conformation of human serum albumin (HSA) and γ-globulins (γ-G) upon adsorption to polystyrene and poly(styrene/acrolein) latexes: Studies by fluorescence spectroscopy, *J. Bioactive Compatible Polym.*, **9**, 282–309.

Morawetz, H. (1965) *Macromolecules in Solution*, John Wiley & Sons, Inc., New York, pp. 348–356.

Ohno, N., Nitta, K., Makino, S., and Sugai, S. (1973) Conformational transition of copolymer of maleic acid and styrene in aqueous solution, *J. Polym. Sci., Polym. Phys. Edn.*, **11**, 413–425.

Okahata, Y., Ozaki, K., and Seki, T. (1984) pH-Sensitive permeability control of polymer-grafted nylon capsule membranes, *J. Chem. Soc., Chem. Commun.*, 519–521.

Okahata, Y., Ariga, K., and Seki, T. (1986a) Redox-sensitive permeation from a capsule membrane grafted with viologen-containing polymers, *J. Chem. Soc., Chem. Commun.*, 73–75.

Okahata, Y., Noguchi, H. and Seki, T. (1986b) Functional capsule membranes (part 23), Thermoselective permeation from a polymer-grafted capsule membrane, *Macromolecules*, **19**, 493–494.

Sakai, H. and Imamura, Y. (1980) The configuration of the adsorbed polymer at the solid–liquid interface as depicted by ESR and IR studies, *Bull. Chem. Soc. Jpn.*, **53**, 1749–1750.

Temerk, Y.M., Ibrahim, M.S., Ahmed, M.E., Ahmed, Z.A., and Kamal, M.M. (1994) Voltametric studies of the adsorption and conformation of polyribocytidylic acid at the mercury-electrode surface, *Bioelectrochem. Bioenerg.*, **32**, 77–82.

Trau, M., Grieser, F., Healy, T.W., and White, L.R. (1992) Investigation by evanescent waves of the charge and conformation of an adsorbed polyelectrolyte at the silica aqueous-solution interface, *Langmuir*, **8**, 2349–2353.

Tsai, M.F. and Moshe, L. (1984) Controlled release by polyelectrolyte microcapsule membranes, *J. Polym. Sci., Polym. Chem. Edn.*, **22**, 2525–2531.

Yanagisawa, M. (1993) Adsorption and configuration of lubricant molecules on overcoat materials, *Wear*, **168**, 167–173.

Xia, J., Dubin, P.L., Izumi, T., Hirata, M., and Kokufuta, E. (1996) Dynamic and electrophoretic light scattering of poly(dimethyldiallylammonium chloride) in salt-free solutions, *J. Polym. Sci., Polym. Phys. Edn.*, **34**, 497–503.

7 Applications of smart hydrogels in separation

Jung Ju Kim and Kinam Park

7.1. Introduction

Separation is an important part of chemical or biological processes for obtaining compounds in pure and/or concentrated forms. Hydrogels have been used widely in separation of biomolecules and others. Recent development in smart hydrogels has made them more useful in separation. Smart hydrogels undergo drastic volume changes in response to small changes in environmental conditions, such as pH, temperature, solvent, salt type, etc. Such a property of smart hydrogels has been applied in the separation of various molecules using relatively simple processes under mild conditions. Temperature-sensitive hydrogels have been most widely used in separation. The smart hydrogels can be prepared in various sizes and shapes as well as various physical states. The relatively mild conditions for separation using smart hydrogels made it useful for the separation of various molecules including proteins which may be easily denatured or degraded during other separation processes. Various applications of hydrogels in separation of biological substances and chemicals are described.

7.2. Description of Hydrogels

Hydrogel is a network of hydrophilic polymers that can hold a large amount of water. A three-dimensional network is formed by crosslinking polymer chains. Crosslinking can be provided by covalent bonds, hydrogen bonding, van der Waals interactions, physical entanglements, or hydrophobic interactions (Kamath and Park, 1993; Park et al., 1993). The key feature of hydrogels is the swelling/deswelling property in aqueous solutions. Hydrogels swell in aqueous solution because hydrophilic polymer chains of the crosslinked network try to dissolve in water. The water-absorbing property of hydrogels has been used in many areas ranging from controlled drug delivery to separation. Hydrogels which swell or deswell (i.e., shrink) upon changes in the environmental condition are known as smart hydrogels (Park and Park, 1999). The property, which makes the smart hydrogels unique, is that the swelling (or shrinking) occurs by only minute changes in the environmental condition. For example, a temperature change of a few degrees is enough to make smart hydrogels swell or shrink. Changes in hydrogel volumes can reach more than a few hundred times. This unique property of smart hydrogels has been exploited in bioseparation. The environmental factors that affect the swelling (or shrinking) of hydrogels are listed in Table 7.1. The most commonly used environmental stimuli are temperature and pH, since the two variables are relatively easy to change. Other stimuli are electricity, ions, solvents, light, and pressure.

Table 7.1 Environmental factors which cause sharp response of hydrogels

Factor	Applications	References
Temperature	Drug delivery	(Bae *et al.*, 1991; Chun and Kim, 1996; Dinarvand and D'Emanuele, 1995; Dong and Hoffman, 1990)
	Separation	(Cussler *et al.*, 1984; Freitas and Cussler, 1987)
	Bioreaction	(Park and Hoffman, 1993b)
	Shape memory	(Hu *et al.*, 1995)
	Artificial muscle	(Kishi *et al.*, 1993)
	Enzyme immobilization	(Shiroya *et al.*, 1995)
pH	Drug delivery	(Bala and Vasudevan, 1982; Brazel and Peppas, 1996; Dong and Hoffman, 1991)
Electric field	Drug delivery	(Kwon *et al.*, 1991; Sawahata *et al.*, 1990)
	Artificial muscle	(Kajiwara and Ross-Murphy, 1992; Osada *et al.*, 1992; Ueoka *et al.*, 1997)
Ions		(Park and Hoffman, 1993a; Starodoubtsev *et al.*, 1995)
Solvents		(Tanaka, 1981)
Light		(Mamada *et al.*, 1990; Suzuki *et al.*, 1996a; Suzuki and Tanaka, 1990; Zhang *et al.*, 1995)
Pressure		(Lee *et al.*, 1990a; Zhong *et al.*, 1996)
Specific molecules	Drug delivery	(Kokufata *et al.*, 1991; Obaidat and Park, 1997; Suzuki *et al.*, 1996b)

For biological processes, smart hydrogels responding to specific biomolecules are also highly useful.

7.3. Applications of Hydrogels in Separation

7.3.1. *Gel Electrophoresis for Separations of Proteins and DNAs*

Crosslinked polyacrylamide and agarose are the most common separation media used in slab gel electrophoresis of proteins and DNA. Choice between crosslinked polyacrylamide and agarose depends on the size of DNAs or proteins and experimental conditions. Crosslinked polyacrylamide hydrogels are used for fine resolution of DNA fragments smaller than 2000 base pairs, while agarose gels are effectively used for separating large DNA fragments ranging from thousands to millions of base pairs. Even though the preparation of polyacrylamide gels is straightforward, the potential of exposing the user to neurotoxic acrylamide monomer is the major drawback in practical applications of polyacrylamide gels. Recently, many new hydrogels have been used in electrophoresis. Of those, temperature-sensitive hydrogels have been most widely used, since they provide a number of advantages over the conventional hydrogels. Temperature-sensitive hydrogels undergo sol-gel phase transition with only a small change in temperature. Thus, they can be easily loaded into a capillary tube by simply lowering the temperature to make a sol state. In addition, the gel-to-sol transition allows easy recovery of the proteins separated after electrophoresis as illustrated in Figure 7.1 (Yoshioka *et al.*, 1994). Crosslinked poly(N-isopropylacrylamide) (PNIAAm) gel was used for electrophoretic separation like normal polyacrylamide gel. After the electrophoresis, separated bands that contained desired substances were excised, and finely crushed. Finally, this gel underwent shrinking and swelling cycles

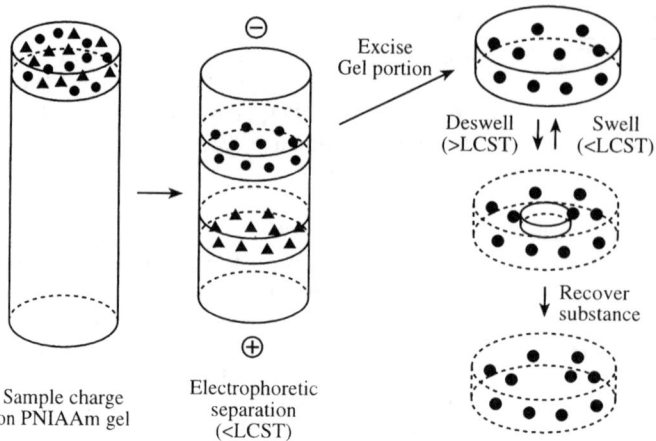

Figure 7.1 Schematic illustration of electrophoretic separation and recovery of substances using a PNIAAm gel. From reference Yoshioka *et al.* (1994).

three times at 37 and 4 °C, respectively. By shrinking the gel at 37 °C, which is above the lower critical solution temperature (LCST), horse heart myoglobin (MW 18,800) and bovine hemoglobin (MW 64,500) were effectively recovered at almost 100% yield. This recovery technique is advantageous for the high recovery yield in addition to the mild operating condition that prevents proteins from the denaturation. Electrophoresis using stimuli-sensitive hydrogels may be applicable to other proteins and DNAs.

Thermoreversible hydrogels are reported to be useful for separation of double stranded DNA fragments (<2000 base pairs) in capillary, tube, and slab electrophoresis (Sassi *et al.*, 1996a). The phase transition behavior of LCST polymers was used to drive a viscosity transition. Sassi *et al.* (1996a) investigated two classes of formulations, gel microsphere suspensions and solutions of uncrosslinked polymers. In their work, separation of DNA fragments with the single-base resolution was achieved for DNA fragments with up to 150 base pairs in the capillary electrophoresis. One of the problems in capillary electrophoresis is the difficulty of loading the polymeric media into capillaries (Gelfi *et al.*, 1995; Grossman, 1994). If there are reversible viscosity transitions without the bulk phase separation or aggregation in suspensions or solutions, such viscosity responsiveness can be used effectively in capillary electrophoresis. Figure 7.2 illustrates the basis for a viscosity transition in a suspension of temperature-sensitive microspheres (Hooper *et al.*, 1997). In a suspension of solid particles, simply adding more particles into solution (Figure 7.2a) increases viscosity. In a suspension of temperature-sensitive PNIAAm particles, viscosity can be increased at a fixed number by swelling microspheres (Figure 7.2b). In addition, the temperature responsiveness enables active control of sieving properties during electrophoresis. Temperature-sensitive polymers, such as poly(N,N'-dimethylacrylamide) or poly(N,N'-diethylacrylamide), can be used for this application. Usually, the resolution is improved as the polymer concentration is increased. The high polymer concentration, however, results in high viscosity making it very difficult to load polymer solution into the capillaries with an inner diameter of usually not more

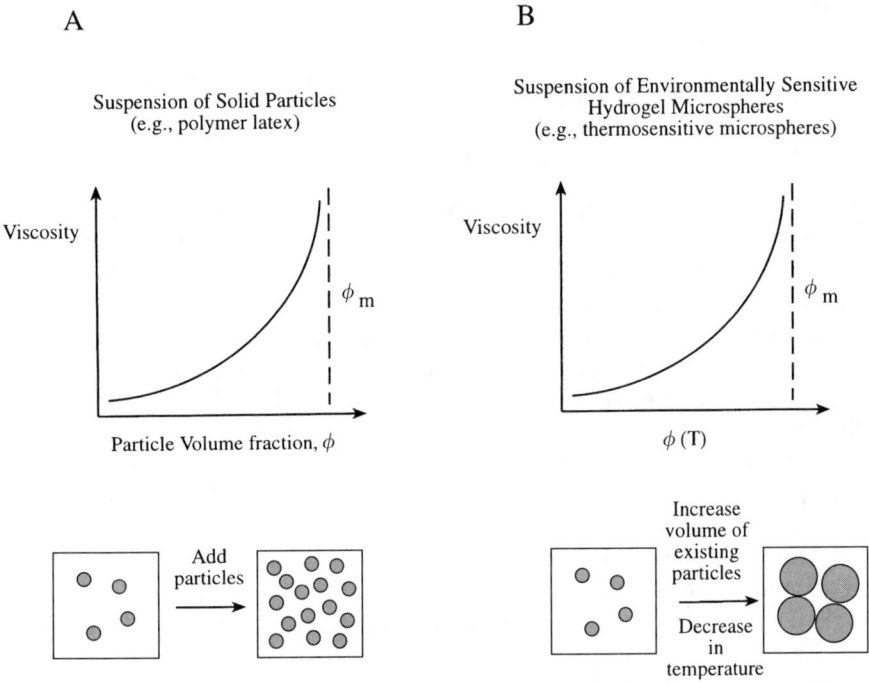

Figure 7.2 Viscosity transitions in particle suspensions. (A) Viscosity increases as the volume fraction, ϕ, increases by adding more particles to the suspension. (B) Viscosity increases as the effective microsphere volume fraction, ϕ, increases by expanding the volume of individual temperature-sensitive microspheres. In both systems, viscosity approaches infinity at some maximum particle packing fraction, ϕ_m. From reference Hooper *et al.* (1997).

than 100 µm. Temperature-sensitive polymers, however, eliminate this particular problem, since viscosity of the solution can be lowered by increasing the temperature.

Poly(ethylene oxide)-poly(propylene oxide)-poly(ethylene oxide) (PEO-PPO-PEO) triblock copolymers have a unique sol-gel transition property in aqueous solution. At low temperatures and low polymer concentrations the polymer is in a unimer state, since both blocks (PEO and PPO) are water-soluble at low temperatures. Increased hydrophobic interaction among the PPO blocks caused by increased temperature leads to micellar formation. As unimers, PEO-PPO-PEO triblock copolymers have relatively low molecular weights. Therefore, such a solution has a relatively low viscosity even at high polymer concentrations (Lenaerts *et al.*, 1987; Prud'homme *et al.*, 1996). However, a gel-like medium with supramolecular crystalline structure can be formed at an appropriate concentration and temperature. By taking advantage of the unique sol-gel property, Wu and coworkers have demonstrated that PEO-PPO-PEO triblock copolymers can be used as a separation medium in capillary electrophoresis (Wu *et al.*, 1997). The block copolymer solution was filled into the capillary tubing by using a microsyringe at 4 °C. Then, the polymer solution became a gel by increasing the temperature to room temperature. After capillary electrophoresis, the gel-like block copolymer separation medium was transformed back into a solution by lowering temperature to 4 °C for easy removal from the capillary tube.

7.3.2. Separation Based on Dewatering Process

7.3.2.1. Dewatering of the Biological Slurries

Large water-absorbability (i.e., high swelling property) of hydrogels can be applied to dewatering process in the treatment of biological slurries. Since most of biological slurries contain large amounts of water, they must be dewatered efficiently before further processing, such as transportation, fermentation, and incineration. Currently, slurry dewatering is done mainly by filtration process. However, the filtration process is inefficient because of the limited filtration area, filter cake, and slow filtration rate. A dewatering process utilizing poly(vinylmethyl ether) (PVME) as a dewatering medium has been proposed to overcome the drawbacks of conventional dewatering methods (Huang *et al.*, 1989). PVME gel is a nonionic gel with a temperature-sensitive swelling property. The equilibrium volume of a spherical PVME gel was related to temperature as shown in Figure 7.3. Photographs of the gel corresponding to 298, 301, and 313 K are also included in the figure. At temperatures above 310 K the gel showed a constant volume, and below 310 K the gel's volume increased with decrease in temperature. In this case, the gel's transition temperature of swelling and shrinking was 310 K. Figure 7.4 shows the reversible volume change of a PVME gel responding to temperature switch from 298 to 321 K. A swollen spherical gel at 298 K (A in Figure 7.4)

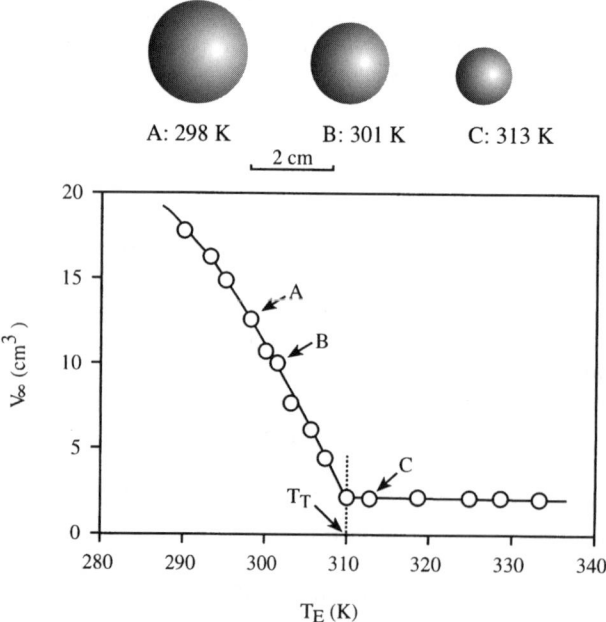

Figure 7.3 Relationship between equilibrium volume V_∞ of one spherical PVME gel and temperature T_E. Photographs show the gels corresponding to 298, 301, and 313 K. From reference Huang *et al.* (1989).

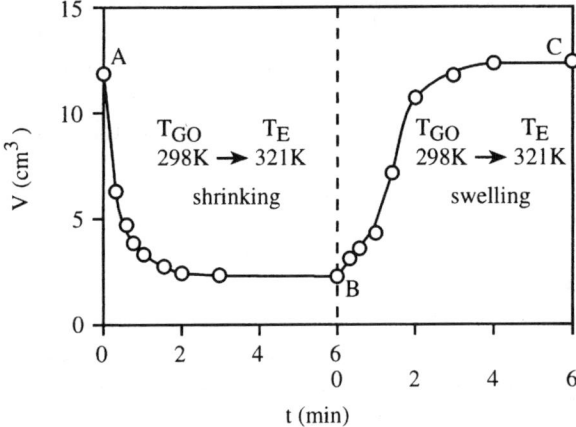

Figure 7.4 Reversible volume change of PVME gel between two different temperatures. From reference Huang *et al.* (1989).

shrunk at 321 K (B in Figure 7.4), and swelled at 298 K again (C in Figure 7.4). The equilibrium volume of gel at C in Figure 7.4 was equal to the initial value at A in Figure 7.4. The change in gel volume according to temperature change was completely reversible and the swelling or shrinking rate was relatively rapid. Porous and nonionic structure of PVME gel is suitable for a medium that requires rapid swelling/shrinking and little effect by ions (Huang *et al.*, 1987). Dewatering of excess activated sludge using spherical PVME gel was calculated to be cost effective. Similar experiments using PNIAAm gel for dewatering fine coal slurries showed similar results (Gehrke *et al.*, 1998). The gels effectively dewatered slurries to around 70 wt% solids without sign of deterioration over a period of 2 months and 20 cycles.

Superabsorbents are also highly useful in dewatering process. It was also demonstrated that initial water content of 30% in the emulsion is reduced to 3% within 10 min by using hydrolyzed starch-*g*-polyacrylonitrile, a superabsorbing hydrophilic polymer, in a semi-continuous separation method (Buwa *et al.*, 1996). A gel-based dewatering of an emulsion would provide an alternate separation method over conventional distillation that requires high-energy consumption. In the presence of surfactant, however, it was seen that rapidly moving tiny water droplets tend to coalesce faster than being absorbed by gel. Hence, the absorption of water from this emulsion occurred much more slowly than in the absence of surfactant.

7.3.2.2. Concentration Using Temperature-sensitive Hydrogels

The temperature-sensitive hydrogels can be allowed to swell in a solution containing molecules to be concentrated. Hydrogels swell by absorbing water, and while water is absorbed, molecules smaller than the pore size of hydrogels can be absorbed into the hydrogel. Molecules larger than the pore size of hydrogels are excluded in the process. After equilibrium swelling is reached, the hydrogels are physically removed from the solution and transferred into another solution. The deswelling (or shrinking) of fully swollen hydrogels can be achieved by a small change in temperature. The shrinking process releases small molecules that were previously absorbed into the hydrogels.

The shrinking process is also called 'collapse' due to a sudden decrease in the hydrogel volume. The collapsed hydrogels are ready to be used again. As water is absorbed into the swelling hydrogel, the volume of aqueous solution in the container is reduced, and this results in concentration of the large molecules remaining in solution. The small molecules absorbed into the hydrogel can be removed from the hydrogel by shrinking the swollen hydrogel through altering the environmental temperature. Since the only change in this process is a small change in temperature, the bioseparation can be achieved relatively easily compared with conventional separation approaches. For temperature-sensitive hydrogels containing ionizable groups (e.g., carboxyl or amine groups), pH can also be used to control the swelling property.

Cussler and coworkers have applied the size selective extraction technology to soy protein isolation process (Trank *et al.*, 1989; Wang *et al.*, 1993). Their process utilizes gel particles made of temperature-sensitive PNIAAm. In the first stage, collapsed PNIAAm hydrogel particles (i.e., those shrunken at temperatures above the LCST) are placed into the defatted protein solution at 5 °C. Approximately 40% of the water is absorbed by the gel as the gel particles swell. The swollen gel particles are removed by centrifugation, and the retentate is diluted with water to the original volume. The water removal process is repeated using additional collapsed gel particles. In the final stage the retentate is not diluted, but is concentrated for spray drying. This alternative process produces 45 kg of protein isolate containing albumins excluded in the acid precipitation process from 100 kg of defatted soybean flakes. Furthermore, it provides more native proteins which may have better flavor and excludes undesirable components such as phytins which are toxic in large amounts.

As shown by the above examples, size selective bioseparation can concentrate the solution at mild conditions without any harmful effect on biomolecules such as proteins. In addition to proteins, enzymes, small solutes, polymer latex particles (Cussler *et al.*, 1984), and even virus (Roepke *et al.*, 1987) can be separated by the size selective separation process. Temperature-sensitive hydrogels do not require any severe changes in other environmental conditions, such as ionic strength, pH, pressure, or shear conditions of the medium during the process (Park and Orozco-Avila, 1992). Since most temperature-sensitive hydrogels have LCST below 50 °C, they require low energy to operate (Galaev and Mattiasson, 1993).

7.3.2.3. Chemically Selective Separation of Mixtures of Organic Solvents

Removal of small amounts of water from organic liquids is an important step in many industrial operations. However, complete removal of water is often complicated by the formation of azeotropes. Typical separation methods, azeotropic or extractive distillation, require high reflux ratio and large number of stages. Those energy intensive and expensive separation processes can be replaced with a new separation method based on chemically-selective polymer gels. Water sorption by polysaccharide adsorbents, such as corn grits, for industrial applications to dehydrate ethanol vapors has been proposed (Ladisch *et al.*, 1984). Polysaccharide adsorbents were regenerated by passing compressed air at low temperatures (100–110 °C) (Bienkowski *et al.*, 1986). A sulfonic acid-type cation-exchange resin was used as sorbent for liquid-phase organics-water separations (Sinegra and Carta, 1987). Sorption equilibrium was determined at 30 °C by means of batch equilibration experiments. The initial hydration of the resin may be interpreted as the result of ion-dipole interactions between water and the resin counterions. Water tends to solvate the counterions and the fixed functional groups as salts are solvated in solutions. Due to chemical crosslinking of the resin, the polymer is not dissolved and the net result at equilibrium is a balance of solvation or osmotic forces, and the elastic forces of the polymeric network. The resin exhibited high selectivity for water and considerable sorption capacity. Selectivity and capacity were dependent upon the ionic form of the resin and the nature of the organic solvent.

7.3.3. Separation through Hydrogel Membrane

Membrane technology for the separation offers a simple and energy efficient process compared to other processes. Separation of products from the substrate during reaction was attempted by immobilization of enzymes to a composite temperature-sensitive membrane (Chen *et al.*, 1998). A composite temperature-sensitive membrane was prepared by using a non-woven polyester support onto which PNIAAm-based hydrogel was cast. α-Amylase was immobilized to the membrane by covalent bonds through reacting with the highly reactive ester groups in N-acryloxysuccinimide (NAS) of the polymer. The composite membranes are temperature-sensitive and can hydrolyze soluble starch and separate it from the hydrolyzed products by stepwise temperature changes. The mechanism of enhanced separation of the products relies on opening/closing of the membrane pores. At temperatures below the LCST of PNIAAm, the polymer membrane provides increased flux of the products by swelling of the temperature-sensitive hydrogel layer. On the other hand, at temperatures above the LCST, a collapsed hydrogel layer blocks the pores and results in decreased flux of the products. Starch hydrolysis with the immobilized enzyme was investigated in two-compartment permeation cells with a composite membrane between the cells. Reaction was carried out by hydrolyzing soluble starch in the donor side and collecting the hydrolyzed products in the receptor side. From a reaction point of view, the reactor should be operated at a high temperature to take advantage of enhanced reaction rates. On the other hand, the temperature should be limited to below LCST for increased transport of the products to the receptor side.

Temperature-sensitive hydrogel membranes can also be used for sequential separation of molecules of different sizes from a mixture (Feil *et al.*, 1991). A crosslinked membrane of poly(N-isopropylacrylamide-co-butylmethacrylate) (95 : 5 mol%) was used to sequentially separate uranine (MW 376), small dextran (MW 4,400), and large dextran (MW 150,000). The mesh size of the hydrogel membrane was controlled by changing the temperature. As temperature is lowered, the mesh size of the membrane increased and larger molecules can diffuse through the membrane. The separation by this method resulted in high purity and recovery of the separated compounds. One limitation of this approach was that it took a long time (100–460 h). It is not only similar to conventional methods such as dialysis and membrane filtration, but also difficult to operate in large scale. Nonaka and coworkers also demonstrated that thin and strong poly(vinyl alcohol-*g*-N-isopropylacrylamide) (PVA-*g*-NIAAm) copolymers could be used as separation membranes (Nonaka *et al.*, 1994). The LCSTs of the copolymers were almost the same as that of PNIAAm hydrogel regardless of the composition of PVA-*g*-NIAAm copolymers and the molecular weight of PVA. The swelling ratio was significantly affected by temperature. This implies that changing the temperature can control the pore size of swollen gel consisting of PVA-*g*-NIAAm copolymers.

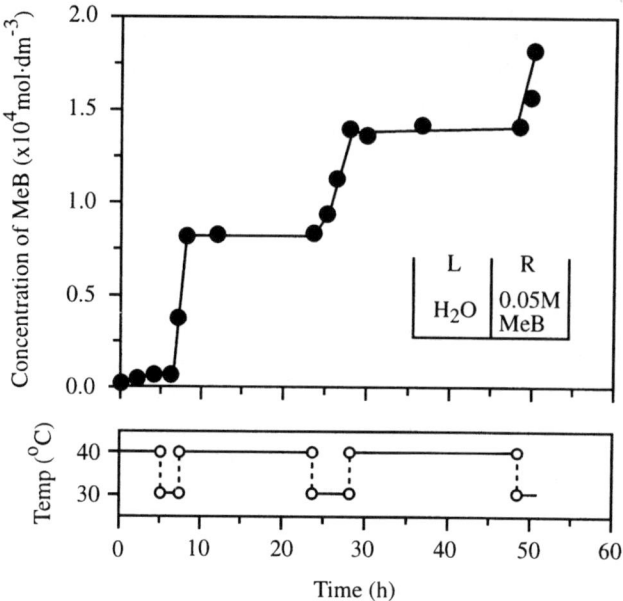

Figure 7.5 Changes in concentration of MeB on the left side of the membrane with stepwise changing of the temperature between 30 and 40 °C. From reference Nonaka *et al.* (1994).

The permeation of methylene blue (MeB) through the membrane increased sharply at 30 °C; however, it was negligible at 40 °C. Figure 7.5 shows that concentration of MeB was discontinuously changed by stepwise changing of the temperature. The concentration of MeB was not changed at 40 °C, but increased sharply at 30 °C.

7.3.3.1. Pervaporation

Pervaporation is recognized as an effective process tool for separation and recovery of liquid mixtures (Fleming, 1992). Pervaporation is characterized by the imposition of a membrane layer between a liquid and a gaseous phase with mass transfer occurring selectively across the membrane to the gas side. The process is termed pervaporation because of permselective evaporation of the liquid molecules. The driving force for permeation in the membrane is achieved by lowering the activity of the permeating components on the permeate side by applying vacuum on the downstream side. The process consists of selective sorption of the components of the liquid mixture into the membrane, diffusion through membrane, and evaporation on the membrane downstream side. The separation can be achieved because the membrane has the ability to transport one component more readily than the other. Since pervaporation involves phase transition from liquid to vapor, it is not essentially an energy saving process. However, pervaporation can offer favorable economics, efficacy, and simplicity. The technique can be integrated easily into distillation and extraction processes and even replace them. Pervaporation could become one of the fundamental methods in membrane separation processes if highly selective and permeable membranes were obtained. Recently, several groups have investigated the pervaporation of water-organic solvent systems using radiation or plasma-grafted composite membranes (Hirotsu, 1987), cationic/anionic interpenetrating polymer networks (Lee *et al.*, 1990b), polyelectrolyte membranes (Reineke *et al.*, 1987), composite membranes prepared from concentrated emulsions (Ruckenstein and Sun, 1995), poly(dimethylsiloxane-co-siloxane) membranes (Lee *et al.*, 1989), and γ-alumina microporous membrane grafted by organosilanes (Alami-Younssi *et al.*, 1998). Sakohara *et al.* have investigated the separation of acetone/water mixtures by thin acrylamide gel membrane prepared in pores of thin ceramic membrane (Sakohara *et al.*, 1990). The gel membrane gave quite large fluxes of water and an extremely high separation factor at 95 mol% of acetone concentration in the upstream. The main reason for such a high separation factor was attributed to the volume change of the gels with high acetone concentration, which blocks the permeation of acetone molecules.

Two types of water–alcohol permselective membranes, i.e., water permselective and alcohol permselective membranes, have been made. The water permselective membrane is much more popular at present because the strength of the hydrogen-bonding interaction between membrane materials and water can be easily modulated using polymeric hydrogels. For the water permselectivity, the membrane should be hydrophilic to sufficiently absorb water from the feed solution. Membranes which contain ionic groups have shown greater permselectivity. However, too much hydrophilicity often leads to defects in mechanical strength, and even dissolution, of the membrane in contact with the aqueous feed solutions. To obtain stronger membranes, crosslinking and/or grafting of polymer molecules onto substrates that have high mechanical strength or composite membranes has been pursued (Ruckenstein and Sun, 1995; Sakohara *et al.*, 1990). In addition, the selectivity and mechanical strength of the membrane (high permeability of specific component) is highly required for efficient operation.

Nakabayashi and coworkers have developed a facilitated transport membrane for CO_2 separation, which can achieve high selectivity (Nakabayashi *et al.*, 1995). The facilitated transport occurs in a membrane containing reversible complexation agents, called carriers (Figure 7.6a). The carrier interacts with dissolved CO_2 in the upstream side of the membrane and forms a CO_2-carrier complex. Then the complex diffuses

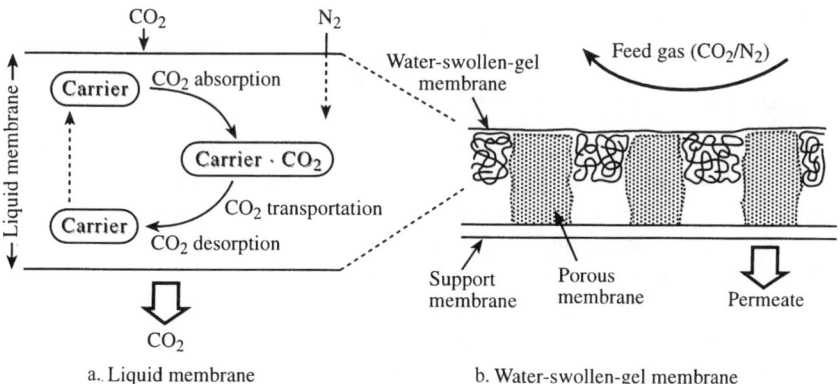

Figure 7.6 Facilitated transport membrane for carbon dioxide separation using a hydrogel membrane. From reference Nakabayashi *et al.* (1995).

across the membrane and releases CO_2 in the downstream side of the membrane. The facilitated transport of CO_2, however, never applied for the industrial use because of weak stability and low permeation rate. To make a stable and highly permeable facilitated transport system, they used a vinyl alcohol/acrylic acid copolymer membrane as a support for the liquid membrane. The stability of a water swollen gel membrane was confirmed over 30 days. Permeability and selectivity substantially increased by adding a crown ether analogue, diaza-15 crown as a complexation agent with potassium ion to K_2CO_3 aqueous solution. The reactivity of carbonate ion which was a carrier of CO_2 was increased by the addition of the complexation additive.

7.3.4. Separation by Adsorption and Desorption

7.3.4.1. Separation of Surfactants

The ionic surfactants in waste water can be easily recovered with polymer flocculants, while it is more difficult to remove nonionic surfactants. Adsorption/desorption phenomena controlled by a slight temperature change can be applied to separate organic substances using temperature-sensitive polymers. An attempt was made to develop a new separation technology by using porous PVME gel to separate organic substances, especially to recover nonionic surfactants in aqueous solution (Ichijo *et al.*, 1994). PVME is one of the well-known temperature-sensitive polymers and undergoes phase transition at around 38 °C. PVME is fully swollen at room temperature, and collapses with temperature increase above the LCST. Elevation of temperature causes the increase in hydrophobicity of the gel. Since PVME gel is dehydrated and becomes hydrophobic with temperature increase, the hydrophobic interaction between PVME and organic substances (nonionic surfactants) also increases with temperature. Poly(oxyethylene nonylphenyl) ether was used as a nonionic surfactant. Poly(oxyethylene nonylphenyl) ether was only slightly adsorbed on PVME gel at a low temperature but was well adsorbed at a high temperature. The adsorbed surfactant was released from the gel into solution by lowering the temperature. A poly(oxyethylene nonylphenyl) ether surfactant was selectively separated from the mixture of those surfactants that have different numbers of oxyethylene unit. This shows that the separation can be accomplished by adsorption/desorption mechanism based on the hydrophobicity of organic substances.

7.3.4.2. Temperature-sensitive Chromatography

Kanazawa *et al.* proposed a selective separation method in chromatography that utilized a temperature-responsive surface with a constant aqueous mobile phase (Kanazawa *et al.*, 1997; Kanazawa *et al.*, 1996). A temperature-responsive semitelechelic copolymer, poly(N-isopropylacrylamide-co-butyl methacrylate), was grafted to the surface of (aminopropyl)silica through the reaction of activated ester-amine coupling. Separation of steroids and proteins (insulin chains A and B, and β-endorphin fragment 1-27) was investigated using the polymer-modified silica as a packing material. As the hydrophobicity of the polymer-modified silica increased, the capacity factors and retention times for steroids increased. The retention times of five steroids and benzene largely depended on temperature as shown in Figure 7.7. Changes in the retention times for hydrophobic steroids such as testosterone were

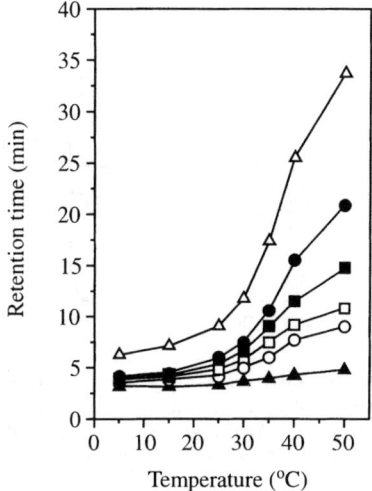

Figure 7.7 Retention times of the five steroids on the temperature-responsive column. ▲, benzene; ○, hydrocortisone; □, prednisolone; ■, dexamethasone; ●, hydrocortisone acetate; and △, testosterone. From reference Kanazawa *et al.* (1996).

larger than those of their hydrophilic counterparts and benzene on the columns packed with PNIAAm-modified silica. Retention times of insulin A and B, and β-endorphin fragment 1-27 were less than 15 min, while those of steroids were less than 30 min at 25 °C. The hydrophobicity of the column was achieved by temperature increase. Only a small change of column temperature, in the range of 5–35 °C, was the major controlling factor in this new chromatographic technique. There was no need to change the aqueous mobile phase. Temperature-responsive chromatography is able to separate solutes without the use of organic solvents. This technology may provide cost-effective separation of proteins and peptides maintaining biological activity in preparatory liquid chromatography.

7.3.5. Separation by Aqueous Two-phase System

Aqueous two-phase systems can be formed when two incompatible polymers or a polymer and an inorganic salt are mixed in water at appropriate concentrations. These systems provide an efficient, mild separation method which is suitable for many biomolecules, such as proteins, nucleic acids, and even cellular particulates (e.g., cells, cell organelles) (Walter and Larsson, 1994). They are also easy to manipulate, reliable in scaling-up, and simple in operation. Unlike the partitioning of soluble materials which distribute according to their solubilities between the two phases, particulates partition between one bulk phase and the interface. The normal or resting position of particulates is the interface because particulates can serve to reduce the interfacial energy. In most cases, the cell partitions between the top phase and the interface as shown in Figure 7.8. The partitioning behavior of cells depends on their surface properties. The most common systems used in cell separation are poly(ethylene glycol) (PEG)/dextran and PEG/salt systems. The phases have high water content and

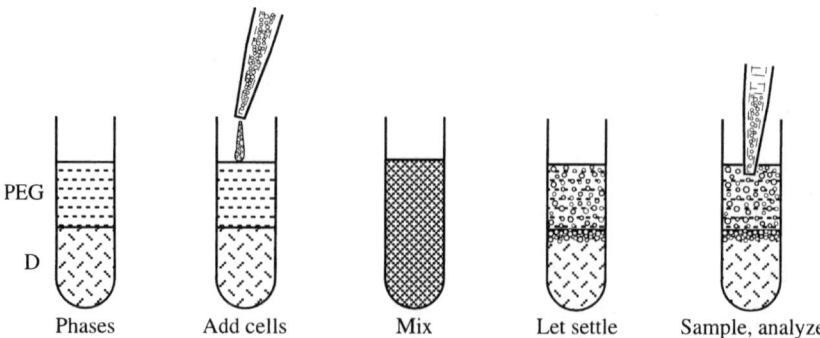

PEG

D

Phases   Add cells   Mix   Let settle   Sample, analyze

Figure 7.8 Schematic presentation of partitioning of cells. PEG and D indicate PEG-rich phase and dextran-rich phase, respectively. Known quantities of cells are added to the phases, mixed and centrifuged to settle. At the end of settling an aliquot is withdrawn from the top phase, and the cell quantity is determined. From reference Walter and Larsson (1994).

low interfacial tension. Relatively high costs of these polymers have prevented wide applications of these systems in biotechnology. This partly comes from difficulties in polymer recycling.

As an alternative, hydroxypropyl starch derivatives with poly(ethylene glycol) (PEG) has been used (Venancio *et al.*, 1996). Various hydroxypropyl starch systems (Reppal® PES) with different degrees of substitution were prepared (Ling *et al.*, 1989). Reppal® PES is commercially available at reasonable cost for large-scale use from carbamyl AB in Kristianstad, Sweden. It is also operated in mild conditions in comparison with PEG/salts system operated at rather high ionic strength, which may cause the denaturation of sensitive biological substances. However, it also has a limitation in polymer recycling and in obtaining the target protein in a polymer-free solution after extraction. Since a 10% (w/w) solution of PEG (MW, 20,000) in water has a cloud point near 120 °C, temperature-induced phase separation cannot be used. Thus, this system may be used only in selected applications due to denaturation or deformation of biomolecules at high temperature (Harris *et al.*, 1991).

More advanced commercial systems have been developed based on PEO-PPO copolymer (Ucon® 50-HB-5100, Union carbide). Ucon® 50-HB-5100 is a non-ionic random copolymer of 50% ethylene oxide (EO) and 50% propylene oxide (PO) with an average molecular weight of 4,000. This polymer allows operation at low temperature due to its low cloud point (less than 60 °C). This provides more acceptable mild operational conditions. In the Ucon/Reppal PES systems, the top phase is enriched with PEO-PPO copolymer. The PEO-PPO copolymer phase is then isolated in a separate container and another phase separation is formed with gradual increase of temperature and addition of a low concentration of salt. In this step the top water phase is almost free of polymer. The target substance is obtained in the top water phase and the polymer can be recycled for another use. Wider applications of this system requires more exploration on the partitioning effect on phases, salts effects on the lowering the cloud point as well as large scale use. Partitioning and recovery of ecdysteroids, effects of ions on partitioning of bovine serum albumin and lysozyme using Ucon/dextran systems have been reported (Alred *et al.*, 1993;

After mixing and phase separation:

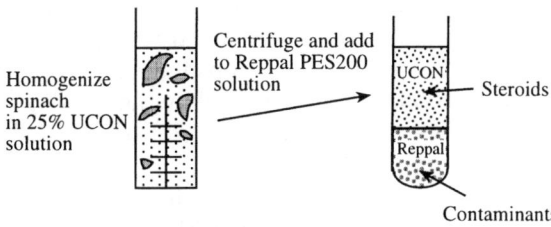

Temperature-induced phase separation:

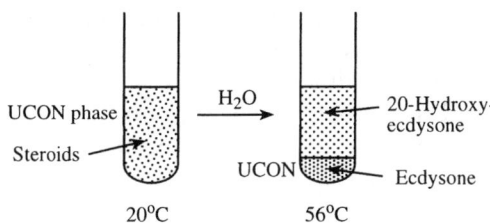

Figure 7.9 Schematic representation of extraction and purification of ecdysteroids from spinach leaves using aqueous two-phase partitioning and temperature-induced phase separation. From reference Modlin *et al.* (1994).

Johansson *et al.*, 1996). Aqueous two-phase partitioning coupled with temperature-induced phase separation was a quick, easy, and inexpensive technique for extracting and purifying ecdysteroids from raw materials (Modlin *et al.*, 1994), amino acids (Li *et al.*, 1997), enzymes (Farkas *et al.*, 1996), and peptides (Johansson *et al.*, 1997). Figure 7.9 shows purification procedure of ecdysteroids from spinach leaves using aqueous two-phase partitioning and temperature-induced phase separation. Ecdysteroids (ecdysone and 20-hydroxyecdysone) are hormones that regulate molting cycles of anthropods. Extraction methods using one or several non-polar solvents and/or hot water are generally used to obtain ecdysteroids from raw materials. After phase separation in the primary aqueous two-phase system the upper Ucon-rich phase is removed, isolated and followed by temperature-induced phase separation with the formation of an Ucon-rich and water-rich phase. A different partitioning result was obtained from the temperature-induced phase separation according to hydrophobicity of ecdysteroids. The greater hydrophobicity of ecdysone caused it to partition more to the Ucon-rich phase.

7.4. Controlling Factors for High Separation Efficiency

7.4.1. *Polymerization Condition*

A reproducible synthesis of PNIAAm hydrogels is difficult to achieve (Gehrke, 1993). The main reason to this problem is that LCST of PNIAAm chains (approximately 32 °C) is close to the usual polymerization temperatures. A local or overall temperature rise in the polymerization reaction may cause a phase separation and formation of heterogeneous structures. Thus, the polymerization techniques significantly affected the properties of PNIAAm hydrogels in the protein separation efficiency (Kayaman *et al.*, 1998). PNIAAm hydrogels formed below 18 °C were homogeneous, whereas those formed at higher temperatures exhibited heterogeneous structures. Visual observations showed that the gels prepared at 18 °C or lower were transparent, whereas those formed at higher temperatures (e.g., at 25 °C) were opaque. Since the turbidity of a gel is a direct result of light scattering from the spatial inhomogeneity of the gel refractive index, it also indicates the water molecules expelled from the PNIAAm network at high temperatures.

Different preparations of PNIAAm hydrogels resulted in different equilibrium swelling ratios. The gels prepared at temperatures higher than 30 °C exhibited a loose structure and increased the equilibrium swelling ratio (Kayaman *et al.*, 1998). Variation of the separation efficiency in concentrating penicillin G acylase was also observed as a function of the gel preparation temperature. Higher water absorption capacity and rapid swelling rate was obtained from the gels prepared by inverse suspension polymerization than by solution polymerization, although the gels were prepared under the same experimental conditions. PVME gels prepared by γ-irradiation is an another example of the effect of different polymerization techniques on the separation efficiency. Those also showed high porosity and fast swelling rate, and thus they were used for dewatering processes of biological slurry (Huang *et al.*, 1989, 1987). The fast response of those gels was ascribed to a macroporous structure caused by phase separation of the polymer solution during irradiation.

7.4.2. Gel Composition

The network homogeneity of polyacrylamide hydrogels is largely affected by the concentration ratio of N,N′-methylenebisacrylamide (BIS, crosslinker) to acrylamide (AAm, monomer). Increase in BIS concentration reduces the effective pore size of the network, while increasing the heterogeneity of the network (Sakohara *et al.*, 1992). The amount of ethanol absorbed into the gel increased with increasing BIS concentration at the composition of 1 and 1.5 M of AAm. It was thought that the heterogeneity of the network rather than the pore size played a dominant role in the penetration of ethanol into the gel. On the other hand, at a composition of 2 M of AAm, the gel prepared with a concentration of 0.08 M of BIS held less ethanol than that prepared with 0.04 M of BIS. Thus, the effective pore size of the network and its homogeneity are expected to be the key factors for the separation of ethanol–water mixtures.

Network defects in BIS-crosslinked polyacrylamide hydrogels increased with higher amount of the crosslinking agent. The crosslinking efficiency was progressively lowered as the concentration of crosslinking agent increased, and the polymer volume fraction at equilibrium swelling reached a plateau at the 7% (w/w) of the crosslinker concentration (Baselga *et al.*, 1987, 1989). The non-random distribution of BIS and the formation of intramolecular cycles were suggested. If a growing radical finds, by cyclization, a pendent vinyl group with very low probability, then the formation of intramolecular cycles has a minor contribution to the formation of non-effective networks. As the crosslinker concentration increases to 7% of the monomer, the probability of forming permanent intramolecular cycles became significant. The efficiency of the crosslinking reaction increases mainly due to the lower cyclization during network formation (Ilavsky and Hrouz, 1983). The homogeneity of gels is one of the important factors which determines the physical properties of a gel. The homogeneous, transparent gel might be obtained even in high monomer concentration by lowering the reaction rate (Tanaka *et al.*, 1988). The low reaction temperature is most effective for this purpose, since the reaction heat may accelerate the gellation locally. The diffusion of reaction heat and the dependence of the reaction rate on the temperature probably dominate the spatial heterogeneity.

7.4.3. Pore Size of Hydrogel

The selectivity in separation can be controlled by changing mesh size of the gel network, and the separation degree is affected by solute shape and size (Ishidao *et al.*, 1997). The mechanism by which the gels separate molecules remains unclear. The most obvious variable is the solute size. The gels clearly separate large molecules from small molecules. The size exclusion effect can be rationalized by imagining the gel as an expanded mesh. Small molecules can easily penetrate the pores in the mesh, but large molecules cannot enter.

The mesh size of the polymer network, ξ, is related to the volume swelling ratio and the unperturbed mean end-to-end distance of the chains between crosslinks as defined by the crosslink density:

$$\xi = \alpha \langle \bar{r}_o^2 \rangle^{1/2} = Q^{1/3} \left[2C_n \left(\frac{\bar{M}_c}{\bar{M}_r} \right) \right]^{1/2} l \tag{1}$$

where α is an expansion factor, which is equal to the cube root of the volume swelling ratio, $\langle \bar{r}_o^2 \rangle$ is the unperturbed mean square end-to-end distance of the chains between crosslinks (Peppas and Barr-Howell, 1986). C_n is the characteristic ratio or rigidity factor for the polymer (obtained as the molar average of the C_n values of the two homopolymers), $\bar{M}_c$ is the molecular weight of polymer chains between crosslinks, and $\bar{M}_r$ is the effective molecular weight of the repeating unit, which was determined by a weighted average of the two monomer molecular weights, based on the copolymer composition. The l term represents the carbon–carbon bond length of 0.154 nm and numeric value of two represents for all vinyl polymers. The degree of crosslink or crosslinking density, ρ_x is defined as:

$$\rho_x = \frac{1}{v \bar{M}_c} \tag{2}$$

where v is the specific volume of polymer. In addition, the theoretical crosslinking density, $\rho_{x,\text{theory}}$ can be calculated from the theoretical molecular weight of polymer chains between crosslinks, $\bar{M}_{c,\text{theory}} = \bar{M}_r / 2X$ by assuming the vinyl groups in the crosslinking agent react quantitatively without crosslinking by impurities or entanglements. The nominal crosslinking ratio, X is defined as the ratio of moles of crosslinking agent to the moles of monomers. $\bar{M}_c$ was measured from swelling studies for poly(2-hydroxyethyl methacrylate) hydrogels (Peppas *et al.*, 1985) and poly(glyceryl methacrylate) hydrogels (Leung and Robinson, 1993), and corresponding mesh sizes were obtained. Simple change of the nominal crosslinking ratio in hydrogel preparation results in obtaining different mesh sizes of hydrogels. As the nominal crosslinking ratio increases, the mesh size and swelling ratio of networks decreases.

The mesh size of hydrogels changed greatly between swollen and collapsed states by changing environmental stimuli, such as temperature and pH, in medium

(Brazel and Peppas, 1995). Pore-size distribution have been measured for cationic acrylamide-based hydrogels differed in extent of crosslinking ratio (Kremer *et al.*, 1994). Measurements were based on the mixed-solute-exclusion method using dextran or poly(ethylene glycol/oxide) as probes. Pore-size distributions shifted to lower pore sizes with increase of crosslinker to monomer ratio. For design of novel gels, it is desirable to predict the size-exclusion curves, especially as a function of polymer volume fraction. Such predictions can provide guidance for synthesis of a gel for particular applications where the size exclusion is important. However, predicting the effect of nominal crosslinking ratio on size-exclusion curves remains an area for continuing research (Sassi *et al.*, 1996b).

7.4.4. Charge Density

Since size-selective separation depends on the relationship between solute size and pore size, any variable which affects pore size will lead to different exclusion behavior. The ionizable monomer content of the gel affects the degree of swelling and exclusion. The exclusion behavior of a polyelectrolyte gel is highly dependent on its interaction with charged solutes. Separation efficiency or selectivity can be affected by thermodynamic partitioning due to charge densities of solutes and hydrogels. An anionic solute like sodium pentachlorophenolate was almost completely excluded by partially hydrolyzed polyacrylamide, a pH-sensitive hydrogel (Gehrke *et al.*, 1986), whereas its separation was negligible for the nonionic PNIAAm hydrogel (Freitas and Cussler, 1987). In the separation of charged solutes using polyelectrolytes, Donnan equilibrium should be considered. When we assume that a gel is anionic with sodium counterions, Donnan equilibrium can be explained as follows (Tanford, 1961):

$$[Na^+]_i [A^-]_i = [Na^+]_o [A^-]_o \tag{3}$$

where $[A^-]_i$ and $[A^-]_o$ are the concentration of monovalent anionic solute inside and outside of the gel, respectively. The sodium ion concentration outside the gel must equal the anion concentration outside because of electroneutrality. The sodium ion concentration inside the gel must balance the anionic charge of $[A^-]_i$ and of the gel itself. Thus, the following relationship is obtained:

$$([RCOO^-] + [A^-]_i)[A^-]_i = [A^-]_o^2 \tag{4}$$

where $[RCOO^-]$ is the charge concentration of gel per volume of the swollen gel. We can solve this quadratic equation to find $[A^-]_i$, thus the concentration of anionic solute inside of the gel which is related to the charge concentration of the gel. The effect of charge on the exclusion of dextran sulfate in 0.5 M NaCl solution was performed using copolymers of N-isopropylacrylamide and 2-acrylamido-2-methyl-1-propanesulfonic acid ($R–SO_3–H^+$) (Vasheghani-Farahani *et al.*, 1992). In this solution, the swelling ratio of copolymers as a function of ionizable monomer content was nearly constant due to high concentration of electrolyte. However, dextran sulfate was much more excluded from an ionic copolymer gel with 10 mol% of $R–SO_3–H^+$ than from nonionic PNIAAm. Donnan equilibrium was responsible for the enhanced exclusion of the negatively charged solute by the anionic copolymer gel.

The ionizable fixed charge groups of polyelectrolyte give rise to important electromechanical and swelling effect on protein transport and separation. Alteration of the electrostatic swelling forces arising from the fixed charge groups of polyelectrolyte will lead to dramatic changes in bulk dimensions and microstructures. Four mechanisms are identified in controlling the solute flux through polyelectrolyte membrane by chemical and electrical control. Those are electromechanical deformation of the membrane, electroosmotic and electrophoretic augmentation of the solute flux within the membrane, and electrostatic partitioning of charged solutes into charged membranes (Grimshaw *et al.*, 1989). Selectivity changes in a particular separation process can be achieved through a combination of those mechanisms. The selective changes can be obtained based on the size and charge of the solutes. Electrically controlled changes were demonstrated by a 21-fold change in the relative flux of labeled bovine serum albumin and ribonuclease A (Grimshaw *et al.*, 1990).

7.5. Future Possibilities of Hydrogels in Bioseparation

There are a number of properties of smart hydrogels to be improved for the enhanced utilization in bioseparation (Galaev and Mattiasson, 1993; Huang *et al.*, 1989; Kim and Park, 1999). Ideal hydrogels for use in concentrating protein solutions should exhibit fast swelling/shrinking rates and high swelling capacities, and the gel should exclude all the protein molecules; that is, the separation efficiency should be 100%. Besides the physicochemical properties of hydrogel, the interaction of biomolecules with hydrogels and separation conditions are important for the successful bioseparation using hydrogels. These improvements are likely to increase the utilization of smart hydrogels in biotechnology. The biotechnology industry is explosively expanding and downstream processing mainly constitutes the costs of biological macromolecular products. Smart hydrogels can be useful in downstream processing because of their smart phase transition in response to small change of environmental stimuli (Park and Park, 1999). New separation technology using physical means (changing the temperature or electric field) will replace the conventional separation methods based on chemical means (changing salt concentration or adding specific eluents). Because the principal chemical composition remains unchanged, a number of other procedures (removing salts and eluents) could be eliminated. We hope to see other various applications of smart hydrogels in separation in the near future. New horizons of separation using smart hydrogels will promote further advances in biotechnology and applications of the biotechnology products.

7.6. Symbols and Abbreviations

ϕ	Volume fraction of particles in suspension
ϕ_m	Maximum volume fraction of particles in suspension
ξ	Mesh size of polymer network
α	Expansion factor
$\langle \bar{r}_o^2 \rangle$	Unperturbed mean square end-to-end distance of the chains between crosslinks
Q	Equilibrium volume swelling ratio
C_n	Characteristic ratio or rigidity factor of the polymer
$\overline{M}_c$	Molecular weight of polymer chains between crosslinks
$\overline{M}_r$	Effective molecular weight of the repeating unit

l	Carbon–carbon bond length (0.154 nm)
ρ_x	Degree of crosslink or crosslinking density
ν	Specific volume of polymer
$\rho_{x,\text{theory}}$	Theoretical crosslinking density
$\overline{M}_{c,\text{theory}}$	Theoretical molecular weight of polymer chains between crosslinks
X	Molar ratio of crosslinking agent to monomers
$[A^-]_i$	Concentration of monovalent anionic solute inside of the gel
$[A^-]_o$	Concentration of monovalent anionic solute outside of the gel
$[RCOO^-]$	Charge concentration of gel per volume of the swollen gel
Bis	N,N'-methylenebisacrylamide
LCST	lower critical solution temperature
MeB	methylene blue
NAS	N-acryloxysuccinimide
PEO	poly(ethylene oxide)
PNIAAm	ploy(N-isopropylacrylamide)
PpO	poly(propylene oxide)
PVA	poly(vinyl alcohol)
PVME	poly(vinylmethyl ether)

7.7. References

Alami-Younssi, S., Kiefer, C., Larbot, A., Persin, M., and Sarrazin, J. (1998) Grafting γ alumina microporous membranes by organosilanes: characterisation by pervaporation, *J. Membr. Sci.*, **143**, 27–36.

Alred, P.A., Tjerneld, F., and Modlin, R.F. (1993) Partitioning of ecdysteroids using temperature-induced phase separation. *J. Chromatogr.*, **628**, 205–214.

Bae, Y.H., Okano, T., and Kim, S.W. (1991) "On-off" thermocontrol of solute transport. II. Solute release from thermosensitive hydrogels, *Pharm. Res.*, **8**, 624–628.

Bala, K. and Vasudevan, P. (1982) pH-sensitive microcapsules for drug release, *J. Pharm. Sci.*, **71**, 960–962.

Baselga, J., Hernandez-Fuentes, I., Pierola, I.F., and Llorente, M.A. (1987) Elastic properties of highly crosslinked polyacrylamide gels, *Macromolecules*, **20**, 3060–3065.

Baselga, J., Llorente, M.A., Hernandez-Fuentes, I., and Pierola, I.F. (1989) Network defects in polyacrylamide gels, *Eur. Polym. J.*, **25**, 471–475.

Bienkowski, P.R., Barthe, A., Voloch, M., Neuman, R.N., and Ladisch, M.R. (1986) Breakthrough behavior of 17.5 mol% water in methanol, ethanol, isopropanol, and t-butanol vapors passed over corn grits, *Biotechnol. Bioeng.*, **28**, 960–964.

Brazel, C.S. and Peppas, N.A. (1995) Synthesis and characterization of thermo- and chemomechanically responsive poly(N-isopropylacrylamido-co-methacrylic acid) hydrogels, *Macromolecules*, **28**, 8016–8020.

Brazel, C.S. and Peppas, N.A. (1996) Pulsatile local delivery of thrombolytic and antithrombotic agents using poly(N-isopropylacrylamide-co-methacrylic acid) hydrogels, *J. Controlled Rel.*, **39**, 57–64.

Buwa, V.V., Lele, A.K., and Badiger, M.V. (1996) Gel-based separation of an *o*-toluidine-water emulsion, *Ind. Eng. Chem. Res.*, **35**, 4182–4184.

Chen, J.P., Sun, Y.M., and Chu, D.H. (1998) Immobilization of alpha-amylase to a composite temperature-sensitive membrane for starch hydrolysis, *Biotechnol. Prog.*, **14**, 473–478.

Chun, S.-W. and Kim, J.-D. (1996) A novel hydrogel-dispersed composite membrane of poly(N-isopropylacrylamide) in a gelatin matrix and its thermally actuated permeation of 4-acetaminophen, *J. Controlled Rel.*, **38**, 39–47.

Cussler, E.L., Stokar, M.R., and Varberg, J.E. (1984) Gels as size selective extraction solvents, *AIChE J.*, **30**, 578–582.

Dinarvand, R. and D'Emanuele, A. (1995) The use of thermosensitive hydrogels for on-off release of molecules, *J. Controlled Rel.*, **36**, 221–227.

Dong, L.-C. and Hoffman, A.S. (1990) Synthesis and application of thermally reversible heterogels for drug delivery, *J. Controlled Rel.*, **13**, 21–31.

Dong, L.-C. and Hoffman, A.S. (1991) A novel approach for preparation of pH-sensitive hydrogels for enteric drug delivery, *J. Controlled Rel.*, **15**, 141–152.

Farkas, T., Stalbrand, H., and Tjerneld, F. (1996) Partitioning of β-mannanase and α-galactosidase from *Aspergillus niger* in Ucon/Reppal aqueous two-phase systems and using temperature-induced phase separation, *Bioseparation*, **6**, 147–157.

Feil, H., Bae, Y.H., Feijen, J., and Kim, S.W. (1991) Molecular separation by thermosensitive hydrogel membranes, *J. Membr. Sci.*, **64**, 283–294.

Fleming, H.L. (1992) Consider membrane pervaporation, *Chem. Eng. Prog.*, **88**, 46–52.

Freitas, R.F.S. and Cussler, E.L. (1987) Temperature sensitive gels as extraction solvents, *Chem. Eng. Sci.*, **42**, 97–103.

Galaev, I.Yu. and Mattiasson, B. (1993) Thermoreactive water-soluble polymers, nonionic surfactants, and hydrogels as reagents in biotechnology, *Enzyme Microb. Technol.*, **15**, 354–366.

Gehrke, S.H. (1993) Synthesis, equilibrium swelling, kinetics, permeability and applications of environmentally responsive gels, *Adv. Polym. Sci.*, **110**, 81–144.

Gehrke, S.H., Andrews, G.P., and Cussler, E.L. (1986) Chemical aspects of gel extraction, *Chem. Eng. Sci.*, **41**, 2153–2160.

Gehrke, S.H., Lyu, L.-H., and Barnthouse, K. (1998) Dewatering fine coal slurries by gel extraction, *Separation Sci. Technol.*, **33**, 1467–1485.

Gelfi, C., Orsi, A., Leoncini, F., and Righetti, P.G. (1995) Fluidified polyacrylamides as molecular sieves in capillary zone electrophoresis of DNA fragments, *J. Chromatogr. A*, **689**, 97–105.

Grimshaw, P.E., Grodzinsky, A.J., Yarmush, M.L., and Yarmush, D.M. (1989) Dynamic membranes for protein transport: chemical and electrical control, *Chem. Eng. Sci.*, **44**, 827–840.

Grimshaw, P.E., Grodzinsky, A.J., Yarmush, M.L., and Yarmush, D.M. (1990) Selective augmentation of macromolecular transport in gels by electrodiffusion and electrokinetics, *Chem. Eng. Sci.*, **45**, 2917–2929.

Grossman P.D. (1994) Electrophoretic separation of DNA sequencing extension products using low-viscosity entangled polymer networks, *J. Chromatogr. A*, **663**, 219–227.

Harris, P.A., Karlström, G., and Tjerneld, F. (1991) Enzyme purification using temperature-induced phase formation, *Bioseparation*, **2**, 237–246.

Hirotsu, T. (1987) Water-ethanol separation by pervaporation through plasma graft polymerized membranes, *J. Appl. Polym. Sci.*, **34**, 1159–1172.

Hooper, H.H., Yu, J., Sassi, A.P., and Soane, D.S. (1997) Viscosity transitions in aqueous suspensions of hydrogel microspheres, *J. Appl. Polym. Sci.*, **63**, 1369–1372.

Hu, Z., Zhang, X., and Li, Y. (1995) Synthesis and applications of modulated polymer gels, *Science*, **269**, 525–527.

Huang, X., Unno, H., Akehata, T., and Hirasa, O. (1987) Analysis of kinetic behavior of temperature-sensitive water-absorbing hydrogel, *J. Chem. Eng. Jpn.*, **20**, 123–128.

Huang, X., Akehata, T., Unno, H., and Hirasa, O. (1989) Dewatering of biological slurry by using water-absorbent polymer gel, *Biotechnol. Bioeng.*, **34**, 102–109.

Ichijo, H., Kishi, R., Hirasa, O., and Takiguchi, Y. (1994) Separation of organic substances with thermo-responsive polymer hydrogel, *Polym. Gels Netw.*, **2**, 315–322.

Ilavsky, M. and Hrouz, J. (1983) Phase transition in swollen gels. 5. Effect of the amount of diluent at network formation on the collapse and mechanical behavior of polyacrylamide networks, *Polym. Bull.*, **9**, 159–166.

Ishidao, T., Sugimoto, H., Onoue, Y., Song, I.S., Iwai, Y., and Arai, Y. (1997) Mesh sizes of poly(N-isopropylacrylamide) gel in aqueous solution, *J. Chem. Eng. Jpn.*, **30**, 162–166.

Johansson, H.O., Lundh, G., Karlström, G., and Tjerneld, F. (1996) Effects of ions on partitioning of serum albumin and lysozyme in aqueous two-phase systems containing ethylene oxide/propylene oxide co-polymers, *Biochim. Biophys. Acta*, **1290**, 289–298.

Johansson, H.O., Karlström, G., and Tjerneld, F. (1997) Temperature-induced phase partitioning of peptides in water solutions of ethylene oxide and propylene oxide random copolymers, *Biochim. Biophys. Acta*, **1335**, 315–325.

Kajiwara, K. and Ross-Murphy, S.B. (1992) Synthetic gels on the move, *Nature*, **355**, 208–209.

Kamath, K.R. and Park, K. (1993) Biodegradable hydrogels in the drug delivery, *Adv. Drug Delivery Rev.*, **11**, 59–84.

Kanazawa, H., Yamamoto, K., Matsushima, Y., Takai, N., Kikuchi, A., Sakurai, Y., and Okano, T. (1996) Temperature-responsive chromatography using poly(N-isopropylacrylamide)-modified silica, *Anal. Chem.*, **68**, 100–105.

Kanazawa, H., Kashiwase, Y., Yamamoto, K., Matsushima, Y., Kikuchi, A., Sakurai, Y., and Okano, T. (1997) Temperature-responsive liquid chromatography. 2. Effects of hydrophobic groups in N-isopropylacrylamide copolymer-modified silica, *Anal. Chem.*, **69**, 823–830.

Kayaman, N., Kazan, D., Erarslan, A., Okay, O., and Baysal, B.M. (1998) Structure and protein separation efficiency of poly(N-isopropylacrylamide) gels: Effect of synthesis conditions, *J. Appl. Polym. Sci.*, **67**, 805–814.

Kim, J.J. and Park, K. (1999) Smart hydrogels for bioseparation, *Bioseparation*, **7**, 177–184.

Kishi, R., Ichijo, H., and Hirasa, O. (1993) Thermo-responsive devices using poly(vinyl methyl ether) hydrogels, *J. Intel. Mater. Syst. Struct.*, **4**, 533–537.

Kokufata, E., Zhang, Y.-Q., and Tanaka, T. (1991) Saccharide-sensitive phase transition of a lectin-loaded gel, *Nature*, **351**, 302–304.

Kremer, M., Pothmann, E., Rossler, T., Baker, J., Yee, A., Blanch, H., and Prausnitz, J.M. (1994) Pore-size distribution of cationic polyacrylamide hydrogels varying in initial monomer concentration and cross-linker/monomer ratio, *Macromolecules*, **27**, 2965–2973.

Kwon, I.C., Bae, Y.H., and Kim, S.W. (1991) Electrically erodible polymer gel for controlled release of drugs, *Nature*, **354**, 291–293.

Ladisch, M.R., Voloch, M., Hong, J., Bienkowski, P., and Tsao, G.T. (1984) Cornmeal adsorber for dehydrating ethanol vapors, *Ind. Eng. Chem. Process Des. Dev.*, **23**, 437–443.

Lee, K.K., Cussler, E.L., Marchetti, M., and McHugh, M.A. (1990a) Pressure-dependent phase transitions in hydrogels, *Chem. Eng. Sci.*, **45**, 766–767.

Lee, Y.K., Tak, T.-M., Lee, D.S., and Kim, S.C. (1990b) Cationic/anionic interpenetrating polymer network membranes for the pervaporation of ethanol–water mixture, *J. Membr. Sci.*, **52**, 157–172.

Lee, Y.T., Iwamoto, K., Sekimoto, H., and Seno, M. (1989) Pervaporation of water-dioxane mixtures with poly(dimethylsiloxane-co-siloxane) membranes prepared by a sol-gel process, *J. Membr. Sci.*, **42**, 169–182.

Lenaerts, V., Triqueneaux, C., Quarton, M., Rieg-Falson, F., and Couvreur, P. (1987) Temperature-dependent rheological behavior of Pluronic F-127 aqueous solutions, *Int. J. Pharm.*, **39**, 121–127.

Leung, B.K.-O. and Robinson, G.B. (1993) The structure of crosslinked poly(glyceryl methacrylate) hydrogel networks, *J. Appl. Polym. Sci.*, **47**, 1207–1214.

Li, M., Zhu, Z.-Q., and Mei, L.-H. (1997) Partitioning of amino acids by aqueous two-phase systems combined with temperature-induced phase formation, *Biotechnol. Prog.*, **13**, 105–108.

Ling, T.G.I., Nilsson, H., and Mattiasson, B. (1989) Reppal PES – a starch derivative for aqueous two-phase systems, *Carbohydr. Polym.*, **11**, 43–54.

Mamada, A., Tanaka, T., Kungwachakun, D., and Irie, M. (1990) Photoinduced phase transition of gels, *Macromolecules*, **23**, 1517–1519.

Modlin, R.F., Alred, P.A., and Tjerneld, F. (1994) Utilization of temperature-induced phase separation for the purification of ecdysone and 20-hydroxyecdysone from spinach, *J. Chromatogr.*, **668**, 229–236.

Nakabayashi, M., Okabe, K., Fujisawa, E., Hirayama, Y., Kazama, S., Matsumiya, N., Takagi, K., Mano, H., Haraya, K., and Kamizawa, C. (1995) Carbon dioxide separation through water-swollen-gel membrane, *Energy Convers. Mgmt.*, **36**, 419–422.

Nonaka, T., Ogata, T., and Kurihara, S. (1994) Preparation of poly(vinyl alcohol)-*graft*-N-isopropylacrylamide copolymer membranes and permeation of solutes through the membranes, *J. Appl. Polym. Sci.*, **52**, 951–957.

Obaidat, A.A. and Park, K. (1997) Characterization of protein release through glucose-sensitive hydrogel membranes, *Biomaterials*, **18**, 801–806.

Osada, Y., Okuzaki, H., and Hori, H. (1992) A polymer gel with electrically driven motility, *Nature*, **355**, 242–244.

Park, C.-H. and Orozco-Avila, I. (1992) Concentrating cellulases from fermented broth using a temperature-sensitive hydrogel, *Biotechnol. Prog.*, **8**, 521–526.

Park, K. and Park, H. (1999) Smart hydrogels. In Salamone, J.C. (ed.), *Concise Polymeric Materials Encyclopedia*, CRC Press, Boca Raton, pp. 1476–1478.

Park, K., Shalaby, W.S.W., and Park, H. (1993) *Biodegradable hydrogels for drug delivery*, Technomic Publishing Co., Lancaster.

Park, T.G. and Hoffman, A.S. (1993a) Sodium chloride-induced phase transition in nonionic poly(N-isopropylacrylamide) gel, *Macromolecules*, **26**, 5045–5048.

Park, T.G. and Hoffman, A.S. (1993b) Thermal cycling effects on the bioreactor performances of immobilized β-galactosidase in temperature-sensitive hydrogel beads, *Enzyme Microb. Technol.*, **15**, 476–482.

Peppas, N.A. and Barr-Howell, B.D. (1986) Characterization of the cross-linked structure of hydrogels, In Peppas, N.A. (ed.), *Hydrogels in medicine and pharmacy*, vol. 1, CRC Press, Boca Raton, pp. 27–56.

Peppas, N.A., Moynihan, H.J., and Lucht, L.M. (1985) The structure of highly crosslinked poly(2-hydroxyethyl methacrylate) hydrogel, *J. Biomed. Mater. Res.*, **19**, 397–411.

Prud'homme, R.K., Wu, G., and Schneider, D.K. (1996) Structure and rheology studies of poly(oxyethylene-oxypropylene-oxyethylene) aqueous solution, *Langmuir*, **12**, 4651–4659.

Reineke, C.E., Jagodzinski, J.A., and Denslow, K.R. (1987) Highly water selective celluosic polyelectrolyte membranes for the pervaporation of alcohol-water mixtures, *J. Membr. Sci.*, **32**, 207–221.

Roepke, D.C., Goyal, S.M., Kelleher, C.J., Halvorson, D.A., Abraham, A.J., Freitas, R.F., and Cussler, E.L. (1987) Use of temperature-sensitive gel for concentration of influenza virus from infected allantoic fluids, *J. Virol. Methods*, **15**, 25–31.

Ruckenstein, E. and Sun, F. (1995) Concentrated emulsion pathway to novel composite polymeric membranes and their use in pervaporation, *Ind. Eng. Chem. Res.*, **34**, 3581–3589.

Sakohara, S., Muramoto, F., Sakata, T., and Asaeda, M. (1990) Separation of acetone/water mixture by thin acrylamide gel membrane prepared in pores of thin ceramic membrane, *J. Chem. Eng. Jpn.*, **23**, 40–45.

Sakohara, S., Maekawa, Y., Tateishi, Y., and Asaeda, M. (1992) Effects of gel composition on separation properties of ethanol/water mixtures by acrylamide gel membranes, *J. Chem. Eng. Jpn.*, **25**, 598–603.

Sassi, A.P., Barron, A., Alonso-Amigo, M.G., Hion, D.Y., Yu, J.S., Soane, D.S., and Hooper, H.H. (1996a) Electrophoresis of DNA in novel thermoreversible matrices, *Electrophoresis*, **17**, 1460–1469.

Sassi, A.P., Blanch, H.W., and Prausnitz, J.M. (1996b) Characterization of size-exclusion effects in highly swollen hydrogels: correlation and prediction, *J. Appl. Polym. Sci.*, **59**, 1337–1346.

Sawahata, K., Hara, M., Yasunaga, H., and Osada, Y. (1990) Electrically controlled drug delivery system using polyelectrolyte gels, *J. Controlled Rel.*, **14**, 253–262.

Shiroya, T., Tamura, N., Yasui, M., Fujimoto, K., and Kawaguchi, H. (1995) Enzyme immobilization on thermosensitive hydrogel microspheres, *Colloids Surf., B*, **4**, 267–274.

Sinegra, J.A. and Carta, G. (1987) Sorption of water from alcohol-water mixtures by cation-exchange resins, *Ind. Eng. Chem. Res.*, **26**, 2437–2441.

Starodoubtsev, S.G., Khokhlov, A.R., Sokolov, E.L., and Chu, B. (1995) Evidence for polyelectrolyte/ionomer behavior in the collapse of polycationic gels, *Macromolecules*, **28**, 3930–3936.

Suzuki, A., Ishii, T., and Maruyama, Y. (1996a) Optical switching in polymer gels, *J. Appl. Phys.*, **80**, 131–136.

Suzuki, A. and Tanaka, T. (1990) Phase transition in polymer gels induced by visible light, *Nature*, **346**, 345–347.

Suzuki, Y., Tomonaga, K., Kumazaki, M., and Nishio, I. (1996b) Change in phase transition behavior of an NIPA gel induced by solvent composition: hydrophobic effect, *Polym. Gels Netw.*, **4**, 129–142.

Tanaka, H., Fukumori, K., and Nishi, T. (1988) Study of chemical gelation dynamics of acrylamide in water by real-time pulsed nuclear magnetic resonance measurements, *J. Chem. Phys.*, **89**, 3363–3372.

Tanaka, T. (1981) Gels, *Sci. Am.*, **244**, 124–138.

Tanford, C. (1961) *Physical Chemistry of Macromolecules*, John Wiley & Sons, New York.

Trank, S.J., Johnson, D.W., and Cussler, E.L. (1989) Isolated soy protein production using temperature-sensitive gels, *Food Technol.*, **43**, 78–83.

Ueoka, Y., Gong, J., and Osada, Y. (1997) Chemomechanical polymer gel with fish-like motion, *J. Intel. Mater. Syst. Struct.*, **8**, 465–471.

Vasheghani-Farahani, E., Cooper, D.G., Vera, J.H., and Weber, M.E. (1992) Concentration of large biomolecules with hydrogels, *Chem. Eng. Sci.*, **47**, 31–40.

Venancio, A., Almeida, C., and Teixeira, J.A. (1996) Enzyme purification with aqueous two-phase systems: comparison between systems composed of pure polymers and systems composed of crude polymers, *J. Chromatogr. B*, **680**, 131–136.

Walter, H. and Larsson, C. (1994) Partitioning procedures and techniques: Cells, Organelles, and membranes, *Methods Enzymol.*, **228**, 42–63.

Wang, K.L., Burban, J.H., and Cussler, E.L. (1993) Hydrogels as separation agents, *Adv. Polym. Sci.*, **110**, 67–79.

Wu, C., Liu, T., Chu, B., Schneider, D.K., and Graziano, V. (1997) Characterization of the PEO-PPO-PEO triblock copolymer and its application as a separation medium in capillary electrophoresis, *Macromolecules*, **30**, 4574–4583.

Yoshioka, H., Mori, Y., and Tsuchida, E. (1994) Crosslinked poly(N-isopropylacrylamide) gel for electrophoretic separation and recovery of substances, *Polym. Adv. Technol.*, **5**, 221–224.

Zhang, X., Li, Y., Hu, Z., and Littler, C.L. (1995) Bending of N-isopropylacrylamide gel under the influence of infrared light, *J. Chem. Phys.*, **102**, 551–555.

Zhong, X., Wang, Y.-X., and Wang, S.-C. (1996) Pressure dependence of the volume phase-transition of temperature-sensitive gels, *Chem. Eng. Sci.*, **51**, 3235–3239.

8 Surfaces coated with smart polymers: Chromatography and cell detachment

Alexander E. Ivanov and Vitali P. Zubov

8.1. Introduction

Polymer-coated solid supports are attracting increasing interest in the field of liquid chromatography and bioseparation. Adsorbents of this type combine the rigidity of silicas, the biocompatibility of soft organic gels and/or chemical stability of bulky polymers. Methods for the preparation of such composite sorbents, intrinsic features of their structure, and peculiar aspects of interaction with solutes were reviewed by Ivanov *et al.* in 1992. Since that time the intensively developing area of composite sorbents spread to the field of smart polymers so that the methods for their grafting onto solid particles have been elaborated and widely practised.

The idea to change the structure and adsorptivity of polymeric grafts by a slight change of temperature or pH is very attractive because a drastic change of the solute interaction with the sorbent may thus be achieved. For example, a temperature-controlled desorption of proteins was reported by several authors (Galaev *et al.*, 1994; Yoshioka *et al.*, 1995; Kanazawa *et al.*, 1997). The method gives a spectacular alternative to the changes in eluent composition, traditionally employed for the same purpose. One may anticipate separation techniques based on controlled permeability of the grafted polymer layer, and/or controlled shielding of the surface adsorption sites by the segments of grafted chains.

Although the development of such smart chromatographic methods seems to be highly promising (a number of relevant sorbents was prepared and characterized), there is still a lack of an integral approach to the comprehension of obtained separations, especially in the context of structure and behavior of smart stationary phases. Indeed, the grafted layers of composite sorbents were very rarely considered as ensembles of interpenetrating chains undergoing the cooperative transitions (Brooks and Mueller, 1996). Changes of the segment density or the layer thickness proceeding in this case were not evaluated for the studied systems or even considered. In the present review, therefore, we are trying not only to update the developing area of smart composite sorbents but also to discuss some basic problems of their structure and behavior. These include:

- theoretical model of coil-globule type transition in the layer of grafted polymer chains;
- polymer–solvent interaction for selected smart polymers;
- colloid-chemical approach to interpretation of chromatography data collected on smart packings.

We believe that a better understanding of the state of sensitive polymeric grafts attached to solid surfaces will facilitate further progress in the chemical design of smart composite materials.

8.2. Theoretical Background

8.2.1. *End-grafted Chains: Coil-globule Type Transitions*

Many smart polymer grafts reported in the literature were synthesized either by covalent attachment of the polymer chains via their end-groups or by graft polymerization of the relevant monomers. Both ways lead to the formation of polymer layers mainly composed of end-grafted chains. Layers of this kind situated on an impermeable surface have been the objects of theoretical investigations for more than 20 years, starting from the pioneering works of Alexander (1977) and de Gennes (1980). Equilibrium microstructures of end-grafted chains, their phase behaviour, interactions and dynamics were reviewed by Halperin *et al.* (1992). In the present review our purpose is to conceive the probable structures of smart stationary phases, so we shall not go into the complicated mathematical models, but instead exploit the conclusions of theoretical studies to comprehend the main features of the object.

Herein we mainly build upon the works of Prof. T.M. Birshtein *et al.* (1983, 1994) devoted to conformational changes in the grafted chains. According to these works, the properties of the polymer brush immersed in the solvent are defined by polymer chain characteristics (molecular weight, stiffness, chain thickness), brush characteristics (shape of the surface, grafting density) and by interaction of polymer segments with one another and with the solvent. Here we shall restrict our attention to the plane surface populated by flexible grafted chains with thickness a equal to the unit length, polymerization degree n and grafting area per chain $a^2\sigma (\sigma > 1)$. The contour length of the chain is, therefore, $L = na$ (Borisov *et al.*, 1988; Zhulina *et al.*, 1991).

Description of structural characteristics of the polymer layers given in the cited studies is based upon the mean-field Flory theory reduced to the presentation of the chain free energy in a layer in the form of a sum of two items: $\Delta F = \Delta F_{el} + \Delta F_{conc}$ where ΔF_{el} is a free energy of elastic stretching and ΔF_{conc} is a free energy of chain units interaction with the environment. As ΔF_{el} and ΔF_{conc} are functions of n, σ, layer height H and Flory polymer–solvent interaction parameter χ, the minimizing ΔF by H yields the asymptotic expressions for equilibrium value of H in a good solvent $(\chi < 1/2)$, θ-solvent $(\chi = 1/2)$ and in poor solvent $(\chi > 1/2)$. These expressions are valid, however, in the circumstances far enough apart from the θ-point.

8.2.1.1. *Isolated Chains*

For the simplest case of isolated chains, the dimensions of grafts coincide with dimensions of the free chains in solution at the same value of χ (Borisov *et al.*, 1988):

$$H \propto \begin{cases} n^{3/5}(1/2 - \chi)^{1/5}, & \chi < 1/2, \\ n^{1/2}, & \chi = 1/2, \\ n^{1/3}(\chi - 1/2)^{-1/3}, & \chi > 1/2. \end{cases}$$

For enabling the calculations of the layer height H also in the intermediate area of $\chi \approx 1/2$ we suggest the following approximate equations:

$$H \approx \begin{cases} k_1 n^{3/5}(1/2 - \chi)^{1/5}, & \chi < 1/2, \\ n^{1/2}, & \chi = 1/2, \\ k_2 n^{1/3}(\chi - 1/2)^{-1/3}, & 3/5 > \chi > 1/2, \\ n^{1/3}(\chi - 1/2)^{-1/3}, & \chi > 3/5. \end{cases}$$

The proportion coefficients $(k_1 = 1.5, k_2 = 0.7)$ were derived from the assumptions that:

$$k_1[n^{3/5}(1 - \chi)^{1/5}]_{\chi=0.49} \approx n^{1/2} \quad \text{for } n = 200;$$
$$k_2[n^{1/3}(\chi - 1/2)^{-1/3}]_{\chi=0.52} \approx n^{1/2} \quad \text{for } n = 200.$$

Thus calculated dependences of H from n obtained for different χ are plotted in Figure 8.1. Obviously, the higher molecular weight grafts enable the layer height to undergo a much bigger absolute increase caused by swelling in a good solvent. The relative increase of H is also higher for longer chains. Figure 8.1 shows that, from a practical point of view, the transition of grafted smart polymer from the collapsed state ($\chi \approx 0.8-0.9$) to the swollen one attained in the vicinity of cloud point (χ slightly above 0.5), does not display the full degree of the brush swelling. Indeed, a further transition to the values of $\chi \approx 0.4$ leads to much more extended conformations of chains exhibiting a higher excluded volume. Such an expedient may assist control of the inner pore volume of the composite sorbents and/or permeability of the polymer layer.

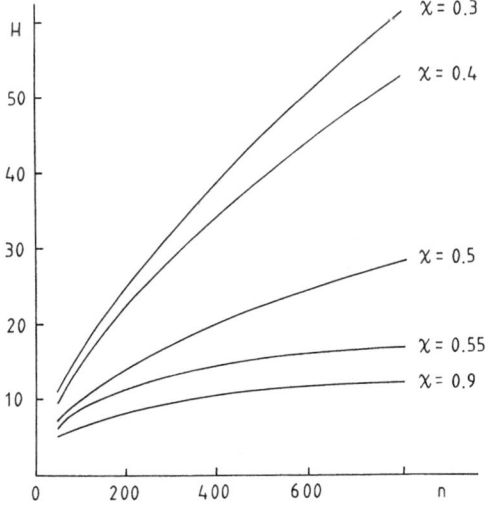

Figure 8.1 Height of a polymer layer H formed by isolated chains as a function of their polymerization degree n at various parameters χ.

8.2.1.2. Overlapping Chains

In θ-conditions, the polymer chains are thought to overlap if $n/\sigma > 1$ (Borisov *et al.*, 1988). For overlapping chains the asymptotic expressions for H/L are the following (Birshtein *et al.*, 1994):

$$H/L \propto \begin{cases} \sigma^{-1/3}(1/2 - \chi)^{1/3}, & \chi < 1/2, \\ \sigma^{-1/2}, & \chi = 1/2, \\ \sigma^{-1}(\chi - 1/2)^{-1}, & \chi > 1/2. \end{cases}$$

Again, to customize them for the needs of crossover calculations, we suggest the following approximate equations:

$$H/L \approx \begin{cases} k_3[\sigma^{-1/3}(1/2 - \chi)^{1/3} + k_4\sigma^{-1/2}], & \chi \leq 0.45, \\ \sigma^{-1/2} + (1 - k_5\sigma)(1/2 - \chi), & 0.6 > \chi > 0.45, \\ k_6[\sigma^{-1}(\chi - 1/2)^{-1} + k_7\sigma^{-1/2}], & \chi \geq 0.6. \end{cases}$$

The proportion coefficients $k_3 = 1.3$, $k_4 = 0.4$, $k_5 = 3.6 \cdot 10^{-3}$, $k_6 = 0.16$, $k_7 = 1.2$ were chosen so that the three branches given by the above functions of $H/L(\chi)$ had the similar slopes at $\chi = 0.45$ and $\chi = 0.6$, and the resultant curve had minimal breaks in these points. The thus calculated dependences of H/L on σ obtained for different χ are plotted in Figure 8.2. Figure 8.3 shows the dependence of H/L from χ for the case of moderately dense grafting ($\sigma = 16$); its character is analogous to the dependence of swelling coefficient $\alpha = H/L : (H/L)_\theta$ from the arbitrary temperature, which accounts for σ and χ presented by Borisov *et al.* (1988). Although three branches given by the above approximate equations do not ideally fit each other at any σ, their combination satisfactorily describes the phase transition for $100 > \sigma > 6$ and $1 > \chi > 0$.

In the layer composed of overlapping chains, the segment interactions lead to stretching of the chains with respect to their Gaussian dimensions not only above but also

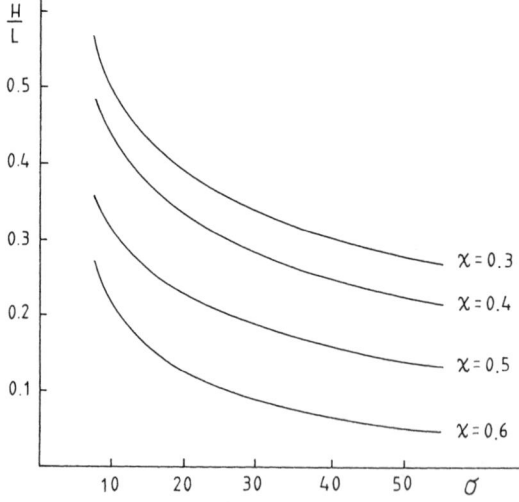

Figure 8.2 Relative height of a polymer layer H/L formed by overlapping chains, as a function of the area per chain σ.

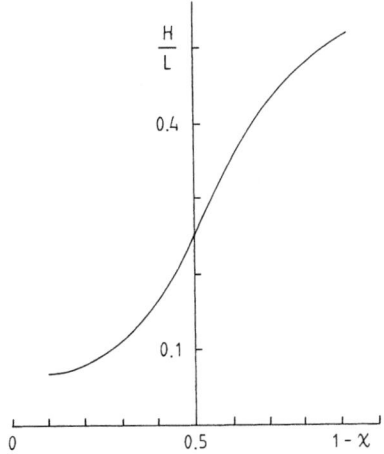

Figure 8.3 Relative height of a polymer layer H/L formed by overlapping chains as a function of χ at constant area per chain $\sigma = 16$.

below the θ-point, i.e. at $\chi > 1/2$ (Borisov *et al.*, 1988; Zhulina *et al.*, 1991). The reason for this is a strong mutual repellency of segments in the conditions of dense grafting. The interchain interactions in the layer make the collapse transition weaker than in an isolated chain; a shift of the transition point to the higher values of χ takes place and the temperature range of transition becomes broader (Zhulina *et al.*, 1991). These effects become more pronounced with the increase in grafting density (see Figure 8.2).

The analysis of structural changes accompanying the collapse of the layer shows that at $\chi \leq 0.5$ the unit density in the layer decreases with the distance from the matrix (x), but remains constant throughout the layer at $\chi \gg 0.5$. The asymptotic expressions for the unit density are the following:

$$
\varphi/\varphi_\theta \approx
\begin{cases}
\dfrac{3}{2}\varphi_\chi[1 - (x/H_\chi)^2], & \chi < 0.5, \\[2mm]
\dfrac{4}{\pi}\varphi_\chi[1 - (x/H_\chi)^2]^{1/2}, & \chi = 0.5, \\[2mm]
\varphi_\chi, & \chi > 0.5.
\end{cases}
$$

where φ_χ is a mean unit density in a layer equal to $na^3/\sigma H_\chi$ and H_χ is a height of the layer at the given χ (Zhulina *et al.*, 1991), see Figure 8.4.

As the Flory parameter χ increases, the collapse of the layer begins from the most diffuse periphery region. The mean unit density increases, whereas the distribution of unit density changes so that a sort of solid surface appears in a poor solvent (Figure 8.4, line 3). It was noted by Milner *et al.* (1988) and Milner (1991) that a brush in solvent with a supposed constant density $\varphi = na^3/\sigma H$ has a characteristic length $\xi \sim a/\varphi$ over which hydrodynamic flow is screened. Beyond this short length ξ, flow does not penetrate into the brush. The length ξ also plays the role of a pore-size in permeable flow (in which a hydrostatic pressure Δp across a brush drives a trickle of fluid of viscosity η at a velocity $v \sim \xi^2 \Delta p/\eta$). Thus, the permeability of a brush may be easily

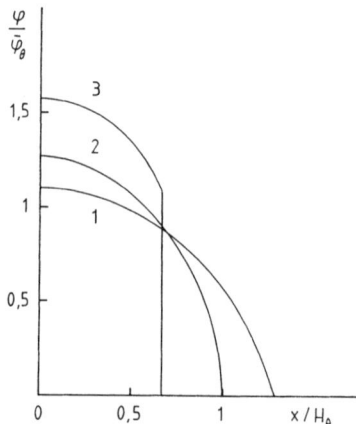

Figure 8.4 Relative unit density φ/φ_θ in a polymer layer as a function of distance from the surface x at various swelling coefficients: $\alpha = 1.3$ (1), $\alpha = 1$ (2), $\alpha = 0.68$ (3) (according to Zhulina *et al.*, 1991).

varied by variation of χ from the higher permeability exhibited at $\chi < 0.5$ to the nearly impermeable state at $\chi \gg 0.5$.

Simultaneously, the collapse is accompanied by the rearrangement of the chain ends initially distributed throughout the layer at $\chi < 0.5$ to the state of being exposed to the periphery of the layer at $\chi \gg 0.5$. Provided the interchain spaces are absorptive or the end-groups display the specific recognition sites, the partition and adsorption of solutes contacting a brush as well as their separation by chromatography might be efficiently controlled in the course of phase transition.

In the conditions of low grafting density and high values of χ the polymer layer may disintegrate into the isolated chains. According to Borisov *et al.* (1988), the borderline condition for this phenomenon is $\sigma > n^{1/2}/\chi - 1/2$. For example, the layer with $\sigma = 36$ and $n = 100$ disintegrates at $\chi > 0.78$ (in this case of $n > \sigma$ the chains do overlap in θ-conditions). This prediction means that structural transitions in the layers of polymer grafts may steeply change the properties of a composite sorbent not only near to the critical solution temperature, but also in quite different conditions of poor swelling.

8.2.2. *Physically and Chemically Adsorbed Chains: Structural Characteristics*

Some of the smart polymer coatings were prepared by means of physical or chemical adsorption of macromolecules to solid supports. In this case transitions of 'coil-globule' type are somewhat restricted compared to the end-grafted chains, because the conformations of coils are already fixed by anchoring their segments to the surface. Nevertheless, the longer loops may reorient in response to the changes in their environment whereas the 'coil-globule' transitions of tails are still possible. Physical adsorption of neutral water-soluble polymers usually results in the coatings being non-stable in aqueous media. In some cases ion-exchange adsorption of polyelectrolytes confers better stability to the composite sorbents. Obviously, the polymeric chains covalently bound to the surface via functions in their units are the most resistant to leakage (Ivanov *et al.*, 1992).

There exists, however, a principal difference between the polymer layers assembled due to either reversible ion exchange or irreversible covalent binding of segments to the surface. This difference was theoretically analyzed in the literature. According to the self-consistent field calculation made by Barford *et al.* (1986, 1987) the surface coupling constant (k_0) was introduced to characterize the polymer–surface interaction. The constant k_0 includes an electrostatic term, thus being $k_0 > 1$ for polyelectrolytes and $k_0 \ll 1$ for neutral polymers. It was found that for the same value of k_0 less polymer can be adsorbed and the polymer profiles are more extended, as a function of adsorbate concentration, in irreversible adsorption. Also, the characteristic length (i.e. the layer thickness) is higher in the latter case (see Figure 8.5, curves a and b). A dynamical Monte Carlo model of irreversible adsorption developed by King and Cosgrove (1993) for neutral polymers predicts the similar structure of the layer, i.e. the moderately dense coating of the surface at short distances combined with some rare long segments exposed to the outer space.

This type of structure is caused by the inability of irreversibly bound chains to minimize their free energy by unfolding and making an exchange of their binding sites, as is the case with reversibly bound chains. Instead, the initially adsorbed amount of polymer repeals the newly coming chains due to the excluded volume of its loops and tails. The latter segments keep on swelling as more polymer is adsorbed. The saturation of the layer is attained when the further adsorption of new chains becomes statistically unfavorable (Cosgrove *et al.*, 1993). For covalent adsorption of neutral polymers (the case of weak attraction but irreversible binding) one can imagine the adsorbed layers as 'rather large diffuse objects' (Barford and Ball, 1987) since their entropy will dominate the attraction to the surface. The relevant segment density profile as a function of the distance from the surface is illustrated in Figure 8.5, curve c.

For the equilibrium adsorption of polyelectrolytes (the case of strong attraction but reversible binding) the adsorbed layers are highly concentrated at the surface, while their density steeply decreases as a function of the distance from the surface (Figure 8.5, curve a). Indeed, it was shown that the layers of polyacrylic acid adsorbed on barium

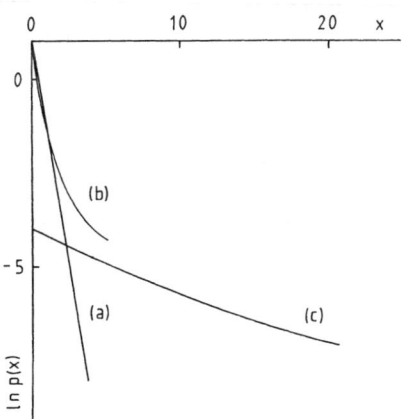

Figure 8.5 Logarithm of unit density $p(x)$ as a function of distance from the surface x as calculated for reversible (a) and irreversible (b,c) polymer adsorption, characterized by strong ($k_0 = 5$; a,b) and weak ($k_0 = 0.1$; c) attraction to the surface (according to Barford *et al.*, 1986).

sulfate crystals from pure water contain up to 90% their carboxyls bound to the surface (Cafe and Robb, 1982) so that almost no swelling may be expected. If a polyelectrolyte adsorbs via irreversible binding of segments some tails appear in the adsorbed layer (curve b).

The outlined differences between the adsorption modes lead to important practical consequences. First, the moderate surface concentration of swollen segments predicted by theory for the neutral chemisorbed chains results in low adsorptivity of macromolecular sorbates like proteins. Many examples of such a passivating effect can be found in literature (Chang *et al.*, 1976; Alpert, 1983; Ivanov *et al.*, 1985; El Rassi and Horvath, 1986). Secondly, the low-density outer part of the chemisorbed polymer layer is very likely composed of isolated tails and long loops. As discussed in Section 8.2.1., these segments undergo the stronger 'coil-globule' transition in a shorter range of χ values compared to overlapping chains. To conclude, the self-assembled structures of neutral chemisorbed polymers and the ensembles of end-grafted chains discussed in Section 8.2.1. may complement each other nicely for the purpose of creating the smart polymer phases.

Lastly, the block copolymers containing polyelectrolyte bound to the smart polymer may also be considered as building blocks of the composite sorbents. The former component provides a 'sticky foot' for electrostatic attachment of a neutral polymer to a solid support. Although such copolymers have not yet been used for the synthesis of smart sorbents, their adsorption to solids can stabilize colloid suspensions, thus being widely studied both theoretically and experimentally (Parsonage *et al.*, 1991).

8.2.3. Polymer–solvent Interaction Parameters for Selected Smart Polymers

In order to evaluate the state of end-grafted or chemisorbed chain ensembles under conditions of chromatography, one should know how the Flory parameter χ depends on temperature and composition of the eluent for the given polymer. The literature data on this point are summarized in this Section.

Based upon examination of poly-N-isopropylacrylamide (PNIPAA) cross-linked gel volumes in water at different temperatures, Hirotsu (1987) obtained the relevant function of χ illustrated in Figure 8.6. Nearly linear dependence of χ on $1/T$ characterizes

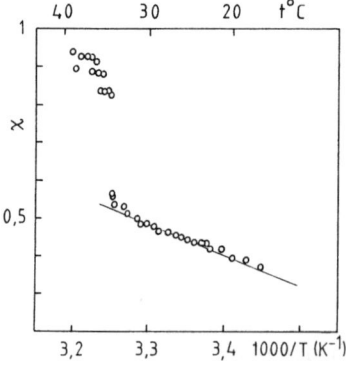

Figure 8.6 Polymer–solvent interaction parameter χ as a function of temperature estimated for poly-N-isopropylacrylamide gel in water (according to Hirotsu, 1987).

the swollen state of the polymer ($\chi < 0.5$), whereas χ changes abruptly at the transition ($\chi > 0.5$). Turning back to Section 8.2.1.1. and Figures 8.1 and 8.2 we see that the extension of PNIPAA brush swelling can be evaluated in the range of temperatures given in Figure 8.6. For example, an extensive brush swelling ($\chi = 0.4$) can be attained for the isolated chains at *ca.* 21 °C, i.e. markedly below LCST.

Further, a stepwise change of χ with temperature was registered also in aqueous solutions with the minor volume fraction of methanol (Hirotsu, 1987). The jumps of χ take place at 35 °C in pure water (see Figure 8.6), at 33 °C in 2% v/v methanol and at 24.5 °C in 16% v/v methanol. With methanol volume fractions of 25% and higher only a gradual growth of χ with temperature was observed.

The above observations make it possible to control adsorptivity of the bonded phase by two independent means: temperature and the eluent composition. For example, a search for the conditions of selective protein adsorption and desorption from PNIPAA-coated packings may thus be achieved: it is known that even low concentrations of alcohols steeply change the protein adsorptivities displayed by neutral polymeric gels (Arakawa and Narhi, 1991).

Kirsh *et al.* (1979) studied the other temperature-responsive, water-soluble polymer: poly-N-vinylcaprolactam (PVCL). Aqueous solution of the polymer becomes opalescent on heating at 34.5 °C whereas its cloud point is equal to 38 °C in 20 mM Tris-HCl (pH 7.3) in the presence of 0.1 M KCl (Galaev *et al.*, 1994). Intrinsic viscosity of PVCL solutions in water [η] strongly depend on temperature: it gradually decreases from 0.25 dl/g at 18 °C to 0.16 dl/g at 33 °C for the sample with $M_n = 27000$ studied (Kirsh *et al.*, 1979).

Although the variation of Flory parameter χ with temperature was not reported for PVCL, one can roughly evaluate its character from the above viscosimetry data. As θ-temperature of PVCL ought to be close to 33 °C, the swelling coefficient α (see Section 8.2.1.2.) can be calculated as a cube root of $[\eta]_{18°C}/[\eta]_{33°C}$, equal to 1.16. Using the approximate equations for the dimensions of isolated chains at $\chi = 0.5$ and $\chi < 0.5$ given in Section 8.2.1.1. it is easy to find out that $\chi(18 °C) \geq 0.45$ at least for $n \geq 30$, what is realistic for the given intrinsic viscosities. Thus, the temperature dependence of χ is apparently weaker for PVCL than for PNIPAA (see Figure 8.6), so an extensive swelling of PVCL coils looks possible at lower temperatures.

8.3. Preparation of Smart Bonded Phases, their Characterization and Use for Liquid Chromatography

8.3.1. *Smart Grafted Packings in Size-exclusion Chromatography*

As an increase of excluded volume is an inherent property of graft polymers under the conditions of swelling, several reports were aimed at characterization of porous packings grafted with smart polymers by means of size-exclusion chromatography. Gewehr *et al.* (1992) prepared two oligomers of poly-N-isopropylacrylamide (PIPA-S and PIPA-L) with carboxyls as the end groups and average-number molecular weights of 1400 and 3400, respectively. The end group of PIPA-S was transformed into N-hydroxysuccinimide ester in the presence of dicyclohexylcarbodiimide. The thus activated oligomer was then coupled to γ-aminopropylsilylated porous glass beads in aqueous medium (pH 9). The carriers with pore diameters from 156 to 408 Å were used. As found by conductometric titration of the silica-bound aminopropyls, 29% to 48% of them were acylated and therefore attached to the oligomer.

When dextrans of various molecular weights were applied to the column packed with the prepared sorbent, their elution times were largely dependent on temperature. Between 25 and 32 °C there was a discontinuous region of the elution behavior, at lower temperatures the elution volumes being smaller. Below the temperature of phase transition the chains of PIPA-S were stretched, and the effective size of pores in the modified glass beads was much smaller than those of the parent glass beads. The swelling of polymer chains constrained the penetration of dextran molecules into the pores. Such a temperature shift was larger in the glass beads with smaller pore size (156 Å). In this case the temperature effect on the elution behavior became the most pronounced so that a change in the resolution of the column below and above the phase transition temperature was very significant (see Figure 8.7).

Hosoya *et al.* (1994) developed the alternative method for preparation of composite sorbents with PNIPAA coating. The monomer and a water-soluble radical initiator were added to the polymerizing mixture of ethylene dimethacrylate and organic porogen suspended in aqueous solution of polyvinylalcohol. At high polymerization temperature of 80 °C, incorporation of hydrophilic PNIPAA onto the polymethacrylate particles may follow one of two different mechanisms. Provided the organic porogen (cyclohexanol) dissolves PNIPAA, the polymer distributes itself in the phase of particles, filling their pores and modifying their internal surfaces via a graft polymerization. In contrast, if the porogen (toluene) does not dissolve PNIPAA, the polymer can not penetrate the inner pores of the beads. In this case, PNIPAA can only graft itself onto the external surface of the beads, which is in contact with the aqueous phase.

High-performance size-exclusion liquid chromatography of dextrans was used to explore how the pore volumes of the composite sorbents change with temperature. Elution volumes of dextrans increase as temperature raises from 30 to 50 °C, which indicates to the shrinking state of the PNIPAA grafts distributed inside the particles,

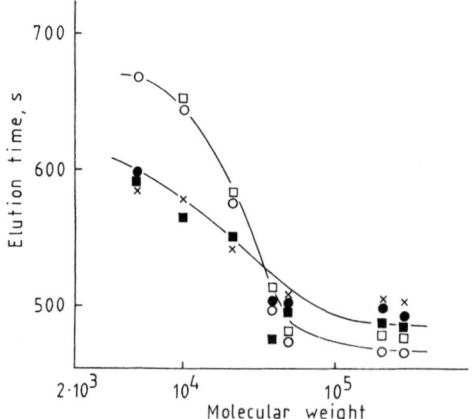

Figure 8.7 Temperature dependence of the elution time of dextran of various molecular weights. Pore size 156 Å, flow rate 0.4 ml/min; (×): 10 °C, (●): 20 °C, (■): 25 °C, (□): 32 °C, (○): 42 °C (according to Gewehr *et al.*, 1992).

the packing being synthesized with cyclohexanol as porogen. The result is generally similar to that obtained by Gewehr *et al.* (1992).

The second type of sorbent grafted with the polymer exclusively on the external surface exhibits a quite the reverse behavior: elution volumes of dextrans tend to decrease with increase of temperature. A similar pattern was also observed with unmodified polymethacrylate beads.

8.3.2. Colloid-chemical Characterization of the Smart Bonded Phases

A systematic study of porous and non-porous inorganic supports chemically coated with PNIPAA and copolymers of NIPAA was performed in the laboratory of Prof. T. Okano (Kanazawa *et al.*, 1998). Aqueous dynamic contact angle measurements were applied to characterize the plane glass surfaces coated with the above polymers (Takei *et al.*, 1994). Both terminally grafted PNIPAA (Ic-120) and multipoint attached copolymer of NIPAA and acrylic acid (IA-3, 97 : 3 mol%) were investigated. The molecular weight of Ic-120 was 13200 and of IA-3 21000. In aqueous 1% solution Ic-120 exhibited its LCST at 32 °C and IA-3 at 34.8 °C.

The coating procedure included the following steps. First, the glass surface was successively treated with solutions of dimethyldichlorosilane and styrene-aminomethylstyrene copolymer (78 : 22 mol%) and dried in a vacuum. Secondly, the terminal carboxyls of Ic-120 or in-chain carboxyls of IA-3 were coupled to the aminoalkyl fuctions of the polystyrene sublayer by means of water-soluble carbodiimide. For both the end-functionalized polymer and the random copolymer, 50% of primary aminogroups of the sublayer were chemically modified.

The advancing contact angle θ measured by the Wilhelmy plate technique in water at 10 °C stabilized after several immersion cycles at 50–55°. This pointed to the much more hydrophilic properties of the coatings compared to those of the polystyrene sublayer $\theta \approx 80°$. According to Young's equation,

$$\gamma_{SL} = \gamma_S - \gamma_L \cos\theta,$$

where γ_S large γ_L are surface free energies of the solid and the liquid, respectively, the high values of contact angle θ correspond to the maximal interfacial free energy γ_{SL}, or maximal hydrophobicity.

Wettability of the PNIPAA-coated plates was also studied at various temperatures ranging from 16 to 36 °C as shown in Figure 8.8. Both the terminally grafted surface and multipoint grafted surface showed hydrophilic properties at lower temperatures. Advancing contact angles of terminally grafted surfaces showed an increase in hydrophobicity at temperature ranging from 20 to 24 °C.

Although the multipoint grafted surface exhibited the advancing contact angle changes above 22 °C as well, the increase in hydrophobicity was weaker. Takei *et al.* (1994) concluded that a multipoint graft conformation constrains the dehydration of polymers and prevents aggregation of their chains. In the present review we mentioned that adsorption of polyelectrolytes results in very dense and thin adsorbed layers (Section 8.2.2., Figure 8.4). In the cited study, multipoint ionic adsorption of IA-3 was followed by the coupling reaction, i.e. the said structure of the layer was fixed by the covalent bonding to the support. Note that a multipoint covalent attachment of a neutral polymer might result in a higher flexibility of chains, larger thickness of the coating and, therefore, less constrained phase transition effects.

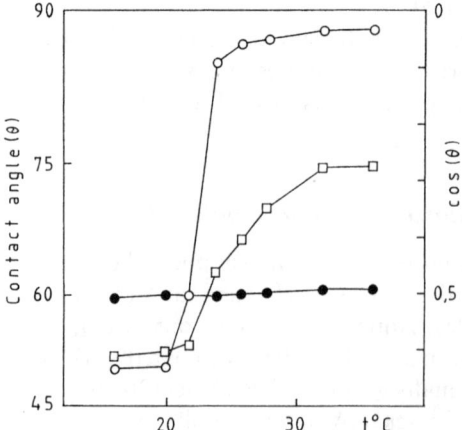

Figure 8.8 Temperature dependence of dynamic contact angle θ changes on PNIPAA-grafted surfaces (○): θ_{adv} (advancing contact angle) for Ic-120 grafted surface; (□): θ_{adv} for IA-3 grafted surface; (●): θ_{rec} (receding contact angle) for Ic-120 grafted surface (according to Takei *et al.*, 1994).

Obviously, the changes in hydrophobicity of both the coatings began at temperatures much lower than LCST. In our opinion, these changes reflect transition of the polymer phases from a largely swollen state ($\chi = 0.3 - 0.4$) to the less swollen although not collapsed state ($\chi \leq 0.5$), see Section 8.2.3. and Figure 8.2.

It follows from Figure 8.8. that end-grafted Ic-120 exhibits unexpectedly high advancing angle and, therefore, high hydrophobicity at elevated temperatures. In fact, even the polystyrene sublayer exhibits somewhat lower advancing contact angle (θ_{adv}). On the other hand, the receding contact angle (θ_{rec}) was nearly independent of temperature for both polymer coatings (*ca.* 60°), see Figure 8.8. Noteworthy, when heterogeneous surfaces are studied by contact angle technique, the advancing angle is quite sensitive to even slight hydrophobicity and rapidly increases with the increase in hydrophobic coverage. Conversely, the receding contact angle is insensitive to changes of hydrophobic coverage in the region of its low values, although even small polar domains appearing on a highly hydrophobic surface will make the angle lower (Dettre and Johnson, 1965). One can not exclude, therefore, that the receding contact angle registered in the above experiments would gradually increase at temperatures above 36 °C.

The interfacial free energy γ_{SL} may be expressed as a combination of γ_S, γ_L their dispersive components γ_S^d and γ_L^d and a non-dispersive (polar) interaction term I_{SL} as follows (Matsunaga and Ikada, 1981):

$$\gamma_{SL} = \gamma_S + \gamma_L - 2(\gamma_S^d \gamma_L^d)^{1/2} - I_{SL}.$$

Judging by variation of the advancing contact angle with temperature (see Figure 8.8) one may presume that I_{SL} becomes very low in the collapsed state of PNIPAA so that γ_{SL} steeply increases. This may be caused by the compact packing of dehydrated polymer chains withdrawn from the contact with water. Chromatographic retention of amphiphilic sorbates resulting from the increase in γ_{SL} is considered in Section 8.3.3.

8.3.3. Hydrophobic-interaction Chromatography Using
PNIPAA-modified Silicas

8.3.3.1. Theories of Hydrophobic Interaction

Many theories have been proposed for hydrophobic-interaction chromatography (HIC). Porath *et al.* (1973) suggested that 'the driving force is the entropy gain arising from structural changes in the water surrounding the interacting hydrophobic groups'. Displacement of the ordered water molecules surrounding the hydrophobic ligands and the proteins leads to an increase in entropy resulting in thermodynamically favorable adsorption (Hjerten, 1977). This theory explains why chromatographic retention in HIC increases with temperature, opposite to normal-phase or reversed-phase chromatography.

The alternative theory (Absolom and Barford, 1988) is based upon a colloid-chemical approach. The change in the Helmholtz free energy due to the process of adsorption (ΔF) can be expressed as follows:

$$\Delta F = \gamma_{SP} - \gamma_{SL} - \gamma_{LP},$$

where γ are the interfacial free energies (or interfacial tensions) between sorbent (S), protein (P) and liquid (L). This theory predicts increasing adsorption of proteins from aqueous solution of salts ($\gamma_L \geq 72\,\text{mJ/m}^2$) on the surfaces with decreasing surface tension (γ_S) the fact approved experimentally by many researchers (Absolom *et al.*, 1987). According to Absolom and Barford (1988), chromatographic retention k' may relate to free energy of adsorption as follows:

$$k' = e^{-\Delta F/RT}(V_S/V_L),$$

where V_S and V_L are volumes of solid and liquid phases available for the sorbate in the column, respectively. As ΔF becomes more negative when γ_{SL} increases, one may expect a rise in k' for amphiphilic substances chromatographed on PNIPAA-bonded phases in the course of the collapse of grafted polymer chains. The growth of k' should be, however, limited by the decrease of V_S/V_L, which is sharp (see Section 8.2.1., Figures 8.1 and 8.2). Moreover, the collapsed layer may become impermeable for large molecules like proteins because of its high density (Section 8.2.1.2., Figure 8.4).

The summarized colloid-chemical theory of HIC appears to be universal because it presumes a surfactant-like behavior of a sorbate, which reduces the free energy of the interface between the sorbent and the contacting liquid. This may be due to either the destruction of the ordered water (see above) or due to stronger water–water interactions compared to water-solute ones (Cooke and Olsen, 1980). The latter reason hardly underlies, however, an increase of hydrophobic adsorption with temperature. Experimental data on the relevant dependences are reviewed in Section 8.3.3.2.

8.3.3.2. PNIPAA-modified Silicas as Chromatographic Sorbents:
The Retention Mechanisms

PNIPAA-modified silica prepared by analogy to the method of Gewehr *et al.* (1992) (see Section 8.3.1.) was tested as a sorbent for HIC of steroids (Kanazawa *et al.*, 1996). Molecular weight of the reactive polymer used was 4400; 73% of aminogroups

(154 µmol/g) chemically attached to 120 Å-pore silica were acylated by the activated ester end-group of the polymer in dioxane. Separation of benzene, hydrocortizone, prednisolone, dexamethasone, hydrocortisone acetate and testosterone carried out with water as an eluent on a column packed with the PNIPAA-modified silica showed steeply increasing chromatographic retention (k') and resolution of the solutes with temperature increased from 5 to 50 °C. Retention times of the more hydrophobic steroids (hydrocortisone acetate, testosterone) were longer. The reference column packed with non-modified silica displayed much shorter retention times decreasing with temperature. It is thus very likely that the mechanism of steroid adsorption to the PNIPAA-silica belongs to the type of hydrophobic interaction, as discussed in Section 8.3.3.1.

It is noteworthy that the retention of steroids also shows large changes above the LCST of PNIPAA: it quickly grows up even between 40 and 50 °C. In our opinion, this may be a consequence of the very high grafting density typical for the chains covalently attached through their reactive end-groups. Indeed, 70% acylated aminopropyls correspond, in our evaluation to the grafting densities of $\sigma \leq 5$. As mentioned in Section 8.2.1.2., the dense grafting causes a shift of phase transition to higher χ and, therefore, temperatures. Moreover, the temperature range of the transition becomes broader.

In the following paper (Kanazawa *et al.*, 1997) the same research group studied a series of temperature-responsive sorbents grafted by random copolymers of NIPAA and butyl methacrylate (BMA) with activated esters as end-groups. LCST of the copolymers varied from 31 to 20 °C depending on the BMA amount in the chains (up to 3.2 mol%). The general pattern of the observed k' temperature dependence for steroids illustrated in Figure 8.9 was similar to that registered with grafted PNIPAA; however, the hydrophobic BMA components dispersed within the hydrated NIPAA sequences enhanced the hydrophobic interaction of the coating polymer with steroids even below the temperature of phase transition. Unlike found with the PNIPAA-modified silica, k' tends to approach a maximum near 45 or 50 °C.

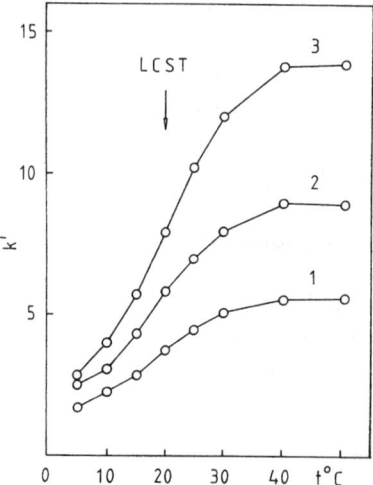

Figure 8.9 Changes in capacity factors for three steroids chromatographed on silica support coated with NIPAA-BMA copolymer; (1): prednisolone, (2): dexamethasone, (3): hydrocortisone acetate (according to Kanazawa *et al.*, 1997).

The latter observation seems to be a far-reaching guide in the studies of hydrophobic adsorption to smart surfaces: it reflects a change in the adsorption mechanism. Zero variation of k' with temperature means that increase in entropy is no longer a driving force of steroid adsorption in certain circumstances. This may be understood as follows.

At the temperature much higher than LCST of the coating copolymer (20 °C) the grafted chains are sufficiently dehydrated and collapsed, i.e. the ordered structure of water around them is broken. Now adsorption of steroids seems to follow the different mechanism based solely on preferential water–water type interactions accompanied by forcing a hydrophobic molecule out of aqueous medium (Cooke and Olsen, 1980). Obviously, the latter process leads to a negative change in enthalpy of the system as water–water interactions (characterized by surface tension of water, γ_L) decrease with temperature. Therefore, one can expect a decrease of k' with temperature for steroids above 50 °C. Another reason for the limiting value of k' attained at higher temperatures may be a drop in V_S proceeding as a result of the chain collapse, as discussed in Section 8.3.3.1.

The actual decrease of k' at temperatures above LCST was reported by Ivanov *et al.* (1997). The authors prepared wide-pore glass (pore diameter of 2000 Å) coated with chemically adsorbed poly(p-nitrophenyl acrylate), $M_w = 46000$ and transformed the bound polymer into copolymer of NIPAA and N(2-hydroxyethyl acrylamide) (7:3) by amidation. LCST of the copolymer estimated in aqueous solution was 41 °C. Chromatographic retention of lysozyme was studied with the copolymer-coated beads as the packing and 1.5 M ammonium sulfate in 0.01 M phosphate buffer (pH 7.5) as eluent. The temperature dependence of elution volume pointed to the maximum near LCST of the copolymer (see Figure 8.10). It is likely that the covalent attachment of the neutral polymer through the reactive groups in the chains led to the coating with moderate grafting density and, therefore, narrow temperature range of the bonded phase transition (see Section 8.2.2.). In fact, the content of monomer units in the layer of chemisorbed polyacrylate was 6 µmol/m². In contrast, the layers of end-grafted

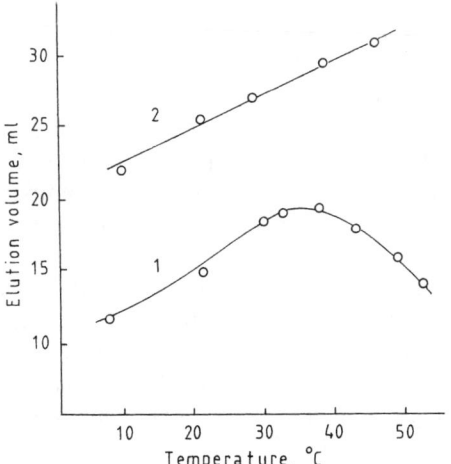

Figure 8.10 Effect of temperature on chromatographic retention of lysozyme on wide-pore glass packing chemically modified with copolymer of NIPAA (1) and poly-N-butylacrylamide (2) (according to Ivanov *et al.*, 1997).

NIPAA and BMA copolymers (Kanazawa *et al.*, 1997) contained as much as *ca.* 25–35 μmol/m² (surface area of 200–300 m²/g can be supposed for 120 Å-pore silica used in the cited study).

8.3.3.3. PNIPAA-coated Sorbents in Separation of Proteins and Peptides

Conventional chromatographic methods in separation of proteins and peptides most often include reversed-phase (RPC) or ion-exchange liquid chromatography. Owing to possible denaturation of proteins under conditions of high-performance RPC, a number of investigations was carried out to elaborate rigid, polymer-coated sorbents for HIC of proteins, which demands no organic solvent for elution and separation of biomolecules. The coating polymers used were polyethylene glycol and polyvinylalcohol (El Rassi and Horvath, 1986), polyvinylpyrrolidone (Kurganov *et al.*, 1994) or poly-N-butylacrylamide (Ivanov *et al.*, 1995). Recently, the relevant synthetic approach was extended to the field of smart polymers grafted or chemisorbed onto silica.

The silica sorbents grafted with copolymers of NIPAA and BMA (Kanazawa *et al.*, 1997) as discussed in Section 8.3.3.2., were studied as packings for HIC of polypeptides. The mixture of insulin chain A, β-endorphin fragment 1-27 and insulin chain B was applied to the column either at 5 or 30 °C and chromatographed under isocratic elution with 0.5 M NaCl (pH 2.1) as shown in Figure 8.11. The polypeptides, consisting of 21–30 amino acid residues, could not be separated at 5 °C (below the LCST of 20 °C). As the column temperature was raised to 30 °C the three polypeptides were well resolved, whereas their retention times markedly increased. The number of relatively hydrophobic amino acid residues in the tested polypeptides (leucine, isoleucine, phenylalanine, tryptophan, tyrosine, valine) increases in the order of their retention times. Thus, hydrophobic interactions are primarily the separating force between the peptides and the polymer-grafted surface.

Porous polymethacrylate beads grafted with PNIPAA as developed by Hosoya *et al.* (1994), see Section 8.3.1., were studied as a packing for separation of serum albumin (BSA), theophyline and barbital (Hosoya *et al.*, 1995). The polymer beads modified with PNIPAA on their external surface excluded BSA from the pore volume so that the protein eluted with the void volume of the column. Conversely, the low molecular weight drugs exhibited a stronger retardation and good resolution, either at 30 or 50 °C; elution was performed by 10% MeCN in 0.02 M phosphate buffer, containing 0.1 M sodium sulfate (pH 7). At 50 °C the peak of BSA became smaller compared to at 30 °C. The authors suggested it was due to a strong retardation of the protein on the hydrophobized surface above LCST.

Temperature-induced desorption of bovine immunoglobuline G (IgG) previously adsorbed to the PNIPAA-grafted silica at elevated 37 °C temperature was reported (Yoshioka *et al.*, 1995). The sorbent was prepared by coupling the PNIPAA oligomer ($M_n = 1600$) containing primary amino end-group to the aminopropylsilica (average pore size 322 Å) activated with glutardialdehyde. IgG (2.5 mg) was brought into contact with 0.15 g PNIPAA-silica batch-wise in 5 ml 0.067 M phosphate buffer solution (pH 7) at 37 °C. The adsorption of protein was 3.4 mg/g support or *ca.* 20% . After lowering the temperature to 24 °C, 2.1 mg protein/g support was desorbed into the supernatant. Adsorption of serum albumin was not detected in the same conditions neither at 37 °C nor at 24 °C.

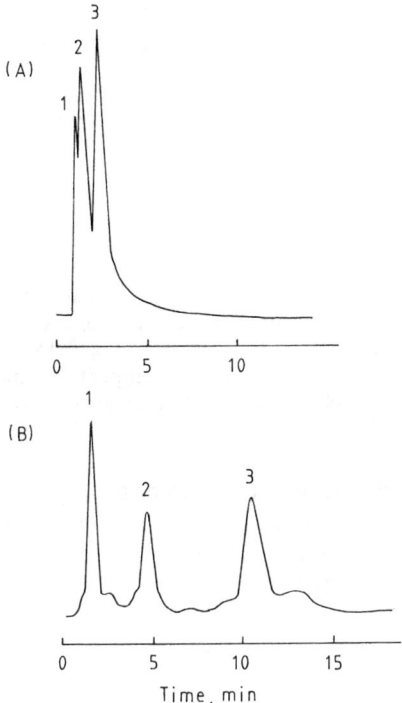

Figure 8.11 Chromatograms of the mixture of insulin chain A (1), β-endorphin fragment 1-27 (2) and insulin chain B (3) on the silica support coated with NIPAA-BMA copolymer; isocratic elution by 0.5 M NaCl, pH 2.1 at (A) 5 °C and (B) 30 °C (according to Kanazawa *et al.*, 1997).

Attempt to detect the temperature-induced desorption of IgG within a column chromatographic technique was made by Ivanov *et al.* (1997). The copolymer-coated sorbent prepared as described in Section 8.3.3.2. was packed into the glass column and equilibrated with 0.067 M phosphate buffer, pH 7. IgG could not be irreversibly adsorbed to the PNIPAA-coated glass, but quantitatively eluted as an asymmetrical peak nearby the total volume of the column, i.e. a weak reversible adsorption took place. The protein could be partially adsorbed from 0.01 M phosphate buffer (pH 7.5) containing 0.15 M NaCl and 1 M (NH$_4$)$_2$SO$_4$ at 45 °C. About 50% of the protein was found in the breakthrough fraction (peak 1 in Figure 8.12). No further desorption was observed when the column was cooled to 30 °C and eluted with the same solution. A small portion of protein desorbed at 17 °C. The adsorbed IgG fraction could be desorbed, however, by 0.01 M phosphate buffer (pH 7.5), containing 0.15 M NaCl at 30 °C or at 17 °C i.e. in the common conditions of hydrophobic-interaction chromatography (HIC). The discrepancy with the results of Yoshioka *et al.* (1995) may arise from different structures of PNIPAA coatings as well as different experimental conditions such as use of batch operation in the cited study.

To sum up the results reviewed in this section we should stress that the interaction of organic solutes with PNIPAA-coated packings principally conform to the usual features of HIC, with a few deviations dealt with a temperature-induced transitions

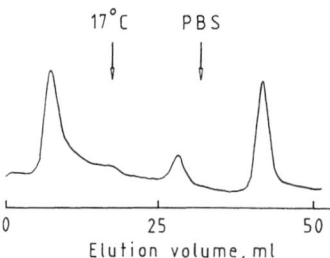

Figure 8.12 Chromatography of immunoglobuline G on wide-pore glass chemically modified with the copolymer of NIPAA according to Ivanov *et al.*, 1997. The
sample applied in 0.01 M phosphate buffer (pH 7.5), containing 0.15 M NaCl
and 1 M $(NH_4)_2SO_4$ at 45 °C. The left arrow indicates a temperature shift, the
right arrow indicates a change of the eluent to 0.01 M phosphate buffer (pH
7.5) containing 0.15 M NaCl.

of the bonded phases. Increase of k' with temperature is an inherent property of HIC
(Wu *et al.*, 1986). Although protein adsorptivity may be principally controlled by a
temperature shift in PNIPAA packings, a quantitative elution of previously adsorbed
proteins in such conditions was not yet demonstrated. In our opinion, a conjugation
of temperature-responsive polymeric grafts with stronger interacting affinity ligands
might be promising for temperature-controlled bioseparations.

8.3.4. Bioaffinity and Ion-exchange Separations Assisted by
Smart Polymeric and Lipid Layers

Galaev *et al.* (1994) reported a temperature-induced separation of proteins on the dye-
affinity agarose packing modified by physically adsorbed poly-N-vinylcaprolactam
(PVCL). Lactate dehydrogenase (LDH) from porcine muscle was bound to the packing
at 40 °C. At this temperature LDH could not be eluted from the column by 0.1 M
KCl. The decrease of temperature to *ca.* 23 °C resulted in LDH elution with 0.1 M
KCl. Presumably, the swollen, mobile chains of PVCL compete with the protein for
binding to the immobilized dye moieties, so that a sort of displacement of the protein
by the polymer takes place at the lower temperature.

Crude porcine muscle extract was applied to the PVCL-coated column at 40 °C and
the foreign proteins were washed out with 0.1 M KCl at 40 °C. The flow was then
interrupted, the column was cooled to room temperature and virtually homogeneous
LDH was eluted with the same buffer. The purification factor was 17 and the recovery
of LDH was 90%. This appeared to be the first reported successful enzyme purification
in which a temperature shift was used as the only eluting factor without changing the
buffer composition.

Loidl-Stahlhofen *et al.* (1996) proposed a different approach to temperature-
controlled chromatography of proteins. They coated a silica gel with a bilayer
of a binary mixture of two lipids: neutral dielaidoyl-phosphatidylcholine (DEPC)
and anionic dimiristoylphosphatidylglycerol (DMPG). The prepared phospholipid
membrane-like coating undergoes its phase transition between 6 and 12 °C. At temperatures above 12 °C a homogeneous charge distribution of molten DEPC and DMPG
mixture takes place, whereas at lower temperatures the domains of frozen individual

phospholipids cause the heterogeneous distribution of charges dealt with enhanced adsorptivity. Positively charged proteins like trypsinogen, cytochrom c and some others adsorbed to the lipid-coated silica at 4 °C and could be eluted by a shift of temperature to 25 °C. Tests of the desorbed proteins proved their high yields (98%–100%) and retained specific activities.

8.4. Effects of Phase Transitions in Polymeric Supports on Attachment, Proliferation and Recovery of Cultured Cells

8.4.1. *Methods for Coating of Polystyrene Dishes with Responsive Polymers and their Characteristics*

Since efficient recovery of cultured cells from PNIPAA-grafted polystyrene dishes was achieved by temperature shift as described by Yamada *et al.* (1990), many studies were made to investigate proliferation, metabolism and temperature-induced detachment of cells brought into contact with the responsive polymer-grafted surfaces. The above area warrants a substantial review; in this chapter we consider the most important effects exerted on cells by the phase transitions in the supporting polymers. Some of the polymeric grafts closely relate to the bonded phases studied by chromatographic techniques, so that their properties will be discussed in combination.

Traditionally, recovery of cells from culture substrates required enzymatic treatment of adhered cells using enzymes like trypsin, pronase, collagenase, etc. These proteolytic enzymes inflict damage on cell surfaces by hydrolysing various membrane-associated protein molecules, resulting in the impairment of cell function. Thus, a technique for mild detachment of cells, which does not require enzymes has long been desired. Cells can adhere and grow more easily on hydrophobic surfaces than on hydrophilic surfaces. The extent of adhesion decreases with increasing substrate surface tension, in accordance with thermodynamic expectations (Absolom *et al.*, 1986). It was, therefore, attractive to culture cells in contact with a hydrophobic polymer, which turns into a hydrophilic one when the cells are needed to be harvested. The hypothesis for the temperature-controlled detachment of cells from PNIPAA-grafted carriers was validated by Yamada *et al.* (1990).

Polystyrene culture dishes were filled with 40 wt.% NIPAA solution in isopropyl alcohol and then irradiated with a 25 Mrad electron beam under vacuum of $4 \cdot 10^{-4}$ Pa to form a PNIPAA coating. Bovine hepatocytes, the cells highly sensitive to enzymatic treatment, were used to study their adhesion to the carriers as well as their subsequent growth and detachment. Dynamic contact angle measurements of the PNIPAA-grafted dish and control polystyrene dish showed 48° and 54° at 37 °C, respectively, and 30° and 54° at 10 °C. The temperature-dependent hydrophilicity of the grafted dish could thus be demonstrated.

The adhesion of hepatocytes to the carriers was not much different, however. The cell number on both PNIPAA-grafted and control dishes after the culture for two days at 37 °C was also nearly identical. For detachment of cultured hepatocytes the dishes were cooled to 4 °C and allowed to stand for 60 min, then the number of detached cells were measured. Nearly 100% of the hepatocytes was detached and recovered from the PNIPAA-grafted dishes by low temperature treatment, while only about 8% of the cells were able to be detached from the control dish.

Hepatocytes recovered from the PNIPAA-grafted dishes were plated onto new control dishes and the number of adhered cells was measured after 18 h of cultivation. As a comparative experiment, hepatocytes recovered by the commonly used trypsin treatment were processed in a similar way. About 73% of the former portion adhered to the new dish, while only 14% of the trypsin-treated cells were able to adhere and the rest of them died due to damage by the enzyme. As found by phase-contrast microscopy, the hepatocytes were detaching from the polymer-coated dishes without loss of their intercellular junctions, i.e. in an assembled state, which is essential for their viability. Thus, the superior subcultures of hepatocytes could be achieved by use of the PNIPAA-grafted dishes. Adhesion and spreading of cells on the plastic dishes physically coated with thin polymerized films of NIPAA were studied by Takezawa *et al.* (1990) and Rollason *et al.* (1993). According to the method of Takezawa *et al.* (1990), 0.5 w/w% aqueous solution of the polymer was poured into a Falcon #3001 culture dish and dried in a clean bench at 10 °C to form the layer of the polymer content of about 100 µg/cm^2. Human dermal fibroblasts did not adhere, however, to the film composed of pure PNIPAA. Effective adhesion and spreading of cells were observed on the mixed polymer films consisting of PNIPAA and collagen (1 : 1, w/w).

Fibroblasts were cultured at 37 °C on the PNIPAA-collagen support until the cells completely covered the surface, after which the ambient temperature was reduced to about 15 °C. After a few minutes, the sheet of fibroblasts started to detach itself from the dish, and in about 15 min the detached cell sheet had floated in the culture medium. During this detaching process, the cells did not scatter but remained linked to each other in a manner apparently similar to the cells detached from the PNIPAA-grafted surfaces and preserved their intercellular junctions (Yamada *et al.*, 1990). When transferred to a hydrophobic plastic dish the cell sheet did not attach itself to the dish and gradually changed into a multicellular spheroid of 0.7–1.1 mm in size.

After multicellular spheroids were cultured for 20 days on hydrophobic plastic dishes, the spheroids were transferred to hydrophilic plastic dishes and were kept in culture. In two days the spheroid adhered to the dish and cells began to migrate and proliferate, indicating that the cells in spheroids sustain capabilities of adhesion and proliferation. The polymer coatings used in the cited studies can hardly be stable, however, because of their solubility in aqueous culture fluids. Although cytotoxicity of soluble PNIPAA was found to be negligibly low (Takezawa *et al.*, 1990), the grafted dishes seem to be advantageous for repeated operations.

NIPAA grafting initiated by UV-irradiation, using benzophenone as a photosensitizer, was reported by Morra and Cassinelli (1995, 1997). The grafted polystyrene surfaces obtained with various irradiation time and NIPAA concentration in the contacting isopropanol solution of the monomer were characterized by ESCA. Elemental composition of the grafted layers almost did not depend on the grafting conditions, being rather close to the composition of PNIPAA. The continuous, anchorage-dependent, contact-inhibited cell line L-929 (mouse fibroblasts) was used to test the cell-supporting and thermally induced cell-recovery behavior of grafted dishes. The cells adhered and grew on PNIPAA-grafted dishes, although the growth was a little slower than on the parent dishes. No obvious trend of attachment and growth rate as a function of NIPAA concentration was found. Cells detached from the surface of modified dishes when maintained at 10 °C for enough time (see Figure 8.13). The quantity of detached cells was lower when grafting was performed from solutions containing

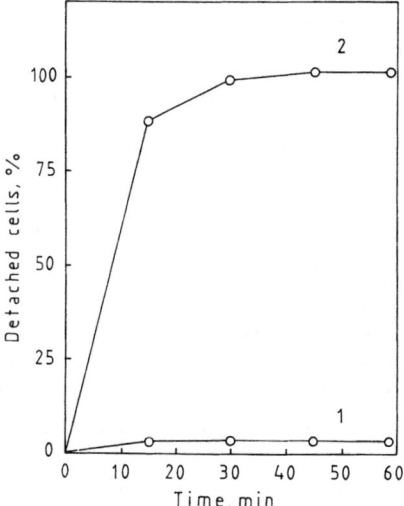

Figure 8.13 Effect of the low temperature treatment on the quantity of the detached mouse fibroblasts. 1: polystyrene dishes, 2: PNIPAA-grafted polystyrene dishes prepared according to Morra and Cassinelli (1995), with 40 wt% NIPAA solution and UV-irradiation time of 10 min.

less than 30% NIPAA, despite ESCA data showing that the surface composition is not effected by the NIPAA concentration.

Several cycles of cooling and heating of a PNIPAA-grafted dish populated by mouse fibroblasts were performed, each cycle taking *ca.* 2 h. The microscopic observation of the number of adhering cells showed that they repeatedly attached and detached as temperature was raised or lowered. This effect of temperature cycling proves that PNIPAA layer remains on the polystyrene surface after the low temperature treatment, and that the hydrophobic–hydrophilic transition repeatedly occurs. Non-consumable temperature-sensitive supports for cell culture might be, therefore, prepared in the described way.

8.4.2. Relationship between the Mechanism of Cell Detachment and Cell Metabolism

Detachment of cultured cells affects not only their adhesion and ability to subculture, but also reflects in a secretion of cell metabolites. On the other hand, metabolic processes themselves are involved in cell detachment, as one may judge by the morphological changes of cells accompanying this process. To investigate the relationship between these phenomena two cell types differing in their sensitivity to proteolytic enzymes were studied as regards their secretion activity changes resulting from detachment and subculture (Okano *et al.*, 1993). The cell types were bovine endothelial cells stable to proteolytic enzymes, and rat hepatocytes, which could not be subcultured after trypsin treatment because of their high sensitivity to the enzyme. Both cell types were cultured on PNIPAA-grafted surfaces at 37 °C and recovered by decreasing temperature. Although endothelial cells recovered either by trypsin treatment or by cooling

Table 8.1 Secretion activities of cells recovered by low temperature treatment (TRS) and enzyme treatment (ERS) from PNIPAA-grafted dishes*

Cells type	Primary system	TRS	ERS
Endothelial cells (ng 6-keto-PGF$_{1\alpha}$/10^5 cells)	0.65 ± 0.05	0.84 ± 0.07	0.16 ± 0.02
Hepatocytes (ng albumin/h 10^5 cells)	264.5 ± 22.9	254.9 ± 31.9	51.9 ± 10.1

*compiled from the data of Okano *et al.* (1993).

from PNIPAA-dishes exhibited almost identical adhesivities and time dependences of proliferation, their abilities to secrete prostacyclin were quite different. Prostacyclin generation, an important function of endothelial cells and a strong indicator of normal methabolism, was *ca*. 5-fold suppressed in the enzyme-treated endothelian cells compared to the cells detached by cooling (see Table 8.1). As discussed in Section 8.4.1., adhesion of hepatocytes previously detached from polystyrene dishes by trypsin treatment was rather low. The similar results were obtained with hepatocytes detached by trypsin treatment from PNIPAA-grafted dishes: only 25% cells was able to adhere to polystyrene dishes. Moreover, the albumin secretion activity was much lower for the enzyme-treated hepatocytes compared to the cells recovered by cooling (see Table 8.1).

It is noteworthy that hepatocytes recovered by cooling retained their native form with numerous bulges and dips. Conversely, the enzyme-treated cells had a smooth outer surface, which might underlie the impairment of their adhesivity. The summarized results confirm that the cells recovered by a temperature shift from PNIPAA-grafted surfaces show an intact structure and maintain normal cell functions. The subsequent investigations were aimed at the solution of the converse problem, namely elucidation the role of cell metabolism in the mechanism of thermally modulated cell detachment.

It was reported by Okano *et al.* (1995) that detachment of rat hepatocytes cultured on PNIPAA-dishes for 2 days at 37°C simultaneously proceeds by only 8%–12% when cooling the plate to the temperature of 4 or 10°C for 30 min. The recovery might be improved up to 80% , however, by the additional incubation of cells at 25°C for 5 min. After 10 min incubation at 25°C 100% cells were detached.

Morphologies of detached cells were observed by optical microscopy during 10 min of additional incubation at 25 °C. Cells changed their shapes from a spread to a round form, and finally they were observed to detach completely from the surface. As the PNIPAA-grafts are still well hydrated at 25 °C, these results suggest that the cell shape changes are accompanying a consumption of cellular metabolic energy. Thus, hydration changes of grafted PNIPAA at the cell-material interface are an important initial stimulus to induce active cell detachment mediated by cellular processes.

Another evidence for the role of intracellular metabolism in the detachment of cells was given by the experiments carried out in the presence of sodium azide (Okano *et al.*, 1995a). Sodium azide is a known inhibitor of cytochrom c oxidase in mitochondria and decreases ATP generation, resulting in the disruption of cellular activities, which require ATP. Presumably, the presence of sodium azide might affect the process of cell detachment, which consumes the cellular metabolic energy. Indeed, after the hepatocytes cultured on PNIPAA-grafted dishes were treated with a solution of sodium azide for 60 min at 37°C, the number of cells detached at 10°C (with additional incubation at 25°C for 5 min) was strongly affected by the azide concentration. Inhibition of cell detachment was registered at the concentrations of 1 mM and higher. These results prove that intracellular processes enabling morphological changes of cells are important prerequisite for their detachment from hydrated surfaces.

8.4.3. Adhesion of Platelets on Naked and Polymer-coated Polystyrene Particles

In contrast to anchorage-dependent cells like hepatocytes and fibroblasts, platelets segregate well in aqueous medium, what enables their interaction with suspended polymeric particles. Adhesion of platelets was followed by measuring the accompanying concentration changes of cytoplasmatic Ca^{2+} in platelets using the fluorescent dye, Fura 2 (Okano *et al.*, 1995b; Takei *et al.*, 1995). It is well-known that platelets are strongly activated in contact with polystyrene surfaces (Grainger *et al.*, 1989). Thus the contact of platelets with suspension of 1 μm polystyrene particles led to an increase of $[Ca^{2+}]$ in platelets (see Figure 8.14). The particles covalently coated with carboxylic end-group polyethylene glycol ($M_n = 5100$) did not interact with platelets, presumably due to effective shielding of polystyrene surface by hydrophilic polymer chains. In turn, thermal transition in the copolymers of NIPAA with N-dimethylacrylamide (4 : 1, LCST = 44.2 °C) caused the interaction of platelets with the copolymer-coated particles at temperatures above 38 °C (see Figure 8.14) with no interaction at lower temperatures. Platelet interaction with these particles appears to be mediated by hydrophilic/hydrophobic changes of the PNIPAA-grafted surface. It is noteworthy that the suspended copolymer-coated particles undergo an aggregation induced by the temperature shift from 37 to 42 °C, i.e. below LCST of the modifying copolymer. Similarly, the polystyrene particles grafted with copolymers IA-3 and Ic-120 (see section 8.3.2.) aggregate at about 22–24 °C, what is also below the LCST's of the copolymers and coincides well with the temperature range of contact angle changes registered in independent experiments with the copolymer-coated plates (see Figure 8.8). The reason

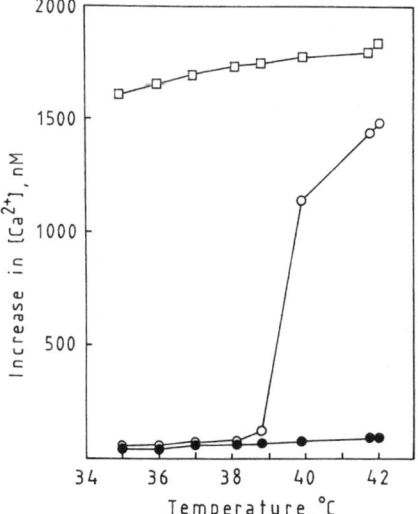

Figure 8.14 Temperature dependence of $[Ca^{2+}]$ increase in platelets in contact with polymer-grafted particles. □ – polystyrene particle; ○ – IDC-20-grafted particles; ● – polyethylene glycol-grafted particles (according to Takei *et al.*, 1995).

for lowering of the phase transition temperature in graft copolymers might be the input of polystyrene core into hydrophobic/hydrophilic balance of the particles. In particular, adsorption of the end-grafted polymer chains to hydrophobic polystyrene carrier may proceed to a larger extent than self-association of free chains at temperatures slightly below LCST.

8.5. Conclusions and Future Trends

Despite many successful applications of smart polymers in bioseparation and bioprocessing the achievements in the chromatographic area have been relatively modest up to now. This is probably due to the fact that the behavior and conformational changes of the responsive grafted macromolecules at the transitions are not yet fully comprehended. In the present chapter the authors tried to draw attention to the existing information in this area.

Better progress has been achieved in the field of cells cultured at smart polymeric surfaces. It should be realized, however, that the state of a cell adhered to the responsive interface is quite different from the state of an adsorbed biomolecule. Whereas the internal energetics of the cell aids its detachment from the repealing surface, a biomolecule may desorb only due to a free energy increase resulting from the phase transition. To amplify such an increase the molecular interactions stronger than hydrophobic ones should be involved in the solute-sorbent association. One can consider novel responsive bioaffinity sorbents based on grafted polymeric phases, ion-exchangers undergoing cooperative transitions in the narrow ranges of pH or ionic strength, and other interactive sorbents. We believe that such smart chromatographic materials will be the subject of intensive research and development.

According to the theoretical analysis reviewed in the present chapter, the most pronounced effects of the responsive polymers related to the conformational changes of their chains are to be observed for the higher molecular weights. For this purpose, creation of functional composite systems incorporating high molecular weight smart polymers capable of controlled interaction with various biological molecules and particles seems to be promising.

8.6. Abbreviations

BMA	butyl methacrylate
BSA	bovine serum albumin
ESCA	electron spectroscopy for chemical analysis
HIC	hydrophobic-interaction chromatography
IA-3	the copolymer of N-isopropylacrylamide and acrylic acid (97:3 mol.%), $M_n = 21000$
Ic-120	poly-N-isopropylacrylamide of $M_n = 13200$ with terminal carboxylic group
LCST	lower critical solution temperature
MeCN	acetonitrile
NIPAA	N-isopropylacrylamide
PIPA-L	the oligomer of N-isopropylacrylamide of $M_n = 3400$ with terminal carboxylic group
PIPA-S	the oligomer of N-isopropylacrylamide of $M_n = 1400$ with terminal carboxylic group

PNIPAA   poly-N-isopropylacrylamide
PVCL   poly-N-vinylcaprolactam
RPC   reversed-phase chromatography

8.7. Symbols

a	polymer chain thickness and unit length
ΔF	free energy of adsorption or free energy of the polymer chain in the polymer brush
H	polymer brush height
I_{SL}	polar interaction term
k_0	surface coupling constant, a measure for polymer–solvent interaction
k'	chromatographic retention
k_1–k_7	proportion coefficients, introduced for crossover calculations of H and H/L in the ranges of different power laws
L	contour length of the polymer chain
n	polymerization degree
$p(x)$	adsorbed polymer density as a function of distance from the surface
V_L	volume of liquid phase available for the sorbate
V_S	volume of solid phase available for the sorbate
α	swelling coefficient
γ_S	surface free energy of solid
γ_L	surface free energy of liquid
γ_{SL}	interfacial free energy
γ^d	dispersive component of surface free energy
η	fluid viscosity
$[\eta]$	intrinsic viscosity of polymer solution
φ	volume unit density
φ_θ	volume unit density in θ-conditions
φ_χ	volume unit density at given χ
θ	contact angle in colloid-chemical measurements
θ-solvent or θ-conditions	theta-solvent or theta-conditions, i.e. the conditions, in which the dimensions of the polymer coil coincide with its unpertubed dimensions.
σ	grafting area per chain
ξ	characteristic length of the polymer brush permeable for the contacting fluid
χ	Flory parameter

8.8. References

Absolom, D.R., Thomson, C., Kruzyk, W., Zingg, W., and Neumann, A.W. (1986) Adhesion of hydrophilic particles (human erythrocytes) to polymer surfaces: effect of pH and ionic strength, *Colloid. Surf.*, **21**, 447–456.

Absolom, D.R., Zingg, W., and Neumann, A.W. (1987) Interaction of proteins on solid-liquid interfaces: contact angle, adsorption and sedimentation volume measurements. In Brash,

D.L. (ed.), *Polymers at Interfaces*, ACS Symposium Series 343, American Chemical Society, Washington, 401–421.

Absolom, D.R. and Barford, R.A. (1988) Determination of surface tension of packings for high-performance liquid chromatography, *Anal. Chem.*, **60**, 210–212.

Alexander, S. (1977) Polymer adsorption on small spheres. A scaling approach, *J. Phys. (Paris)*, **38**, 977–981.

Alpert, A. (1983) High-performance hydrophobic-interaction chromatography of proteins on a series of poly(alkyl aspartamide)-silicas, *J. Chromatogr.*, **359**, 85–97.

Arakawa, T. and Narhi, L.O. (1991) Solvent modulation in hydrophobic-interaction chromatography, *Biotechnol. Appl. Biochem.*, **13**, 151–162.

Barford, W., Ball, R.C., and Nex, C.M.M. (1986) A non-equilibrium configuration theory of polyelectrolyte adsorption, *J. Chem. Soc., Faradey Trans. I*, **82**, 3233–3244.

Barford, W. and Ball, R.C (1987) Towards a complete configuration theory of non-equilibrium polymer adsorption, *J. Chem. Soc., Faradey Trans. I*, **83**, 2515–2523.

Birshtein, T.M. and Zhulina, E.B. (1983) Conformations of polymer chains grafted to unpearmeable plane surface, *Vysokomol. Soedin. A*, **25**, 1862–1873.

Birshtein, T.M. and Lyatskaya, Yu.V. (1994) Polymer brush in a mixed solvent, *Colloid. Surf. A*, **86**, 77–83.

Borisov, O.V., Zhulina, E.B., and Birshtein, T.M. (1988) Diagram of the state and collapse of grafted chains layers, *Vysokomol. Soedin. A*, **30**, 767–773.

Brooks, D.E. and Mueller, W. (1996) Size-exclusion phases and repulsive protein-polymer interaction/recognition, *J. Mol. Rec.*, **9**, 697–700.

Cafe, M.C. and Robb, I.D. (1982) The adsorption of polyelectrolytes on barium sulfate crystals, *J. Colloid Interface Sci.*, **86**, 411–421.

Chang, S.H., Gooding, K.M., and Regnier, F.E. (1976) High-performance liquid chromatography of proteins, *J. Chromatogr.*, **125**, 103–114.

Cooke, N.H.C. and Olsen, K. (1980) Some modern concepts in reversed-phase liquid chromatography on chemically bonded alkyl stationary phases, *J. Chromatogr. Sci.*, **18**, 512–524.

Cosgrove, T., Patel, A., Semlyen, J.A., Webster, J.R.P., and Zarbaksh, A.(1993) A study of chemisorption of poly(hydrogen methylsiloxane) using neutron reflectometry and small-angle neutron scattering, *Langmuir*, **9**, 2326–2329.

Dettre, R.H. and Johnson, R.E. (1965) Contact angle hysteresis. IV. Contact angle measurements on heterogeneous surfaces, *J. Phys. Chem.*, **69**, 1507–1515.

El Rassi, Z. and Horvath, Cs. (1986) Hydrophobic interaction chromatography of t-RNA's and proteins, *J. Liq. Chromatogr.*, **9**, 3245–3268.

Galaev, I.Yu., Warrol, C., and Mattiasson, B. (1994) Temperature-induced displacement of proteins from dye-affinity columns using an immobilized polymeric displacer, *J. Chromatogr. A*, **684**, 37–43.

de Gennes, P.D. (1980) Conformations of polymers attached to an interface, *Macromolecules*, **13**, 1069–1075.

Gewehr, M., Nakamura, K., and Ise, N. (1992) Gel-permeation chromatography using porous glass beads modified with temperature-responsive polymers, *Makromol. Chem.*, **193**, 249–256.

Grainger, D.W., Nojiri, C., Okano, T., and Kim, S.W. (1989) *In vitro* and *ex vivo* platelet interactions with hydrophilic-hydrophobic poly(ethylene oxide)-polystyrene multiblock copolymers, *J. Biomed. Mat. Res.*, **25**, 979–1005.

Halperin, A., Tirrel, M., and Lodge, T.P. (1991) Tethered chains in polymer microstructures, *Adv. Polymer Sci.*, **100**, 31–71.

Hirotsu, S. (1987) Phase transition of a polymer gel in pure and mixed solvent media, *J. Phys. Soc. Japan*, **56** (1987) 233–242.

Hjertén, S. (1977) Fractionation of proteins by hydrophobic-interaction chromatography, with reference to serum proteins. In *Proceedings Intl. Workshop on Technology for Protein Separation and Improvement of Blood Plasma Fractionation*, Reston, Virginia, 410–421.

Hosoya, K., Sawada, E., Kimata, K., Araki, T., Tanaka, N., and Frechet, J.M.J. (1994) *In situ* surface-selective modification of uniform size macroporous polymer particles with temperature-responsive poly-N-isopropylacrylamide, *Macromolecules*, 27, 3973–3976.

Hosoya, K., Kimata, K., Araki, T., Tanaka, N., and Frechet, J.M.J. (1995) Temperature-controlled high-performance liquid chromatography using a uniformly sized temperature-responsive polymer-based packing material, *Anal. Chem.*, 67, 1907–1911.

Ivanov, A.E., Zhigis, L.S., Chekhovskykh, E.A., Reshetov, P.D., and Zubov, V.P. (1985) Carbonylchloride-containing compositional matrices for immobilization of biospecific ligands, *Bioorgan. Khim.* (in Russian), 11, 1527–1532.

Ivanov, A.E., Saburov, V.V., and Zubov, V.P. (1992) Polymer-coated adsorbents for separation of biopolymers and particles, *Adv. Polymer Sci.*, 104, 135–176.

Ivanov, A.E., Zhigis, L.S., Rapoport, E.M., Lisyutina, O.E., and Zubov, V.P. (1995) Characterization of weak-hydrophobic composite sorbents and their application to the isolation of bacterial lectin. *J. Chromatogr. B*, 664, 219–223.

Ivanov, A.E., Zhigis, L.S., Kurganova, E.V., and Zubov, V.P. (1997) Effect of temperature upon the chromatography of proteins on porous glass chemically coated with N-isopropylacrylamide copolymer, *J. Chromatogr. A*, 776, 75–80.

Kanazawa, H., Yamamoto, K., Matsushima, Y., Takai, N., Kikuchi, A., Sakurai, Y., and Okano, T. (1996) Temperature-responsive chromatography using poly(N-isopropyl-acrylamide)-modified silica, *Anal. Chem.*, 68, 100–105.

Kanazawa, H., Kashiwase, Y., Yamamoto, K., Matsushima, Y., Kikuchi, A., Sakurai, Y., and Okano, T. (1997) Temperature-responsive liquid chromatography. 2. Effect of hydrophobic groups in N-isopropylacrylamide copolymer-modified silica, *Anal. Chem.*, 69, 823–830.

Kanazawa, H., Matsushima, Y., and Okano, T. (1998) Temperature-responsive chromatography, *Trends Anal. Chem.*, 17, 435–440.

King, S.T. and Cosgrove, T. (1993) A dynamical Monte-Carlo model of polymer adsorption, *Macromolecules*, 26, 5414–5422.

Kirsh, Yu.E., Sus', T.A., Karaputadze, T.M., Kobyakov, V.V. Sinitsyna, L.A., and Ostrovski, S.A. (1979) Peculiarities of complex formation and conformational transitions of poly-N-vinyllactams in aqueous solutions, *Vysokomolek. Soedin. A* (in Russian), 21, 2734–2739.

Kurganov, A., Puchkova, Yu., Davankov, V., and Eisenbeiss, P. (1994) Polyvinylpyrrolidone-coated silica packings for chromatography of proteins and peptides, *J. Chromatogr. A*, 663, 163–174.

Loidl-Stahlhofen, A., Kaufmann, S., Braunschweig, T., and Bayerl, T.M. (1996) The thermodynamic control of protein binding to lipid bilayers for protein chromatography, *Nature Biotechnology*, 14, 999–1002.

Matsunaga, T. and Ikada, Y. (1981) Dispersive component of surface free energy of hydrophilic polymers, *J. Colloid Interface Sci.*, 84, 8–13.

Milner, S.T., Witten, T.A., and Cates, M.E. (1988) Theory of grafted polymer brush, *Macromolecules*, 21, 2610–2619.

Milner, S.T. (1991) Polymer brushes, *Science*, 251, 905–914.

Morra, M. and Cassinelli, C. (1995) Thermal recovery of cells cultured on poly(N-isopropylacrylamide) surface-grafted polystyrene dishes, *Polymer Preprints (Am. Chem. Soc., Div. Polym. Chem.)*, 36, 55–56.

Morra, M. and Cassinelli, C. (1997) Thermal recovery of cells cultured on poly(N-isopropylacrylamide) surface-grafted polystyrene dishes. In Ratner, B.D., Castner D.G. (eds.), *Surface Modification of Polymer Biomaterials (Proc. Am. Chem. Soc. Div. Polym. Chem. Int. Symp.)* Plenum, New York, pp. 175–181.

Okano, T., Yamada, N., Sakai, H., and Sakurai, Y. (1993) A novel recovery system for cultured cells using plasma-treated polystyrene dishes grafted with poly(N-isopropylacrylamide), *J. Biomed. Mat. Res.*, **27**, 1243–1251.

Okano, T., Yamada, N., Okuhara, M., Sakai, H., and Sakurai, Y. (1995a) Mechanism of cell detachment from temperature-modulated, hydrophilic-hydrophobic polymer surfaces, *Biomaterials*, **16**, 297–303.

Okano, T., Kikuchi, A., Sakurai, Y., Takei, Y., and Ogata, N. (1995b) Temperature-responsive poly(N-isopropylacrylamide) as a modulator for alteration of hydrophilic/hydrophobic surface properties to control activation/inactivation of platelets, *J. Control. Release*, **36**, 125–133.

Parsonage, E., Tirrel, M., Watanabe, H., and Nuzzo, R.G. (1991) Adsorption of poly(2-vinylpyridine) – poly(styrene) block copolymers from toluene solution, *Macromolecules*, **24**, 1987–1995.

Porath, J., Sundberg, L., Fornstedt, N., and Olson, I. (1973) Salting-out in amphiphilic gels as a new approach to hydrophobic adsorption, *Nature*, **245**, 465–466.

Rollason, G., Davies, J.E., and Sefton, M.V. (1993) Preliminary report on cell culture on a thermally reversible copolymer, *Biomaterials*, **14**, 153–155.

Takei, Y.G., Aoki, T., Sanui, K., Ogata, N., Sakurai, Y., and Okano, T. (1994) Dynamic contact angle measurement of temperature-responsive surface properties for poly(N-isopropylacrylamide) grafted surfaces, *Macromolecules*, **27**, 6163–6166.

Takei, Y.G., Aoki, T., Sanui, K., Ogata, N., Sakurai, Y., and Okano, T. (1995) Temperature-modulated platelet and lymphocyte interactions with poly(N-isopropylacrylamide)-grafted surfaces, *Biomaterials*, **16**, 667–673.

Takezawa, T., Mori, Y., and Yoshizato, K. (1990) Cell culture on a thermo-responsive polymer surface, *Bio/Technology*, **8**, 854–856.

Wu, S.-L., Benedek, K., and Karger, B.L. (1986) Thermal behavior of proteins in high-performance hydrophobic-interaction chromatography. On-line spectroscopic and chromatographic characterization, *J. Chromatogr.*, **359**, 3–17.

Yamada, N., Okano, T., Sakai, H., Karikusa, F., Sawasaki, Y., and Sakurai, Y. (1990) Thermo-responsive polymeric surfaces; control of attachment and detacment of cultured cells, *Makromol. Chem., Rapid Commun.*, **11**, 571–576.

Yoshioka, H., Mikami, M., Nakai, T., and Mori, Y. (1995) Preparation of poly(N-isopropylacrylamide)-grafted silica gel and its temperature-dependent interaction with proteins, *Polym. Adv. Technol.*, **6**, 418–420.

Zhulina, E.B., Borisov, O.V., Pryamitsyn, V.A., and Birshtein, T.M. (1991) Coil-globule type transitions in polymers. 1. Collapse of layers of grafted polymer chains, *Macromolecules*, **24**, 140–149.

9 Smart latexes for bioseparation and bioprocessing

Haruma Kawaguchi and Keiji Fujimoto

9.1. Introduction

A 'thermosensitive polymer' is defined as a polymer which exhibits discontinuous response at the critical temperature. Many polymers such as poly(acrylamide derivative)s, polyethyleneglycol-containing polymers, and cellulose derivatives satisfy this criterion (Kawaguchi, 1999). Among them, poly(N-isopropylacrylamide) (PNIPAM) (I in Scheme 1) is the most representative thermosensitive polymer, having its transition temperature or the lower critical solution temperature (LCST) at 32 °C (Schild, 1992).

When microspheres are composed of thermosensitive polymers, they can be used for various purposes such as an adsorbent, a transducer, a coagulant, etc., whose performance is controlled with temperature. One can prepare several kinds of thermosensitive microspheres such as microgels, crosslinked shell-carrying microspheres, and hairy shell-carrying microspheres, shown in Figure 9.1. They are prepared by emulsifier-free emulsion polymerization (Kawaguchi *et al.*, 1986), precipitation polymerization (Pelton and Chibante, 1986), dispersion polymerization, etc. Each polymerization includes some modifications. Seeded polymerization is also one of the useful methods to prepare thermosensitive shell-carrying microspheres.

In order to expand the uses of thermosensitive microspheres, they are sometimes hybridized with biospecific compounds. The resulting hybrid microspheres can be used for several biomedical applications. In this chapter the versatility of thermosensitive microspheres in their biomedical applications is discussed focusing on PNIPAM microspheres.

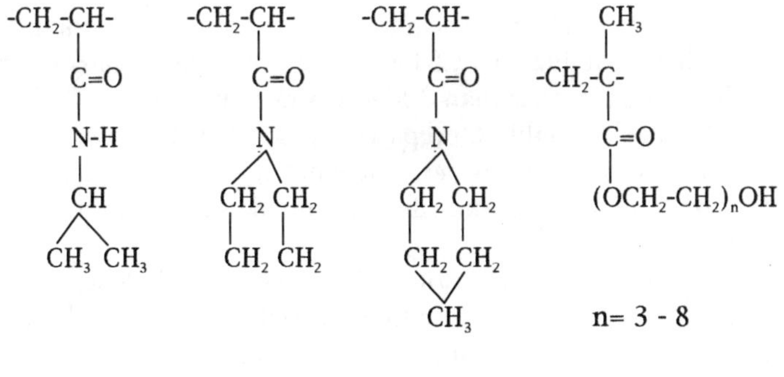

Scheme 1 Repeating units of thermosensitive polymers cited in this review.

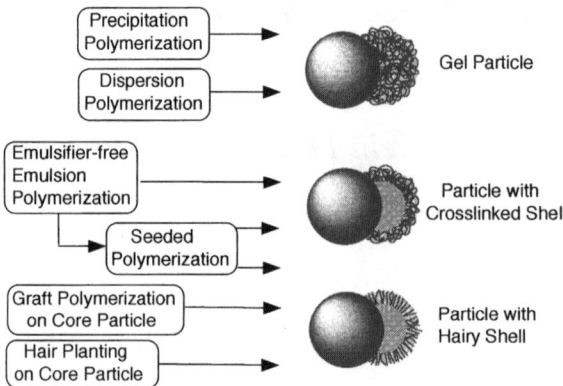

Figure 9.1 Cross-section of thermosensitive particles and methods to prepare them.

9.2. Preparation of Thermosensitive Microspheres

9.2.1. Precipitation and Dispersion Polymerizations to Form Thermosensitive Microgel Particle

NIPAM monomer and many other acrylamide derivatives are soluble in hot water but their polymers are not. Therefore, polymers precipitate from the aqueous solution when they are polymerized in the aqueous solution at the temperature above the LCSTs of the polymers. If aggregation is prevented among the precipitated polymers, they can form stable microspheres. The first trial based on this strategy was carried out by Pelton and Chibante in 1986. They polymerized NIPAM and methylenebis-acrylamide (MB) in an aqueous solution at 70 °C and obtained monodisperse microgel particles of diameter about 500 nm. The particle is swollen in cold water and the hydrodynamic diameter became 1 μm. As no stabilizer was added in this precipitation polymerization, ionic initiator fragment must contribute to electrostatic stabilization. In addition, PNI-PAM precipitated at 70 °C seemed to play a role of self-stabilizer due to its remaining hydrophilicity. The polymerization temperature is crucial for the formation of stable thermosensitive microgel particles and must be at least 30 °C higher than the LCST of polymer. This condition may restrict the application of the precipitation polymerization of other monomers which give higher LCST polymers. Another disadvantage of this method is that a solid content higher than 2.5% was not available.

Much smaller microgel particles could be prepared by an aqueous polymerization of NIPAM and MB with an aid of an emulsifier. The emulsifier might give the nuclei of the particles and stabilize them but the emulsifier remained as a contaminant in the product.

In another aqueous dispersion polymerization of NIPAM, polyethyleneglycol was used as a stabilizer (Topp *et al.*, 1997). The resulting particle was core-shell or core-corona particle whose shell was composed of polyethyleneglycol bound to PNIPAM and core was thermosensitive. The volume of PNIPAM core was proportional to the amount of NIPAM charged. PNIPAM particles thus obtained were employed as a drug carrier for the controlled release (Topp *et al.*, 1998). For these purposes, the inner structure and its homogeneity might become a subject of consideration because almost all of the reports concerning it suggested inhomogeneity of particle interior (Hirose *et al.*, 1987).

9.2.2. *Emulsifier-free Emulsion Polymerization to Form Thermosensitive Shell-carrying Microsphere*

If the thermosensitive microspheres express their function at the surfaces or surface layers, the interior (core) of microspheres is not necessary to be thermosensitive and only the shell of core-shell particles is requested to be thermosensitive. The core-shell particles are preferable to microgel particles in terms of the easy preparation and processing. Emulsifier-free emulsion polymerization is the most applicable method to prepare such core-shell particles.

Almost the same time as Pelton's trial, emulsifier-free emulsion copolymerizations of hydrophobic monomer with acrylamide derivatives were carried out to prepare thermosensitive shell-carrying microspheres (Kawaguchi *et al.*, 1986, 1987, 1993). The acrylamide derivative monomers used were acryloyl pyrrolidine (APr, II in Scheme 1) or acryloyl piperidine (APp, III in Scheme 1) to coat the polystyrene core with the thermosensitive shell. The core-shell structure is spontaneously built up during the copolymerization because the more hydrophilic monomer units, APr or APp in this case, were concentrated to the surface to decrease the polymer-water interfacial energy. An example of the polymerization recipe was; APr or APp x g ($x = 1 - 10$), styrene (St) $(20 - x)$ g, water 120 g, potassium persulfate 0.27 g. The polymerization was carried out at 70 °C under nitrogen. The resulting particles were thermosensitive, indicating reversible swelling (hydrodynamic diameter (HD) 320 nm) and deswelling (HD 280 nm). The transition temperatures of poly-APr and poly-APp shell particles were 55 and 5 °C, respectively. A series of microspheres having the intermediate LCSTs between 55 and 5 °C were prepared by the copolymerization of APr and APp with the appropriate monomer ratios (Hoshino *et al.*, 1987b). Figure 9.2 shows the boundary conditions between stable and unstable dispersions, and the upright zone of each curve indicates the flocculation zone in terms of temperature and ionic strength. When we look at the standing part of each curve from the x-axis, we can see the critical flocculation temperature (CFT) at each ionic strength. Because the flocculation results from the collapse of PNIPAM shell which, in swollen state, contributes to dispersion stabilization, CFT was regarded to be almost the same to the LCST. Therefore, the fact that the CFT becomes lower with increasing NaCl concentration means that NaCl is effective in decreasing the LCST gradually.

With increasing APr fraction, the microsphere had a higher LCST, and lower critical flocculation concentration (CFC, the minimum concentration of KCl to flocculate the dispersion) at higher temperature than LCST, that is, lower level of horizontal part of each curve (CFC$_{min}$). The lower CFC$_{min}$ of high APr particle, than that of high APp particle, was attributed to less deswelling, which restricted the exposure of ionic group to the surface of particle so that the particle had low inter-particle electro-repulsive force and low resistance against salting-out. The figure shows us the conditions to recover the thermosensitive particles from the dispersion medium in terms of temperature and ionic strength.

Under the condition marked with a star, both of APr-rich and APp-rich particles are stably dispersed in the aqueous medium but with different mechanism. The former keeps their stable state by steric stabilization effect whereas the latter does by

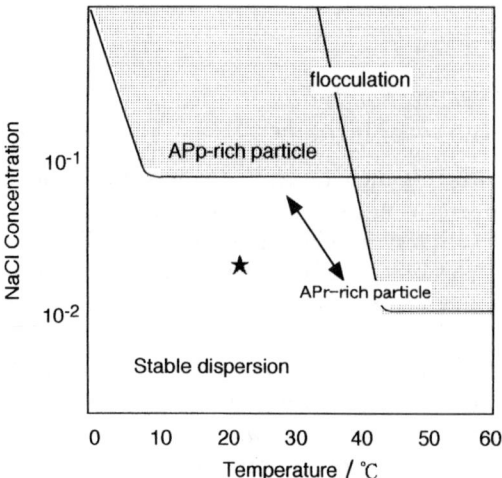

Figure 9.2 Boundary conditions between stable and unstable dispersions for APr- and APp-rich particle systems in terms of temperature and NaCl concentration. ★: a condition for both of APr- and APp-rich particles are stable in the dispersions with different mechanisms.

electrorepulsive effect. Therefore, the former is flocculated by raising temperature but adding more salt is ineffective in flocculation, and vice versa for the latter.

Other soap-free emulsion polymerizations to form thermosensitive particles include polymerizations of St, NIPAM and dimethylaminoethyl methacrylate (DMAEMA) (Duracher *et al.*, 1998), and St and vinyl acetamide (Chen *et al.*, 1997).

The thermosensitivity may be not so high for the core-shell particles prepared by such one-step polymerization. Two-shot polymerization was performed when the thermosensitive shell of the particle prepared by the one-short polymerization is not thick enough (Ohshima *et al.*, 1993). The two-shot polymerization includes the following steps (1) seed particle formation by emulsifier-free emulsion copolymerization of hydrophobic monomer with a small amount of thermosensitive monomer; and (2) seeded copolymerization of thermosensitive monomer with crosslinker. The thermosensitive monomer in step (1) was helpful to invite the thermosensitive monomer in step (2) to the particle and crosslinker charged in step (2) was effective in thickening the shell.

Dingenouts *et al.* (1998) prepared core-shell particles by the same method and studied their thermosensitivity. They emphasized that the thermosensitivity significantly depended on spatial constraint by the core particle surface because the constraint decreased the maximum degree of swelling below LCST but, on the other hand, prevented a full collapse above LCST.

9.2.3. Hairy Particle Formation

Core-shell particles were also prepared by the emulsifier-free emulsion copolymerization of hydrophobic monomer with thermosensitive oligomer-carrying macromonomer, e.g., polyethyleneglycol methacrylate (IV in Scheme 1). The cloud point of the particle dispersion was dependent on the chain length of oligoethylene-glycol (Hoshino *et al.*, 1987a).

The macromonomer units distributed to the surface of particles because they gave low surface energy. Lately PNIPAM-carrying macromonomer was used for the similar purpose (Takeuchi *et al.*, 1993). In this case, PNIPAM exists on the surface of particles as hair or brush. PNIPAM-including macro-azo-initiator was used in an emulsion polymerization to form thermosensitive shell-carrying particles (Kitano and Kawabata, 1995).

When emulsion polymerization of St was carried out in an aqueous solution of PNIPAM at room temperature, St was grafted on a part of PNIPAM chains and the product served as a stabilizer located on the surface of particles (Zhu and Napper, 1994). This is another method to prepare PNIPAM hair-carrying particles although the grafting efficiency is not good.

Hairy or brushy shell-carrying particles were prepared by graft polymerization on a core particle. For example, graft polymerization of NIPAM was carried out by redox initiating system of ceric ion and glycol unit. The latter belonged to glycerol methacrylate on the core particle so that the redox system efficiently forms radicals on the surface and hairs grow up from the surface (Matsuoka *et al.*, 1999). As the polymerization proceeds, the number of hairs on the surface increased but the length of hair, or the degree of polymerization, was not affected. NIPAM homopolymer hair took a loosely coiled conformation at room temperature. On the contrary, the hair of PNIPAM including 0.02% poly(acrylic acid) had a lightly extended conformation due to intra- and inter-particle electrorepulsive force as well as electrorepulsive force between the particle surface and hair chains. Only the acrylic acid-carrying PNIPAM shell particle exhibited discontinuous transition at the LCST as shown in Figure 9.3.

The above-mentioned method to prepare hairy particles is a so-called growth method. There is another method to prepare hairy particles that is called 'hair transplant.' In the method, a preformed polymer having a functional group at the chain end is bound with a preformed particle. For example, carboxyl end group-carrying PNIPAM was chemically bound to the aminated surface of particles (Yasui *et al.*, 1997). If excess amount of PNIPAM chains having two carboxyl groups at α- and ω-positions are incubated with the aminated particles, the PNIPAM hairs have unreacted carboxyl groups at the outermost layer, which can be used for immobilization of functional compounds such as enzymes and antibodies, etc.

9.3. Protein Adsorption on Thermosensitive Particles

9.3.1. Protein Concentration and Separation

PNIPAM gel beads have been used for protein concentration and separation (Sassi *et al.*, 1992). The technology is based on two mechanisms in terms of the affinity of protein to the polymer beads. Proteins that have no affinity towards PNIPAM can be concentrated in their aqueous solution because water is selectively absorbed into the polymer beads (Freitas and Cussler, 1987). Specific proteins which have affinity towards PNIPAM can be separated from others by being selectively absorbed into the polymer. To adjust the affinity between protein and PNIPAM, in some cases, ionic groups were incorporated into PNIPAM and the pH for the operation was controlled. Comprehensive studies on protein separation using charged thermosensitive macrogel

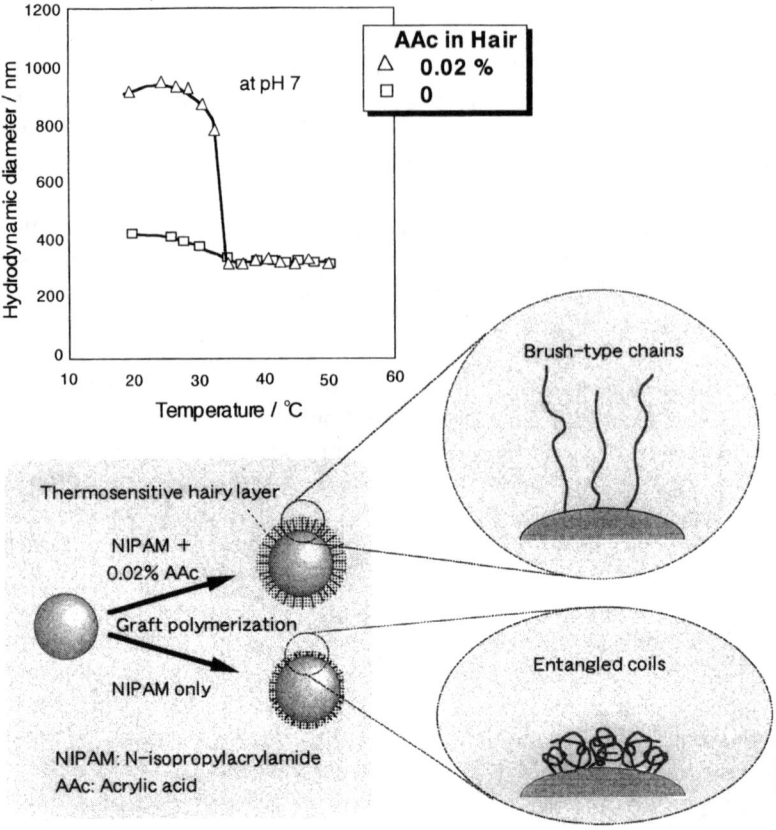

Figure 9.3 Thermosenstivity of latex shells composed of poly-NIPAM or poly(NIPAM-co-acrylic acid).

pointed out that the factors decisive for the partition coefficient are size exclusion (steric effect), electrostatics, and short-range interactions such as hydrophobicity. The higher swelling favours permeation of protein but hampers hydrophobic interaction between protein and gel and the partition coefficient is decided by the balance of these conflicting factors and the electrostatic interaction between protein and gel (Sassi *et al.*, 1996).

After either operation, the PNIPAM gel beads are heated to be collapsed so that they can be re-used.

Protein adsorption and desorption were carried out using submicron-sized thermosensitive particles in place of beads. Protein concentration decreased when submicron particles were added into an aqueous solution of proteins, and the decrease was attributed to adsorption of protein at the surface of the particles, but not to absorption inside the particles. It was because the surface/volume ratio of the submicron particles was quite large and the amount of absorbed proteins was neglected. Tight network structure of the gel microspheres is supposed to assist in increasing the ratio of adsorption/absorption.

The smaller particles have more various advantages, that is, large specific surface area, high diffusion rate, and short inter-particle surface-to-surface distance, all of which promote the encounter of particles with proteins in the dispersion. The smaller particles do not need pore structure inside them and have a plain surface which offers uniform adsorption condition to the proteins. The small size did not give any inconvenience in recovering process if the recovery is carried out by temperature- or ionic strength-induced aggregation. Another promising recovery method is a use of magnetic force, which is applicable to magnetic particles. PNIPAM particles containing magnetite were efficiently separated from the medium with a magnet after thermal coagulation (Kondo *et al.*, 1994).

9.3.2. *Temperature-dependence of Protein Adsorption on PNIPAM Particles*

Adsorption of any protein onto PNIPAM microspheres is more or less sensitive to temperature. It would be attributed mainly to the change in hydrophilicity/hydrophobicity of the surface with temperature because proteins are adsorbed on hydrophobic surfaces preferably in general and thus adsorbed more on PNIPAM microspheres at a temperature above 32 °C than below it. Therefore, the proteins once adsorbed on PNIPAM microspheres at a high temperature could be desorbed more or less when the system is cooled down to room temperature as shown in Table 9.1. This was not the case when polystyrene microspheres were used as an adsorbent, on which a large amount of proteins were adsorbed regardless of the temperature. Polystyrene microspheres did not release the once-adsorbed proteins even when the temperature was decreased (Fujmoto *et al.*, 1993; Kawaguchi and Fujmoto, 1999).

In contract to the adsorption onto polystyrene particle, the adsorption of globulin onto PNIPAM particle was temperature dependent as mentioned above. About 40% of globulin once-adsorbed was desorbed when the system was cooled down through the LCST of PNIPAM.

More significant thermosensitivity was observed in albumin adsorption on PNIPAM particle. Albumin was adsorbed appreciably at 37 °C and most of the protein was desorbed on cooling. In contrast, the hydrophilic protein lysozyme was adsorbed in a small amount at 37 °C, even though the electrostatic force between the surface and protein was favourable for the adsorption, and desorbed a little when cooled. Because the PNIPAM surface is much softer than the polystyrene surface even above the transition temperature, an enzyme desorbed from PNIPAM microsphere retained higher activity than that from polystyrene microspheres.

Table 9.1 Protein absorption at 37 °C and desorption at 25 °C on latex particles

Particle*	Protein (P)	IEP of P	Amount (A) of P absorbed at 37 °C (g/g particle)	Amount (B) of P desorbed at 25 °C** (g/g particle)	B/A
PSt	IgG	6.8	8.5	0.1	
PNIPAM	IgG	6.8	2.4	1.0	0.42
PNIPAM	HSA	5	1.9	1.5	0.79
PNIPAM	Lyz	11	0.40	0.06	0.15
PNIPAM	MG	7	1.7	1.1	0.65

* Both particles have negative charges.
** Desorbed when cooled to 25 °C after 30 min incubation at 37 °C.
Particle: PSt Polystyrene, PNIPAM poly(N-isopropylacrylamide).
Protein: IgG Immuno-globulin G, HSA Human serum albumin, Lyz Lysozyme, MG Mioglobin.
Adsorption and desorption were carried out at pH 7.0.

It should be mentioned that there is another factor to strongly control the protein adsorption on thermosensitive particles at different temperatures. That is the electro-static force which comes from the difference between the surface potentials of PNIPAM microspheres and protein. Therefore the adsorption and desorption were affected by pH, ionic strength, and temperature. Lowering the temperature decreased the charge density in the PNIPAM shell and, consequently, the apparent surface potential as well.

9.4. Thermal Behavior of Enzyme Carrier

9.4.1. Diffusion of Substrate through PNIPAM Layer

An enzyme, trypsin, was immobilized onto different kinds of PNIPAM shell particles and the temperature dependence of the activity of immobilized enzyme was studied. The enzyme activity is affected by the diffusivity of substrate through PNIPAM layer as well as by the immobilized mode and density of enzyme. The temperature dependence of the diffusion of low molecular weight materials through PNIPAM layer has been studied by measuring the reaction rates of reduction and oxidation between oxidant (or reductant) in the core and reductant (or oxidant) in the medium as a function of temperature. For example, the rate was examined at which ascorbic acid in water dif-fused through PNIPAM shell layer to reach ubiquinone in the core (Kato et al., 1994). The diffusion rate was high at low temperature and decreased with increasing temper-ature. This could be explained by the temperature-dependent density and/or viscosity of PNIPAM layer. Namely, above the transition temperature, the dense PNIPAM layer rejected the permeation of water-soluble reductant or oxidant.

In addition, the rise in temperature brought about hydrophobization of PNIPAM layer and consequently decreased the solubility of water-soluble molecules in PNIPAM.

In the system of trypsin immobilized to crosslinked PNIPAM microsphere, the activ-ity of trypsin was measured as a function of temperature (Shiroya et al., 1995). The relative enzyme activity decreased with increasing temperature through the LCST of PNIPAM. This was attributed to the network densing and hydrophobization effect mentioned above.

9.4.2. Activity of Enzyme Immobilized at PNIPAM Chain End

The situation was drastically changed when the enzyme carrier had different archi-tectures. PNIPAM bearing carboxyl group(s) at the chain ends(s) was prepared using solution polymerization and chemically bound to latex particle. Trypsin was immo-bilized at the other end of some PNIPAM chains. Two kinds of PNIPAM chains, trypsin-carrying and free-end chains, coexisting on the microspheres, had different transition temperature as shown in Figure 9.4A (Yasui et al., 1997). The conceiv-able reasons for the rise in transition temperature of PNIPAM chain by immobilizing trypsin are a decrease in the entropy of chain due to suppressed flexibility and an increase in the hydrophilicity of chain induced by hydrophilic protein immobilization.

The difference in the transition temperature between trypsin-carrying and free-end chains gave a unique character to the hybrid particle. Free chains collapsed at lower temperature than trypsin-carrying chains did and so trypsin was exposed toward the aqueous medium above the transition temperature of free-end PNIPAM chain as shown in Figure 9.4B. Trypsin could encounter the substrate more easily under

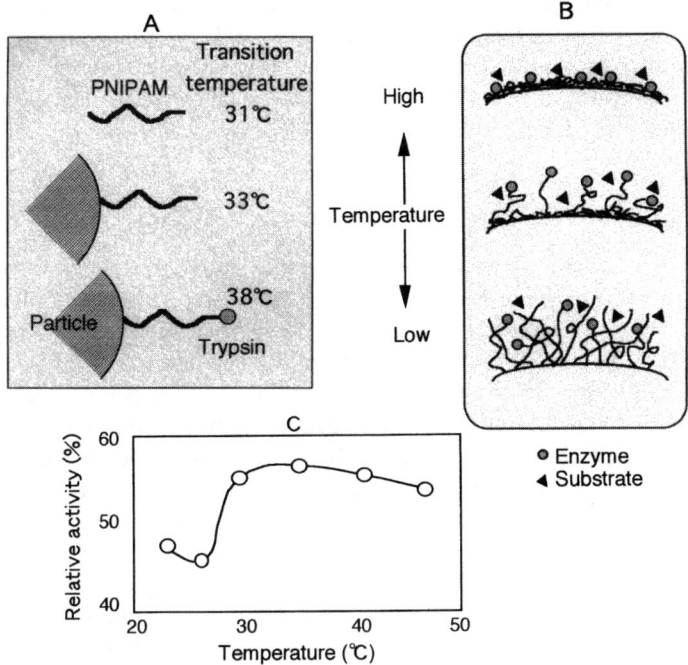

Figure 9.4 Thermosensitive activity of enzyme immobilized at the end of hair on particle. A: Transition temperatures of PNIPAM chains under different conditions.
B: Conformations of PNIPAM at different temperatures. C: Relative activity of
immobilized enzyme as a function of temperature.

such conditions and exhibited high enzymatic activity. The temperature response of
the hybrid particle is shown in Figure 9.4C. Some explanation might be required on
the temperature at which enzymatic activity started to increase because the temperature was slightly lower than the transition temperature of free-end PNIPAM chain.
This temperature shift was supposed to result from n-clustering of free-end PNIPAM
chains (Zhu *et al.*, 1993) which starts at about 10 degrees lower than the LCST.

The above-mentioned, unique temperature dependence of enzyme activity could
be observed when the following conditions were satisfied; the particle surface was
masked with a sufficient amount of PNIPAM chains (both of trypsin-carrying and
free-end chains), and the substrate had a molecular size larger than the critical one.
These requirements are logical because the phenomenon was based on the diffusion of
substrate. As to the latter, the critical molecular weight seemed to be around several
thousands in the present system. One can use this system for a temperature-dependent
selective enzymatic reaction in substrate mixtures.

9.5. Cell Activation by PNIPAM Particles

Cells are activated when contacting with external materials. For example, granulocytes play an important role to protect the body from viruses and toxic materials by
activating themselves when they contact with the foreign materials. Or more precisely,
the cells recognize some kinds of stimuli from the foreign materials and start to defeat

them. The stimuli include the different features of foreign materials such as the molecular structure, shape, hydrophobicty, electrostatic force, etc. Because thermosensitive particles can change these features with temperature, the study on the mechanism and kinetics of cell activation by thermosensitive particles as a function of temperature is challengeable. In this section, three modes of PNIPAM particle systems are discussed as the stimulant or activator for granulocytes and neutrophil-like cells.

9.5.1. Cell Activation in PNIPAM Particle Dispersion

When the cells were stimulated by external materials, they consume oxygen dissolved in the medium and produce active oxygen species. Therefore, ability of polymeric particles to activate cells can be assessed by measuring the amount of oxygen consumed or active oxygen species produced by the cells which were incubated in a dispersion of polymeric particles (Achiha et al., 1994). For example, the oxygen consumption by granulocyte contacting particles was measured according to the following method: 100 μl of the particle dispersion (solid content: 10 wt%) was added to 2 ml of the granulocyte suspension (cell number: 1×10^7 cells/ml) that was pre-incubated for 2 min at 25 or 37 °C. The oxygen concentration in the suspension was measured with an oxygen electrode at 0.65 V. In order to evaluate the normal respiration of granulocytes, water was injected instead of the dispersion of microspheres.

Active oxygen species produced by cells in contact with particles were determined from the intensity of chemiluminescence that was emitted by the reaction between luminol and active oxygen species, which were dissolved in the reaction medium. This measurement could be done by using a very small amount of cells, to say, 0.2 ml of cell dispersion of 2×10^5 cells/ml concentration. The experiment was done in a 1.8 ml phosphate buffered saline containing 0.5 mM luminol. The chemiluminescence was measured with a chemiluminescence photometer at two temperatures, one of which was above and the other was below the LCST of PNIPAM.

The measurements were carried out in three different particle systems, that is, polystyrene, crosslinked polyacrylamide and PNIPAM particles with diameters about 250 nm, which were the representative of highly hydrophobic, highly hydrophilic and highly hydrophilicity-changeable particles, respectively. When granulocytes were mixed with polystyrene particle dispersion, the cells extensively phagocytized the particles and consumed a large amount of oxygen regardless of temperature in the range of 25 to 37 °C. On the contrary, polyacrylamide particles were not phagocytized and the cells consumed a little amount of oxygen even at 37 °C. The slight amount of active oxygen production in the system was attributed to the normal respiration of granulocytes, but not to the stimulus by the particles.

The interaction of PNIPAM particles with granulocyte was rather similar to that of polyacrylamide particle and granulocyte. The microspheres were almost inert at room temperature and gave a slight but non-neglectable stimulus to cells at 37 °C, judging from the amount of oxygen consumption and active oxygen production. It would be worth mentioning that no phagocytosis was detected even at 37 °C in PNIPAM particle system. This means that oxygen consumption and active oxygen production are, at least in this case, not directly related with phagocytosis. In other words, contact of particles with cells can cause oxygen consumption and production of active oxygen species. The different response of cells to PNIPAM particles between two temperatures, above and below the LCST, can be attributed to the differences in hydrophobicity and/or hardness of the particles between two temperatures. The decrease in hydro-dynamic size with increasing temperature seemed not to be the factor to induce the cell activation. The effects of hydrophobicity and hardness of particles was also the reason for the much less oxygen consumption in PNIPAM system at 37 °C than that in polystyrene system. That is, PNIPAM particles are still more hydrophilic and softer at 37 °C than polystyrene particles.

Oxygen consumption measurement was also carried out in above-mentioned three systems in which the temperature of the systems was suddenly changed from room temperature to 37 °C. The amount of consumed oxygen vs. time curve smoothy shifted from the low-temperature ones to the high temperature one and no excess response was observed in every system.

It would be worth mentioning that the amount of active oxygen produced by cells when stimulated with PNIPAM particles depended on the medium. All of the above-mentioned results were obtained in phosphate buffered saline while different results were obtained in serum. In serum, selective adsorption of proteins altered the property of particle surface and hence affected the cell–particle interaction.

9.5.2. Cell Activation by 2-dimensional Particle Assembly

Two kinds of PNIPAM particles having different diameters were prepared by soap-free emulsion polymerization followed by seeded polymerization and deposited on a plate in order to produce a 2-dimensional array of the particles on the surface. Once the plate of particles assembly was dried, the particles were never detached even when the plate was again shaken in water. The micro-patterned plate thus obtained was placed on the bottom of a dish and cell dispersions were poured into the dish. The objective of this experiment was to study (a) the effects of surface morphology of materials on cell activation, and (b) the effect of dynamic motion of surface on cell activation. The dynamic motion was induced on the particle assembling surface due to the volume phase transition of PNIPAM particles in response to raid changes in temperature from below to above the LCST of PNIPAM.

For the objective (a), polystyrene particle assemblies of different particle sizes, 650, 960, 1,050 and 1,230 nm, were also prepared. The dimensional assemblies were obtained by spilling particle dispersions on a plate, incubating for a proper period of time and then treating the plat with a spinner. The four polystyrene particle systems gave densely regulated assemblies although the incubation conditions were differ-ent in each system. The extent of cell activation was measured by luminol-induced chemiluminescence.

Among four kinds of polystyrene particle-assembled surfaces, the micropatterned surface which was composed at 1,050 nm particles gave the largest stimulus to the cells. The pitch of particles might match the size of contact point on the cell. The same was true in the PNIPAM particle-assembled surfaces, that is, the surface having a pitch of *ca*. 1 μm gave the largest stimulus to cells, although much less active oxygen was produced by the PNIPAM particles-assembled surface than the polystyrene particles-assembled surface (Miyaki *et al.*, 1999).

Cells were more activated at 37 °C than at 25 °C as observed in PNIPAM particle dis-persion system. Differing from floating particles, the fixed particles were found to give a stronger stimulus to the cells when the cells settled down one by one to the assembly surface (Fujimoto *et al.*, 1997). The temperature dependence of the response of cells in contact with dimensional PNIPAM particle assembly was again attributed to the

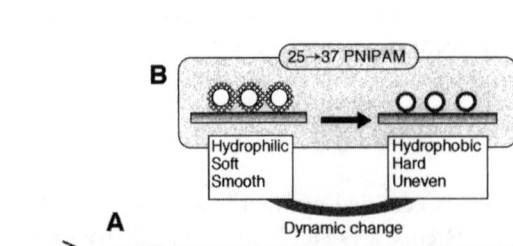

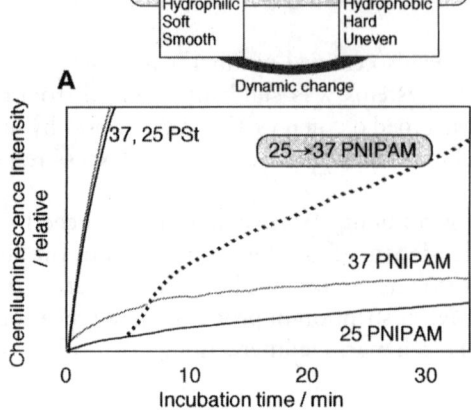

Figure 9.5 Cell activation by polymer particles in dispersion. A: Active oxygen production by cells stimulated under different conditions. B: Schematic illustration of surface covered with PNIPAM particles at different temperatures. Substrate PSt: Polystyrene PNIPAM: 2-dimentional assembly of PNIPAM-shell particles.

structural difference in the surface between two temperatures, a soft and hydrophilic surface at 25 °C and, a hard uneven and hydrophobic one at 37 °C.

In contrast with the response of cells to the static stimulation, that to dynamic stimulation was quite unique. It was observed when the temperature of the system was suddenly changed from 25 to 37 °C after 4 min incubation at 25 °C. Under these conditions, the cells released chemiluminescence as shown with dotted lines in Figure 9.5A, which did not approach the curve for 37 °C but passed over it. A sudden cooling from 37 to 25 °C also brought about the excessive response of the cells. Namely, in this case, cooling did not decrease the chemiluminescence but increased it. The temperature change-induced stimulation was a function of the rate of temperature change and a very slow rise in temperature resulted in no excess response. These effects were attributed to the stimulus by dynamic motion of PNIPAM shell of microspheres resulting from PNIPAM's volume phase transition as shown in Figure 9.5B. Therefore, such phenomenon was not observed on a flat PNIPAM surface, which was prepared by grafting PNIPAM onto a substrate surface, or on an assembly of non-thermosensitive polystyrene microspheres.

This section is concluded by describing that two- (and three-) dimensional assembly of thermosenstitive particles can be a novel cell activator exhibiting a unique temperature effect.

9.5.3. Cell Activation by PNIPAM Particles with the Aid of Ligand/Receptor Interaction

Reactive thermosensitive particles were prepared by precipitation polymerization of NIPAM with a small amount of acrylamide (5% of NIPAM) and methylenebisacrylamide. The resulting thermosensitive copolymer gel particle was coded as NAM particle. The diameter of a dry NAM particle was 300 nm. The hydrodynamic diameter of a NAM particle changed significantly with temperature through the transition temperature. The transition temperature was almost the same to the LCST of PNIPAM despite the coexistence of acrylamide units. The particles were aminated by the Hoffman reaction (Kawaguchi *et al.*, 1984) and then a tetrapeptide, arginine–glycine–aspartic acid–serine (RGDS), was bound chemically to the particles. The hybrid particles thus obtained were coded as NAM-RGDS and their ability to control cell functions was investigated.

9.5.3.1. Cell-adhesive Peptide

RGDS is a key peptide in cell-adhesive proteins. It is usually buried in the protein molecule itself but exposed in response to the change in environmental conditions such as the calcium ion concentration. Then RGDS gets a chance to bind the receptor called integlin on the cell surface. The integlin consist of two membrane proteins, one of which has a specific site for RGDS binding and the other helps to hold the integlin molecule conformation suitable for the adhesion of RGDS-carrying protein. The protein adhesion onto cells mediates then the transduction of signals to the cytoskelton. Oxygen consumption and active oxygen production by the cell are the output of this event, and we can estimate the extent of cell response from the amount of oxygen consumed or/and of active oxygen produced. (In the tetrapeptide, RGD is essential but S is not and exchangeable with other amino acid.) Such transduction does not occur when soluble RGDS adheres a cell. The whole body of protein molecule seems to play a certain role in the transduction.

Immobilization of RGDS onto aminated NAM particles did not affect the particle's thermal property. Namely, the transition temperature of RGDS-carrying particles was the same as that of the NAM particles. The hybrid particles were incubated with cells at different temperatures to examine the quality and quantity of the stimulus which the hybrid particles give to the cells.

9.5.3.2. Interaction between RGDS-carrying PNIPAM Particle and Cell

When cells were added into a buffered dispersion of NAM-RGDS particles at room temperature, they started to release active oxygen. On the other hand, the cell exhibited only a little response when incubated with NAM particles, like as in PNIPAM particle dispersion. Therefore, the active oxygen production in NAM-RGDS system can be attributed to a RGDS-integlin interaction. This view was confirmed by an experiment in which the above-mentioned procedure was carried out in the presence of free RGDS which does not cause cell activation. When present, free RGDS competed with NAM-RGDS in binding integlin. With increasing free RGDS concentration, therefore, more integlin was masked with free RGDS and the production of active oxygen decreased (Figure 9.6) (Kisara, 1998).

When cells were mixed with NAM-RGDS particles in a medium without Mg^{2+} ions, no active oxygen was produced. This is another evidence for the contribution of RGDS-integlin interaction to the active oxygen production Mg^{2+} ions are crucial for keeping the integlin conformation in its active state and the lack of Mg^{2+} ions prevents the RGDS-integlin reaction.

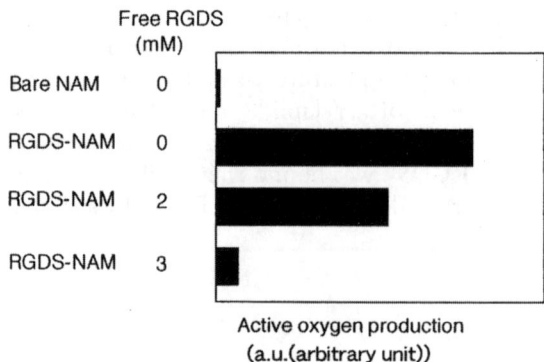

Figure 9.6 Cell activation with RGDS-carrying PNIPAM particles in the absence and presence of free RGDS. NAM: PNIPAM gel particle containing a small amount of
acrylamide.

If temperature is suddenly raised in the dispersion in which NAM-RGDS particles were attached to cells, the cell surface must suffer severe turbulence because the sudden collapse of PNIPAM (carrier of RGDS) forces the RGDS-integlin complex to move. A quick change in temperature to 37 °C after 20 min incubation at 25 °C resulted in an excessive response of the cells. The phenomenon suggested that sudden conformational change of RGDS-carrying PNIPAM chains activated cells by inducing the formation of integlin cluster or giving mechanical stress to the cell skeleton. Although it is unknown which mechanism works, it is obvious that RGDS-carrying thermosensitive particles constitute a convenient tool for cell activation in cell dispersion.

9.6. Conclusions

Several kinds of thermosensitive polymer particles with different architectures were obtained by heterogeneous polymerizations. Some of them were further modified or hybridized with biopolymers or their analogs. Their temperature sensitive properties such as swellability, hydrophilicity, apparent surface charge, dispersion stability, etc. were put into account in bioprocessing, which includes protein adsorption, enzyme reaction, cell activation, and others. The particles were employed for such applications in the form of dispersions or 2- or 3-dimensional particle assemblies.

9.7. Abbreviations

APr	Acryloyl pyrrolidine
APp	Acryloyl piperidine
CFC	Critical flocculation concentration
CFT	Critical flocculation temperature
DMAEMA	Dimethyleminoethyl methacrylate
HD	hydrodynamic diameter
LCST	Lower critical solution temperature
MB	Methylenebisacrylamide
NAM	N-isopropylacrylamide-acrylamide-methylenebisacrylamide

NIPAM   *N*-isopropylacrylamide
PNIPAM   Poly-NIPAM
RGDS   Arginine–glycine–aspartic acid–serine
St   styrene

9.8. References

Achiha, K., Ojima, R., Kasuya, Y., Fujimoto, K., and Kawaguchi, H. (1994) Interactions between temperature-sensitive hydrogel microspheres and granulocytes, *Polym. Adv. Tech.*, 6, 534–540.

Chen, M.-Q., Kishida, A., and Akashi, M. (1997) Graft copolymers having hydrophobic backbone and hydrophilic branches. XI. Preparation and thermosensitivite properties of polystyrene microspheres having poly(N-isopropylacrylamide) branches on their surfaces, *J. Polym. Sci., Polym. Chem. Ed.*, 34, 2213–2220.

Dingenouts, N., Norhausen, Ch., and Ballauff, M. (1998) Observation of the volume transition in thermosensitive core-shell latex particles by small-angle X-ray scattering, *Macromolecules*, 31, 8912–8917.

Duracher, D., Sauzedde, F., Elaissari, A., Perrin, A., and Pichot, C. (1998) Cationic amino-containing N-isopropylacrylamide-styrene copolymer latex particles: 1 Particle size and morphology vs. polymerization process, *Colloid Polym. Sci.*, 276, 219–231.

Freitas, R.F.S. and Cussler, E.L. (1987) Temperature sensitive gels as extraction solvents, *Chem. Eng. Sci.*, 42, 97–103.

Fujimoto, K., Mizuhara, Y., Tamura, N., and Kawaguchi, H. (1993) Interactions between thermosensitive hydrogel microspheres and proteins, *Intl. Mat. Systems Str.*, 4, 184–189.

Fujimoto, K., Takahashi, T., Miyaki, M., and Kawaguchi, H. (1997) Cell activation by the micropatterened surface with setting particles, *J. Biomater Sci. Polym. Edn.*, 8, 879–891.

Hirose, Y., Amiya, T., Hirokawa, Y., and Tanaka, T. (1987) Phase transition of submicron gel beads, *Macromolecules*, 20, 1342–1344.

Hoshino, F., Sakai, M., Kawaguchi, H., and Ohtsuka, Y. (1987a) Soap-free latices of polyoxyethylene chain-binding particles, *Polymer J.*, 19, 383–389.

Hoshino, F., Kawaguchi, H., and Ohtsuka, Y. (1987b) N-Substituted acrylamide-styrene copolymer latices III. Morphology of latex particles, *Polymer*, 19, 1157–1164.

Kato, T., Fujimoto, K., and Kawaguchi, H. (1994) Permeation control by thermosensitive shell layer of submicron microspheres, *Polymer Gels Networks*, 2, 307–313.

Kawaguchi, H., Hashino, H., Amagasa, H., and Ohtsuka, Y. (1984) Modifications of a polymer latex, *J. Colloid Interface Sci.*, 97, 465–475.

Kawaguchi, H., Hoshino, F., and Ohtsuka, Y. (1986) Soap-free emulsion copolymerization of styrene with acrylopyrrolidine and features of resulting latices, *Makromol. Chem., Rapid Comun.*, 7, 109–114.

Kawaguchi, H., Fujimoto, T., Hoshino, F., and Ohtsuka, Y. (1987) N-substituted acrylamide-styrene copolymer latices. 2. Polymerization behavior and thermo-sensitive stability of latices, *Polymer J.*, 19, 241–248.

Kawaguchi, H., Fujimoto, K., Saito, M., Kawasaki, T., and Urakami, Y. (1993) Preparation and modification of monodisperse hydrogel microspheres, *Polymer International*, 30, 225–231.

Kawaguchi, H. (1999) Thermosensitive hydrogel microspheres. In Arshady, R. (ed.), *Microspheres Microcapsules & Liposomes*, Citus Books, The MML Series, pp. 237–252.

Kawaguchi, H. and Fujimoto, K. (1999) Smart latexes for bioseparation, *Bioseparation*, 7, 253–258.

Kisara, K. (1998) Cell activation by thermosensitive polymer particles carrying cell adhesive peptide, Master's thesis, Keio University, Yokohama Japan.

Kitano, H. and Kawabata, J. (1995) Temperature-responsive polymer microspheres prepared with a macroinitiator, *Macromol. Chem. Phys.*, 197, 1721–1729.

Kondo, A., Kamura, H., and Higashitani, K. (1994) Development and application of thermo-sensitive magnetic immunomicrospheres for antibody purification, *Appl. Microbiol. Biotechnol.*, **41**, 99–105.

Matsuoka, M., Fujimoto, K., and Kawaguchi, H. (1999) Monodisperse microspheres exhibiting discontinuous response to temperature change, *Polym. Gels Networks*, **6**, 319–332.

Miyaki, M., Fujimoto, K., and Kawaguchi, H. (1999) Cell response to micropatterned surfaces produced with polymeric microspheres, *Colloids Surfaces A. Physicochem. Eng. Asp.*, **153**, 603–608.

Ohshima, H., Makino, K., Kato, T., Fujimoto, K., Kondo, T., and Kawaguchi, H. (1993) Electrophoretic mobility of latex particles covered with temperature-sensitive hydrogel layers, *J. Colloid Interface Sci.*, **159**, 512–514.

Pelton R.H. and Chibante P. (1986) Preparation of aqueous latices with N-isopropylacrylamide, *Colloids Surfaces*, **20**, 247–256.

Sassi, A.P., Blanch, H.W., and Prausnitz, J.M. (1992) Crosslinked gels as water absorbents in separations, In Sloane, D.S. (ed.), *Polymer Applications for Biotechnology*, Prentice Hall, Englewood Cliffs, pp. 244–275.

Sassi, A.P., Shaw, A.J., Han, S.M., Blanch, H.W., and Prausnitz, J.M. (1996) Partitioning of proteins and small biomolecules in temperature- and pH-sensitive hydrogels, *Polymer*, **37**, 2151–2164.

Schild, H.G. (1992) Poly (N-isopropylacrylamide) experiment, theory and application, *Progress Polym. Sci.*, **17**, 163–250.

Shiroya, T., Tamura, N., Yasui, M., Fujimoto, K., and Kawaguchi, H. (1995) Enzyme immobilization on thermosensitive hydrogel microspheres, *Colloids Surfaces B. Biointerfaces*, **4**, 267–274.

Takeuchi, S., Oike, M., Kowitz, C., Shimasaki, C., Hasegawa, K., and Kitano, H. (1993) Microspheres prepared with a temperature-responsive macromonomer, *Makromol. Chem.*, **194**, 551–558.

Topp, M.D.C., Dijkstra, P.J., Talsma H., and Feijen J. (1997) Thermosensitive micelle-forming block copolymers of poly(ethylene glycol) and poly (N-isopropylacrylamide), *Macromolecules*, **30**, 8518–8520.

Topp, M.D.C., Hamse, I.M., Dikstra, P.J., Talsma, H., and Feijen, J. (1998) Thermosensitive micelles for drug delivery, *Polym. Repr.*, **39**, 176–177.

Yasui, M., Shiroya, T., Fujimoto K., and Kawaguchi, H. (1997) Activity of enzymes immobilized on microspheres with thermosensitive hairs, *Colloids and Surfaces B: Biointerfaces*, **8**, 311–319.

Zhu, P.W. and Napper, D.H. (1994) Experimental observation of coil-to-globule type transitions at interfaces, *J. Colloid Interface Sci.*, **164**, 489–494.

10 Application of water soluble polymers and their complexes for immunoanalytical purposes

Boris B. Dzantiev, Anatoliy V. Zherdev, and Elena V. Yazynina

10.1. Introduction

During the eighties and nineties immunoanalytical approaches became widely applied to the determination of rather different chemical substances and biological objects, from low molecular weight compounds (metal ions, pesticides, hormones, etc.) to virus particles and whole microorganisms (Tijssen, 1985; Gosling, 1990; Wild, 1994). Nowadays the immunoassay is an irreplaceable part of diagnostics and control in medicine, food and environmental monitoring, biotechnology, etc. (Gee *et al.*, 1994; Hock, 1996; Drummer, 1998; Mitchell *et al.*, 1998; Shiba, 1998; Watson, 1998).

This success of the method was due to the combination of two factors: highly specific recognition of different substances by antibody molecules, and labeling of the detected immune complexes by compounds that may be quantitatively measured at extremely low concentrations. Enzymes, fluorochromes, and isotopes are the most widely used immunoassay labels. High sensitivity and speed of assay can thus be achieved. There are many immunoassay formats, differing in the labels used, sequence of the assay steps and molecular structure of detected complexes. A description of the current variety in the assay formats may be found in the reviews of Sugiura and Mizuoka (1995) and Hock (1996).

Although the use of smart polymers in the immunoassay is not so widespread, this direction of investigation is extremely promising. The pertinent studies in this field along with other possibilities for the practical use of the smart polymers were first summarized by Galaev and Mattiasson (1993). Potential implementations of the polymers are conditioned by the possibility of reaching fast changes in the polymer's microstructure triggered by small changes in the characteristics of the medium in which they are used (e.g., its pH, temperature, ionic strength, presence of specific chemicals, light, electric or magnetic field). These microscopic changes may entail different easily-detected changes at the macroscopic level, such as precipitate formation. Nowadays these properties of the smart polymers are successfully used in biochemistry (immo-bilized biocatalysts, size-selective and downstream separation), medicine (controlled drug release, site-directed drug delivery), and other fields (Galaev *et al.*, 1996). The new immunoassay systems that will be described in the present review are also based on the peculiarities of the water-soluble polymers and their ability to change structurally under the influence of the above mentioned factors.

As indicated above, the main tasks in the elaboration of a new immunoassay for-mat are reaching higher specificity and/or sensitivity as well as reducing the assay time. Accordingly, the following directions may be outlined for the immunoanalytical applications of smart polymers:

- use of polymer carriers for effective separation of detected immunoreactants from the reaction mixture;
- use of polymers that can be directly detected as labels of immunoreactants;
- modification of immunoreactants by polymers in order to change properties of the test solution for more effective immune recognition of the corresponding antigens;
- polymeric modification of enzymes or other labels for improving their properties (activity, stability, etc.).

Concrete approaches for realizations of these ideas will be described in the course of the chapter. Typically we will analyze the assay principle, properties of the polymers used, specific features of the assay protocols for different tested substances, achieved results and perspectives of the assay. Some of the given examples are not consistently classified as pertaining to 'smart' properties of the applied polymers, although drastic changes of smart polymers may give additional advantages for these approaches. We shall first describe in detail the techniques that are based on typical realignments of the smart polymer carriers, and thereafter characterize briefly other cases of applications of different water-soluble polymers.

10.2. Immunoassays with Separation through Polyelectrolyte Reaction

10.2.1. *Homogeneous Assays with the Use of Polyelectrolyte Pairs*

The interaction between water-soluble linear polyelectrolytes was proposed as a means for the separation of reactants during immunoanalysis (Dzantiev *et al.*, 1988, 1989, 1990). Poly(methacrylic acid) (further named PMA) was used in these systems as the polyanion, and poly(N-ethyl-4-vinylpyridinium) (PEVP) – as the polycation. The possibility of applying poly(N-N'-dimethyldiallylammonium) as the polycation in the pair with PMA polyanion was also demonstrated.

Interaction between polyanion and polycation molecules leads to the formation of polyelectrolyte complexes (PEC). These complexes combine two properties that might appear at first sight to be mutually exclusive, i.e. their rather high stability and lability (Izumrudov *et al.*, 1991). Due to the cooperative character of multisite polycation-polyanion binding the PEC becomes extremely stable with respect to dissociation. Practically irreversible reaction with an 'infinite' binding constant takes place in some range of pH and ionic strength. The theory of such systems was propounded by Papisov and Litmanovich (1989). Addition of salt in excess (for example, more than 0.5–1 M of NaCl) destroys bonds between the polymers and thus allows the processing individual compounds of the precipitate (Kabanov and Zezin, 1984; Kabanov, 1994; Pergushov *et al.*, 1995).

The solubility of PEC depends significantly on the polycation/polyanion ratio. An insoluble PEC is formed on addition of polyelectrolytes in the stoichiometric ratio 1 : 1 of negatively and positively charged monomeric units (Kabanov and Zezin, 1984). If the ratio differs essentially from this value, the polycation-polyanion reaction gives soluble nonstoichometric complexes, but they can bind the deficient polymer at a later time. Addition of the deficient polyelectrolyte to the soluble PEC leads to the transfer of ionene chains to the added polyelectrolyte chains and, as a result, to the rearrangement of PEC, as Tsuchiba *et al.* (1972) described for the reaction between aromatic ionenes and poly(styrene sulfonate). More recently it was shown that soluble PEC could be obtained from a wide variety of oppositely charged polyions (Kharenko *et al.*, 1979; Tsuchida and Abe, 1986).

These properties of the polyelectrolytes ensure their use as carriers for immunoreactants. In this way immunoassay reactions are carried out in solution, and then the insoluble precipitate is formed for the separation of immunoreactants.

To study this approach, the interaction between antibodies covalently attached to the PMA and peroxidase-labeled antigens, namely, bacterial α-amylase and human immunoglobulin G (IgG), was examined by Dzantiev *et al.* (1988). Incubation of the immunoreactants during 5 min proved to be enough to reach equilibrium. Subsequent addition of PEVP led to the separation of bound and unbound peroxidase-labeled antigen molecules. The formation of a precipitate requires no more than 1 min; it can be easily separated by centrifuging or filtration. Redissolution of the precipitate can be achieved, for example, in a 1 M NaCl solution.

These preliminary data made it possible to design an immunoassay technique based on the use of polyelectrolytes (Dzantiev *et al.*, 1995). The principle of the assay is shown in Figure 10.1. Solutions containing the antigen to be determined and the enzyme-labeled antigen are added to the solution of antibody-PMA conjugate (obtained by carbodiimide/succinimide technique). Reactions of the antibodies with the antigen and with the labeled antigen proceed simultaneously and rather rapidly. After the formation of soluble immune complexes containing PMA, the PEVP solution is added. As a result, specifically bound molecules of the antigen-enzyme conjugate are included in the formed insoluble precipitate. The enzymatic activity can be measured either in the solution (unbound label) or in the redissolved pellet (bound label). In both cases the determined activity offers a measure of the antigen concentration in the analyzed sample.

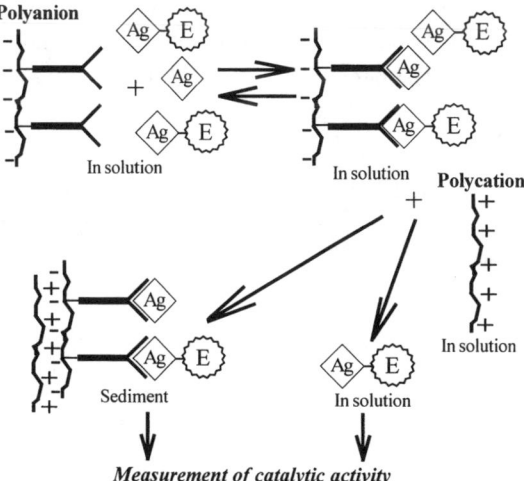

Figure 10.1 Scheme of a homogeneous polyelectrolyte immunoassay; Ag, antigen; E, enzyme label.

Another proposed immunoassay technique is based on the formation of a soluble PEC from PMA-antibody conjugate and PEVP, carrying out the immune reaction and separation of the immune complexes through precipitation of the soluble PEC by addition of Ca^{2+} ions. It was found that the optimal value of initial ratio between monomeric units of the polyanion and the polycation is $3:1$. The PEC formed under these conditions have good solubility and can be separated effectively by Ca^{2+} (Dzantiev et al., 1995).

Absence of significant non-specific binding between the carriers and compounds of the tested samples is an extremely important requirement for correct determination of the antigen. At first glance, the polyelectrolytes would seem unsuitable carriers for assaying biological substances, because different proteins and other macromolecules can interact electrostatically with oppositely charged polyelectrolytes. However, a more detailed analysis demonstrates that non-specific binding in the system is practically negligible (see Figure 10.2). A complete displacement of some model proteins (α-chymotrypsin and penicillin amidase) from PEC was studied in detail by Margolin et al. (1981, 1982, 1985), who described how to use the water-soluble polyelectrolytes as carriers for enzymes immobilization. The observed exclusion of not covalently bound proteins by polyelectrolyte molecules can be explained by the fact that linear polyions have an essentially higher density of charges as compared with the surface of the protein globules. When PEVP is added to a reaction mixture containing PMA and the protein, electrostatically sorbed amphoteric proteins cannot compete successfully with the polycation and therefore are squeezed out of the PEC particles (Izumrudov, 1996).

The proposed immunoassay technique was realized for a number of antigens differing in their physico-chemical properties (molecular weight, charge, isoelectric point, etc.). The achieved detection limits are given in Table 10.1.

These levels are principally equal to those of the traditional microplate enzyme-linked immunosorbent assay (ELISA) with the use of the same antibody preparations. However, the polyelectrolytes allow a significant reduction in the time of the assay. The total duration of the proposed assay does not exceed 15 min (if the activity is measured in the supernatant after separation of PEC). ELISA technique at equilibrium conditions needs at least 1–1.5 h to be carried out, and reduction of this time (kinetic regime) leads to a decrease in its sensitivity (Yazynina et al., 1999b).

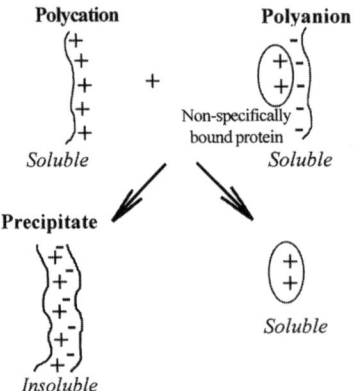

Figure 10.2 Replacement of non-specifically bound proteins from polyelectrolyte carriers.

Table 10.1 Sensitivities of enzyme immunoassays based on the polycation–polyanion interaction

Compound	Detection limit, ng/ml
Testosterone	0.04
Insulin	0.5
Surface antigen of hepatitis B virus (HBsAg)	1
Potato X virus	5
Human IgG	10
Bacillary α-amylase	10

The obtained results can be summarized as follows: the technique described above combines two important properties of the homogeneous and heterogeneous assays. Firstly, carrying out the immune reaction in solution (as in the case of homogeneous assay) permits to reduce the assay duration. Secondly, polyelectrolyte separation allows sensitive measurement of bound or unbound labels (as in the case of heterogeneous assay).

It should be noted that conjugation of the polyelectrolytes with antibodies in combination with separation of immunoreactants by the electrostatic binding of the polyelectrolytes may be applied also for other purposes. Thus, Dainiak *et al.* (1998a and b) used this system as a simple model of chaperon action when the inactive forms of enzyme (glyceraldehyde-3-phosphate dehydrogenase) are removed from the reaction media preventing aggregation. The obtained results are comprehensively described by V.A. Izumrudov in this volume.

10.2.2. *Homogeneous Assays with the Use of a Polyelectrolyte – Metal Ion Pair*

An analogous immunoassay technique was proposed by Marshall (1985). The author used polyanion carriers that can change their solubility depending on the cations present in solution. Alginic, pectic, celluronic and poly(acrylic) acids, carrageenan, ethylene-maleic anhydride copolymer, and carboxymethylcellulose were studied. These substances (both acids and sodium salts) are water soluble but may be rendered insoluble, for example, by pH lowering or by addition of certain metal ions such as calcium. The insolubilized polymeric substance may be redissolved by pH raising or by the addition of metal chelating agents (citrate ion, ethylenediamine tetraacetic acid etc.). Alginic acid was found to be the preferred polymeric substance. Antibodies are attached to the carrier either directly (with the use of either cyanogen bromide or a carbodiimide reagent such as dicyclohexylcarbodiimide or 1-ethyl-3-(3-dimethylamino-propyl) carbodiimide), or through an additional binding agent (second antibody, protein A or avidin). During the assay the antigen to be tested competes with the enzyme-labeled (for example, glucose oxidase-labeled) antigen for binding with polymer-modified antibodies. The cited examples of antigens assayed by the present scheme are theophylline, human chorionic gonadotropin, heroin, phenobarbital, digoxin, testosterone, aspirin, and folic acid. The author notes that there are no special requirements as regards the properties of the tested antigen, but the preferable range of its molecular weight is 100–50,000 Da. The described assay includes 30 min incubation for immunochemical reaction and followed by a rapid (1–2 min) separation of the polymer.

10.2.3. Microplate Semi-homogeneous Assays

The main limitation of the above approaches is that they may necessitate the use of additional devices for the separation of the polymeric precipitate from the reaction mixture. Centrifugation or filtration procedures increase complexity of the assay and require special equipment. To avoid these disadvantages, a combination of the poly-electrolyte approach with the standard ELISA format was proposed by Yazynina *et al.* (1999a and b), the scheme of the assay is given in Figure 10.3.

As evident from the figure, one more modification is made in the assay protocol. Antibody-polyanion conjugate is changed to a pair of reactants – non-modified anti-bodies and a polyanion conjugate with staphylococcal protein A. The latter protein is well known in immunochemistry because of its ability to interact specifically with anti-body molecules without loss of their antigen-binding activity (Forsgren and Sjöquist, 1969; Langone, 1982). The application of this pair of reactants allows one to use antibodies against different antigens without any chemical modification (and even use of whole antisera).

The proposed microplate polyelectrolyte ELISA format includes the following steps:

(1) carrying out immune interactions in a solution that contains a tracer antigen, antigen-peroxidase conjugate, antibodies, and polyanion-protein A conjugate;
(2) binding of the polyanion-containing complexes formed with the polycation that was immobilized in microplate wells;
(3) washing of the wells; and
(4) optical measurement of peroxidase activity using a substrate solution.

Kinetic and concentration dependencies of the assay have been studied and 90% saturation of binding sites of the immobilized polycation was found to be reached in 8 min, whereas 2 min of incubation proved to be sufficient for binding of more than 50% of the sites. Although both the polyelectrolyte ELISA and a standard one

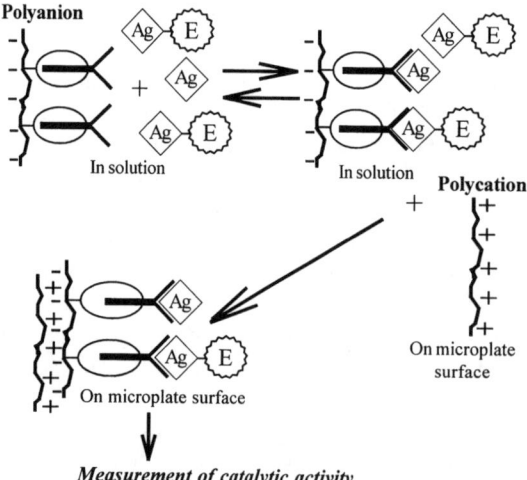

Figure 10.3 Scheme of a microplate semi-homogeneous polyelectrolyte immunoassay; Ag, antigen; E, enzyme label.

include steps of heterogeneous interaction, the polycation-polyanion binding is a significantly faster process as compared with the antigen-antibody one. This is connected with the absence of necessity of strong orientation of the interacting molecules. It was shown that the number of polymeric complexes formed increased significantly within a narrow range of polyanion/polycation ratio that provides maximum formation of large-size aggregates of the polymers. With the polyelectrolyte ELISA, immune complexes were detected in two short steps of 8 and 5 min, whereas the traditional format requires 60 min incubations to saturate binding sites of the immobilized antibodies.

The polyelectrolyte ELISA was utilized for detection of the herbicides simazine and atrazine. Sensitivity of simazine detection is 0.5 ng/ml, that of atrazine – 0.03 ng/ml. Total assay duration including the stage of enzymatic reaction is 30–40 min, while for the traditional assay this value is equal to 100–120 min (even if the interaction between protein A and antibodies is interpreted as a preliminary stage that is carried out before analysis of the samples). The proposed assay was characterized by high reproducibility. Coefficients of variance for four parallel tests range from 1.4% to 9.3%, as expected for typical ELISA tests.

The technique was successfully applied to the detection of simazine in different liquid media, namely, drinking water, orange juice, and milk. Recovery of simazine varied from 95% to 135%, making this a suitable quantitative method.

The proposed polyelectrolyte technique permits a significant increase in speed of the resulting ELISA without loss of sensitivity. By using the polyanion-protein A conjugate, the described assay does not require the preparation and purification of specific antibody derivatives. Therefore the same scheme can be easily adapted for other analytes.

10.2.4. *Immunofiltration Assays*

There are many practical tasks in analytical chemistry where a high precision of determination is not needed for making a decision. The real aim of such analysis is to establish some critical concentration level. Immunochemical techniques are intensively applied for solving such tasks (Hock, 1996; Parija, 1998). Immunofiltration and immunochromatography are two main kinds of the assay formats that are proposed for qualitative immunochemical tests.

Polyelectrolyte interaction was combined with immunofiltration membrane immunoassay by Dzantiev *et al.* (1994a and b, 1997) and Yazynina *et al.* (1999b). The principle of the proposed technique is illustrated in Figure 10.4. It is close to the polyelectrolyte microplate ELISA described above. After incubation the same reaction mixture is filtered through a porous membrane with adsorbed polycation. (The membrane is placed in a special holder with microwells.) The polyanion reacts with the immobilized polycation almost instantly, and the immune complex bound to the polycation becomes immobilized on the membrane surface. All the other components of the reaction mixture are removed through the membrane into layers of filter paper by washing. Finally, the membrane is immersed in the peroxidase substrate solution or the solution is filtered through the membrane, giving an insoluble coloured product. In case the antigen is absent in the tested sample, only polyanion–protein A : antibodies : antigen-peroxidase complexes bind to the membrane, the amount of immobilized peroxidase is high, and therefore the spots formed are dark. Decrease of the spot intensity (to pale) points to the antigen presence in the sample.

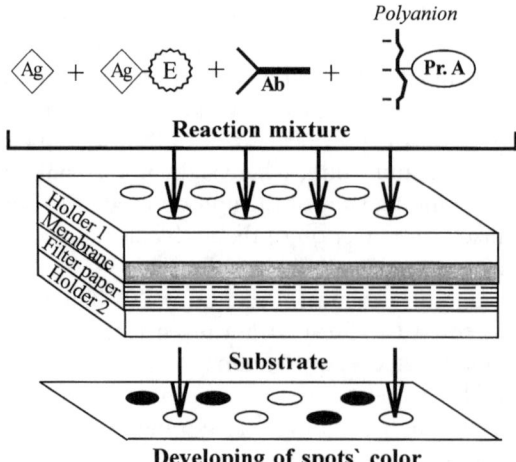

Figure 10.4 Scheme of a filtration immunoassay and device for the assaying; Ag, antigen; Ab, antibody; Pr. A, staphylococcal protein A; E, enzyme label.

The technique described above was realised for the determination of pesticide simazine, the drug methamphetamine and detection of the hormone testosterone. The obtained limits of antigen detection with instrumental measurement of spot density are close to those of traditional ELISA (1 ng/ml for simazine, 0.1 ng/ml for testosterone).

Visual detection in this dot blot immunoassay was typically limited to distinguishing 'positive' and 'negative' samples by comparison with some cut-off level. The sensitivity of such tests corresponds to a midpoint of the competitive curve and was about one order lower. For our model antigens it was equal to 10 ng/ml in the case of simazine, and 1 ng/ml in the case of testosterone.

The assay is highly reproducible. Coefficients of variance for four parallel tests on the same membrane ranged from 3.4% to 5.1%.

The described approach for detection of haptens is based on a competitive scheme of immunoassay. For multivalent antigens it is reasonable to use two-site ('sandwich') immunoassay schemes. Such technique was proposed for determination of hepatitis B surface antigen (HBsAg) (Dzantiev *et al.*, 1996). It proved to be more than six times as rapid as the well-known ELISA for this substance (30 min instead of 3–4 h). The detection limit for HBsAg was 1 ng/ml.

The advantage of the polyelectrolyte-based visual tests is their speed. Total analysis time was 20 min and mainly limited by handling procedures. The developed system permits detection without any special equipment. In dealing with real environmental samples under field conditions, measurement errors are reduced by using volumes of several hundred microliters over smaller volumes. The polyelectrolyte technique facilitates a rapid concentration of these volumes to a small spot with a high signal/noise ratio.

10.2.5. *Application of Polyelectrolytes for Viruses Immunoassay*

As we have underlined above, the high affinity of polyelectrolyte interaction allows one to avoid non-specific binding of polyelectrolytes with protein antigens or compounds of the tested samples. Working with virus particles leads to a significant growth of the antigen-polyelectrolyte direct binding because of the high density of charged groups on the surface of the virus particle. However, this potential disadvantage can be turned to advantage by creating a new immunoassay format that combines polyelectrolyte and immune recognition of the detected object.

The study of model virus-polyelectrolyte systems indicated that under certain conditions a soluble virus-PEVP complex efficiently interacted with PMA and formed an insoluble PEC that quantitatively entrapped the huge particles of the virus (Dzantiev *et al.*, 1990; Blintsov *et al.*, 1995).

The principle of the relevant assay is as follows. The enzyme-labeled antibodies interact with the virus in solution, the virus particles bind PEVP molecules (soluble product), and the formed complexes are precipitated by PMA (insoluble product). The enzymatic activity determined in the PEC pellet after its dissolution in 0.5 M NaCl is proportional to the concentration of the assayed virus in the tested substance. It was shown that compounds of the plant sap or its extract did not affect the specificity and sensitivity of the assay. The method allows one to detect phytoviruses (potato X virus, tobacco mosaic virus, and potato leaf curl virus) at concentrations of 5–10 ng/ml. The immunochemical part of the assay may be carried out in 5 min, and the total assay time is about 20 min.

10.3. Immunoassays with Generation of a Polymeric Solid Phase

10.3.1. *Induction of Carriers Polymerization*

A.S. Hoffman and co-workers developed two novel immunoassays in which the solid phase is generated *in situ* after the specific binding reaction has occurred, thereby enhancing reaction kinetics and minimizing the opportunities for non-specific binding.

In the first system the capture antibody is conjugated to an organic monomer, polymerization of which leads to formation of insoluble polymer particles (Auditore-Hargreaves *et al.*, 1987; Nowinski and Hoffman, 1987, 1989; Thomas *et al.*, 1988). The assay, named as *de novo* polymerization, involves two antibody preparations – an antibody conjugated with a polymerizable organic monomer, and a fluorophore-labeled antibody (Figure 10.5). This assay can be performed in any of several configurations. These can include competitive, sandwich and non-competitive immunoassay configuration. The order of addition of reagents can also be varied. Assays with simultaneous addition of reactants are most preferred because they require the least number of manipulations. All the configurations, however, offer significant advantages over prior art immunoassays in that incubation times are shortened and washing steps are eliminated. After the specific binding between antigen and antibody has taken place, polymerization is initiated by a reaction involving free radicals. This results in formation of insoluble polymer particles, the label content of which is directly proportional to the amount of antigen in the sample.

Auditore-Hargreaves *et al.* (1987) compared a variety of monomers, such as vinyl benzoate, styrene sulfonic acid, and isothiocyanatophenylacetylene. The most effective incorporation of the label was found for vinyl benzoate. The polymer particles formed by this reaction were polydisperse, with smaller sizes predominating. The size distribution is not static, but changed as a function of time after the initiation of polymerization. If left long enough, the particles would eventually coalesce into a single

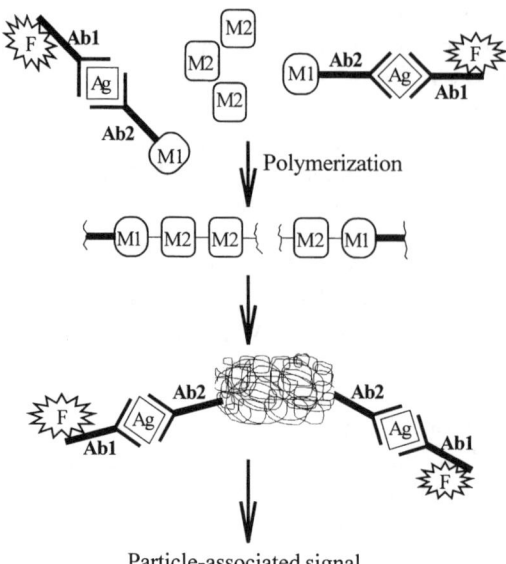

Particle-associated signal

Figure 10.5 Scheme of a *de novo* polymerization immunoassay: Ag, antigen; Ab1, Ab2, two kinds of antibodies; F, fluorescent label; M1, conjugated monomer; M2, free monomer.

large bead that could not be readily dispersed. The sensitivity of the developed assay for IgG is about 5 ng/ml $(2 \cdot 10^{-11}$ M).

Multi-analyte system was also proposed for simultaneous determination of IgG and IgM in a single sample, using phycoerythrin-labeled antibody to report IgM molecules and a fluoresceinated antibody to report IgG. The presence of serum compound in quantities up to 20% does not influence the assay results.

A wider range of carriers was described by Nowinski and Hoffman (1989). The possibility to use acrylic acid, methacrylic acid, acryloyl chloride, methacryloyl chloride, glycidyl acrylate or methacrylate, glycerol acrylate or methacrylate, allylamine, allyl chloride, hydroxy lower alkyl-acrylates (e.g., 2-hydroxyethyl methacrylate or 2-hydroxypropyl methacrylate), and amino lower alkyl-acrylates (e.g., 2-aminoethyl methacrylate) as specific monomers was demonstrated. Preferred monomers are those which are soluble in water or water/polar organic solvent mixtures. Homopolymerization of the monomer/reactant conjugate itself or copolymerization with non-derivatized monomers is initiated by generation of free radicals. The free radicals may be generated by oxidation–reduction initiation, photochemical initiation, ionizing radiation or thermal initiation. An advantage of both oxidation–reduction initiation and photochemical initiation is production of free radicals at reasonable rates at relatively low temperatures that are acceptable for protein stability (such as 20–37 °C). Types of oxidation–reduction initiators which may be used include:

(1) peroxides in combination with a reducing agent, e.g. hydrogen peroxide with ferrous ion, or benzoyl peroxide with N,N-dialkylaniline or toluidine; and
(2) persulfates in combination with a reducing agent, such as sodium metabisulfite or sodium thiosulfate.

Photoinitiated polymerization may also be used by employing photoinitiators, such as azodiisobutyronitrile, azodiisobutyroamide, benzoin methyl ether, riboflavin, thiazine dyes such as methylene blue and eosin, and transition metals such as ferric chloride, in combination with ultraviolet and/or visible light irradiation of the reaction system.

It should be noted that possible labels may include enzymes, fluorophores, radioisotopes, chemiluminescent materials, dye particles, etc. In general, however, fluorophores are preferred. Some suitable fluorophores include fluorescein, rhodamine, phycoerythrin, and Nile blue.

The detection of human IgM with fluorescent label was described by Nowinski and Hoffman (1989) as an example of an application of the proposed assay technique. The obtained calibration curve was linear in the range of antigen concentrations from 0.5 to 5.0 μg/ml.

10.3.2. Carriers Having Temperature-dependent Solubility

The second system proposed by A.S. Hoffman and co-workers is based on the application of polymers with temperature-dependent solubility (Auditore-Hargreaves *et al.*, 1987; Monji and Hoffman, 1987; Monji *et al.*, 1988). The assay principle is given in Figure 10.6. Capture antibody is conjugated to a polymer with appropriate properties. Specific binding is conducted below the critical solution temperature of the polymer, which is then separated from solution by increasing the temperature. The incorporation of labeled antibody into the precipitated polymer is directly proportional to the antigen concentration. The study was focused primarily on acrylamide and N-alkyl acrylamide polymers and copolymers. Polymers with a lower critical solution temperature provide a dual advantage: the efficiency of washing is maximized while dilution of the label and sample is minimized. Ideally, the polymer is precipitated initially from a relatively small volume, to minimize dilution of the sample, and is redissolved in subsequent steps in a relatively large volume, to promote washing. Signal development can be carried out in either phase – when the polymer is precipitated or when

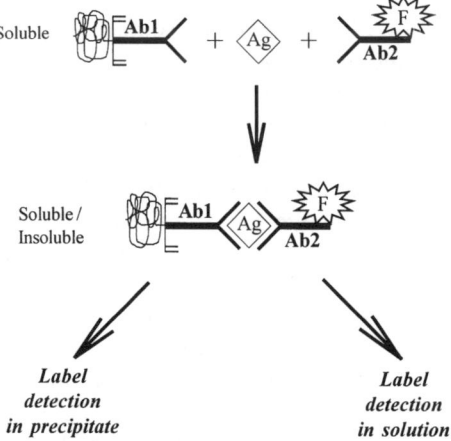

Figure 10.6 Scheme of a thermal precipitation immunoassay: Ag, antigen; Ab1, Ab2, two kinds of antibodies; F, fluorescent label.

it is in solution. The maximum concentration effect is achieved when signal development is carried out in the solid phase, which represents the smallest possible volume the polymer can occupy. However, if the signal is generated enzymatically, it may be desirable to carry out signal development in solution, because the rate of diffusion of the enzyme substrate into the precipitate may be rate-limiting.

The homopolymer of N-isopropyl acrylamide exhibits a lower critical solution temperature of 31 °C. Its copolymers with acrylamide or N-alkyl (ethyl or methyl) acrylamides can be essentially tailored to any lower critical solution temperature desired, simply by adjusting the relative proportions of the monomers from which they are synthesized. To be useful in an immunoassay, a polymer must have a critical temperature that is unaffected by the presence of substances commonly encountered in biological fluids. A wide number of tested substances (serum, urine, ions, anticoagulants, ionic, non-ionic, and zwitterionic detergents, chelating agents) did not influence the critical solution temperature of the polymer over the range of concentrations in which they might be expected to be found in biological samples. Furthermore, this temperature was unaffected by conjugation of antibody to the polymer backbone. The size of the particles that formed above the critical solution temperature was dependent on the degree of antibody substitution.

In addition to being thermally precipitable, poly(N-isopropyl acrylamide) and its copolymers thereof can be precipitated from solution by adding ammonium sulfate to 14%–20% of saturation, depending on the composition of the polymer. Because immunoglobulins do not precipitate at these salt concentrations, this is a convenient method of separating antibody-conjugated polymer from free antibody during the preparation of reactants for assay.

The developed system was applied for enzyme immunoassay of HBsAg and *Chlamydia trachomatis* (Auditore-Hargreaves *et al.*, 1987), enzyme and fluorescent immunoassay of IgG (Monji and Hoffman, 1987). The proposed and traditional ELISA systems for HBsAg are a favorable comparison, both in terms of sensitivity and performance time. The sensitivity of the polymer-based HBsAg assay was approximately 0.5 ng/ml for chromogenic substrate (tetramethylbenzidine), and approximately 0.25 ng/ml for fluorogenic substrate (p-hydroxyphenylpropionic acid). The sensitivities of the developed competitive enzyme immunoassay for mouse IgG, an antigen capture fluorescence immunoassay for human IgG, and a secondary antibody sandwich enzyme immunoassay for rabbit anti-mouse IgG are in the range of typical non-isotopic heterogeneous assays. Studies of the *Ch. trachomatis* systems had also shown equal sensitivity and essential reduction of the assay time for the polymer-based assay.

The conjugate of protein A with poly(N-isopropylacrylamide) was synthesized and utilized in thermally-induced separation of human immunogammaglobulin (Chen and Hoffman, 1990). The principle of this affinity precipitation procedure is presented in Figure 10.7. The obtained universal conjugate may be also applied for immunoassays of different compounds.

Takei *et al.* (1994) analyzed the properties of the intermolecular conjugate formed by IgG and poly(N-isopropylacrylamide). It was shown that the polymer content of the conjugate was 20–30 wt% for the case of 6,100 Da polymer preparation to demonstrate specific antigen binding activity most effectively and to reduce Fc-dependent binding with protein A. Fluorescein isothiocyanate-labeled human serum albumin was a model antigen. A peculiarity of the proposed conjugation technique is the use of semitelechelic poly(N-isopropylacrylamide), containing a carboxylic group at one end

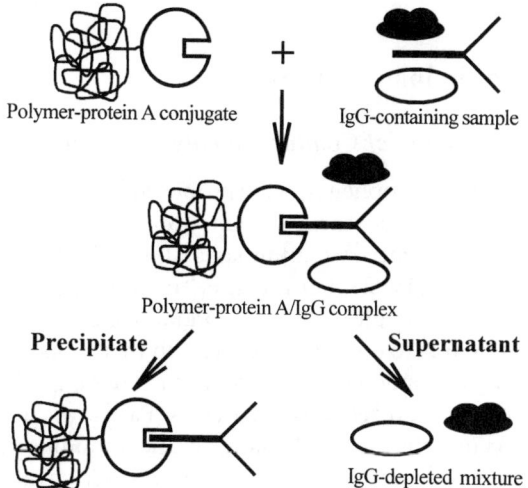

Figure 10.7 Scheme of an affinity precipitation procedure with the use of thermal precipitation.

of the polymer. Owing to this, the synthesis of polymer-protein conjugate excludes formation of multi-point bonds, that may lead to non-stable solubility and denaturation of biomolecules.

Membrane immunoassay format was developed on the basis of the described properties of the room-temperature-precipitable polymers (Monji and Cole, 1993a and b; Monji *et al.*, 1994). An activated terpolymer consisting of N-isopropylacrylamide/N-n-butylacrylamide/N-acryloxysuccinimide was prepared and conjugated to an antibody. The conjugate was evaluated in a novel membrane-based immunoassay which utilizes the especially strong physical attachment of the polymer to cellulose acetate to bind and concentrate the polymer-attached protein onto the membrane. One advantage of this membrane is that it has low non-specific protein binding properties, which can be further minimized by treatment with bovine serum albumin. In addition, it has been determined that there is minimal interference by proteins and mild detergents in binding of the complex to this membrane. Antibodies to HIV were the main analyte of interest. It was noted that different N-alkylacrylamines, N-arylamines, alkyl acrylates and combinations thereof can be the polymers used (Monji and Cole, 1993a). The labels include enzymes, fluorophores, radioisotopes, luminescent groups, particles, etc. Particularly preferred among the luminescers are chemiluminescent structures, such as 3-(2′-spiroadamantane)-4-methoxy-4-(3′-phosphoryloxy)phenyl-1,2-dioxetane (AMPPD), 3-(2′-spiroadamantane)-4-methoxy-4-(3″-β-D-galactopyranosyloxy)phenyl-1,2-dioxetane (AMPGD), luminol, isoluminol, and firefly luciferin.

Liu *et al.* (1995) studied the retention of antibodies labeled by poly(N-isopropylacrylamide) on cellulose acetate/nitrate membrane. The polymer can adhere quickly and tightly to the membrane either below (less efficiently) or above (more efficiently) the lower critical solution temperature, and the retention increased over 30-fold when compared with the unconjugated antibody. These characteristics were used to develop polymer-enzyme-linked immunoassay for HBsAg. The system can detect as little as 1 ng/ml of the antigen. High speed and low non-specific background of the assay were noted.

10.4. Other Polymers Applications in Immunoassays

10.4.1. Immunoassays with Direct Detection of Conductive Polymeric Labels

Carboxylic and acetic acid-substituted poly(thiophenes) were proposed by Englebienne and Weiland (1994, 1996a) as labels for immunoreactants. This approach was named as water-soluble conductive polymer homogeneous immunoassay (SOPHIA).

It was shown earlier by Huang and Lee (1992) that the antigen–antibody interaction can induce a local pH change near the complex. Besides, conductive polymers are extensively conjugated macromolecules able to conduct electricity in their doped state and having an UV-visible spectrum which undergoes important chromatic modifications when subjected to pH changes or to oxido-reductive processes (Patil *et al.*, 1988).

When antigen–antibody binding occurs, the local pH change induces modifications in the absorbance at a characteristic wavelength of a conductive polymer covalently linked to either the antigen or the antibody. Consequently, the extent of tracer binding can be directly monitored by photometry during incubation.

The described principle was realized in competitive immunoassays for human C-reactive protein and human serum albumin. Turbidity changes were measured at 340 nm. The equilibrium of tracer–antibody binding was reached between 20 and 30 min of incubation. The limit of detection of the compounds named above is 0.4 and 1.5 µg/ml, accordingly, and the range of linearity is up to 50 and 250 µg/ml. These parameters are comparable with the ones of analogous immunoassay formats, such as immunoluminescent assay, competitive enzyme immunoassay, and immunoturbidimetry (Englebienne and Weiland, 1996a).

Further the same technique was applied for determination of theophylline (Englebienne and Weiland, 1996b; Englebienne, 1999). The results obtained are representative of the wide analytical range (nanomolar up to higher micromolar) to which such homogeneous assay techniques can be applied. Although some homogeneous systems, such as scintillation proximity or flow-cytometric fluorescence immunoassays can quantitate analytes in the picomolar and even femtomolar range, but they require many manipulations, use complex and costly reagents, and depend on a specific instrumentation (Gosling, 1990).

10.4.2. Immunoassays with Enzymes Creating Detectable Polymers

Oster and Davis (1990, 1991a and b) applied the polymerization reaction at the label detection step of enzyme immunoassay. The label of the formed immune complexes generates a compound that induces polymerization of a monomer solution into further detected polymer molecules. For example, glucose oxidase that was linked to an antibody (and in this way to the detected immune complexes) generates hydrogen peroxide; the latter substance in the presence of ascorbic acid, a reductant, yields free radicals capable of initiating vinyl polymerization. The xanthine oxidase reduction of oxygen to hydrogen peroxide (in the presence of xanthine) generates an intermediate capable of reducing ferric to ferrous ions; the latter ion serves as a reductant in a hydrogen peroxide redox system for vinyl polymerization. Other applied labels in analogous systems were iron chelate (ethylenediamine tetraacetic acid) and horseradish peroxidase (with dihydroxyfumaric acid as a source to generate free radicals that initiate vinyl polymerization).

The intended method of detection of the polymer would be determined by the choice of the monomer. The simplest approach is visual detection of the formed precipitate. Use of fluorescent monomers (such as europium (II) acrylate) or agglutination of suspension of latex particles by formed polymer molecules can be indicated as the most interesting approaches. The following monomers among others can be employed singularly or in combination in the assay: acrylamide, N-octyl acrylamide, methacrylamide, N-methyl-acrylamide, acrylic acid, methacrylic acid, hydroxymethyl acrylamide, methylene bisacrylamide, acrylonitrile, methyl acrylate (and higher esters), ethylene glycol methacrylate, propylene glycol methacrylate, acrylamido propane sulfonic acid, vinyl pyrrolidone, vinyl pyridine. Still further, synthetic and natural polymers to which vinyl groups can be coupled and which are capable of forming addition polymers can be used.

The redox catalyst system serves as a highly efficient source of initiating molecules at temperatures that protect against thermal denaturation of sensitive biological substances. Redox catalysis is also characterized by high yield of polymer, high yields of high molecular weight polymer and by a reduced induction time.

The approach described above can be also applied for the assays based on nucleic acids hybridization.

10.4.3. Modification of Immunoreactants by Polymers to Change Solubility

Development of enzyme immunoassay is a rather complicated task when water-insoluble compounds have to be determined. It is connected with destruction of the reaction centers of enzyme label and antibody in organic solvents. If the antibody displays activity in an organic solvent, enzyme immunoassay could be used directly without extra steps to remove the organic solvent used to extract the antigen from the samples. So approaches for increase of antibody stability in different solvents are of especial interest nowadays.

It is well known that proteins become soluble in an organic solvent after chemical modification of their amino groups with an amphiphilic polymer (Abuchowski et al., 1977; Ferjancic et al., 1990). This approach was successfully applied, for example, by Inada et al. (1984) to provide lipase activity in organic solvent. Sasaki et al. (1997) used poly(ethylene glycol), activated with cyanuric chloride, for modification of antibodies. New enzyme immunoassay of pesticide atrazine was developed with the obtained reactants. Most of the antibody amino groups (namely 72%) were modified with poly(ethylene glycol). By demonstrating that the binding constant of the modified antibody was equal to that of the native antibody, it was shown that modification had no influence on the binding affinity of the anti-atrazine antibody in the water phase. The modified antibody was soluble both in toluene and chloroform at concentrations of up to 26 and 270 μg/ml, respectively, following centrifugation at 5800 g. On the other hand, the native antibody did not show any solubility in these two solvents. Enzyme immunoassay of atrazine in toluene with the modified antibody demonstrated 0.5 ppm sensitivity, which is about ten times better than the traditional ELISA.

Kiselev et al. (1999) used modification by polybrene to increase stability of antibodies in ethanol/water and isopropanol/water systems. The obtained growth of the stability and affinity to antigen allowed to develop high-sensitive magnetic beads-based immunoassay for the immunodepressant drug cyclosporin A, being a hydrophobic compound. The method allows cyclosporin A determination in a medium with a higher content of ethanol compared to conventional immunochemical techniques. Monitoring of the drug in ethanol extracts from a patient's whole blood without many-fold dilution with aqueous buffer is possible.

The obtained data opens the way to the development of immunoassays for other hydrophobic substances in organic solvents.

10.4.4. Modification of Immunoreactants by Polymers for Two-phase Separation

Another idea connected with the changes of immunoreactant's solubility by means of its chemical modification by a polymer was realized by B. Mattiasson and co-workers (Mattiasson and Ling, 1980, 1983; Mattiasson *et al.*, 1981, 1983; Ling *et al.*, 1982; Ling and Mattiasson, 1983, 1984). The described partition affinity ligand assay (PALA) combines enzyme immunoassay with partition in aqueous two-phase systems to give a separation stage. At a typical immunochemical stage the antigen (native and bound with the enzyme label) reacts with antibodies. When binding is complete the bound and unbound fractions of antigen are separated by partitioning in a two-phase system. Proper partitioning of the antibody can be achieved by choosing a proper phase system or if necessary by modifying the surface structure of the antibody to make it partition exclusively to one of the phases. Such modification has been most effective by the covalent attachment of a polymer such as poly(ethylene glycol) or poly(propylene glycol). Modification of immunoreactants by hydrophilic substituents also may be used. The desired degree of hydrophobicity/hydrophilicity may be achieved by using either a great number of small substituents or a low number of big substituents. The two-phase system of liquid may be a system known *per se* for the fractionation of mixtures of e.g. high molecular weight substances of different physical characteristics by means of two immiscible aqueous and (usually) polymeric liquids. Examples of such systems are: dextran/water-soluble copolymer of sucrose and epichlorohydrin (ficoll)/water, dextran/hydroxypropyldextran/water, poly(ethylene glycol)/dextran sulfate/water, dextran/poly(ethylene glycol)/water, poly(propylene glycol)/poly(ethylene glycol)/water, poly(propylene glycol)/poly(vinyl alcohol)/water, poly(propylene glycol)/poly(vinyl pyrrolidone)/water, etc. Such systems containing at least two water-soluble polymers may also contain an addition of salt or organic solvent. Other groups of aqueous two-phase liquid systems which may be used are at least one polymer/at least one salt/water and at least one polymer/at least one organic solvent/water. The polymer may, for example, be poly(ethylene glycol), poly(propylene glycol), poly(vinyl pyrrolidone) or a polysaccharide. The salt may be an inorganic salt or an organic salt which is soluble in water, for example a sulfate, a phosphate or a chloride. The organic liquid should be a water-soluble solvent which does not affect the properties of any of the reactants taking part in the biospecific reactions. Examples of organic solvents which may be used are propylalcohol, glycerol and 2-butoxyethanol.

Mattiasson and Ling (1982) described three techniques for IgE detection in serum which are based on reactants modification by poly(ethylene glycol) (molecular weight 2000 Da) and the following separation with the use of magnesium sulfate/poly(ethylene glycol)/water separation system. IgE can be quantitatively determined by a competitive radioimmunoassay, a sandwich radioimmunoassay, and a sandwich enzymeimmunoassay.

Mattiasson *et al.* (1981) used this approach for the quantification of bacterial cells, namely *Staphylococcus aureus*. The immunoassay was developed both in direct and competitive formats. Using this method, cell numbers in the region 10^5–10^7 can be quantified. An assay takes 40–90 min. Streptococci were the second analyzed cellular antigen (Ling *et al.*, 1982).

Regularities of the PALA technique were analyzed in detail during the development of β2-microglobulin immunoassay (Ling and Mattiasson, 1983). In order to get efficient separation, the antibodies were modified to favor their partition in a different phase from that of antigen. However, such modification significantly decreased binding capacity of antibodies. This was overcome by using antibodies bound to previously modified staphylococci, which had proper partitioning behavior. Alternatively, antibodies conjugated with biotin could be used in combination with modified avidin. A method for the evaluation of data from separation immunoassays was described in this paper.

The necessary binding strength of the polymeric ligand was found by mathematical calculations and the model was verified experimentally by Ling and Mattiasson (1984).

The same approach can be applied for different biospecific affine systems. Protein A, complement factor, lectins, enzyme inhibitors, bioreceptors may be used as binding reactants.

10.4.5. Modification of Immunoreactants by Polymers to Amplify Detected Signals

During the seventies polymeric carriers were proposed for the combination of one immunoreactant molecule with a significant number of hapten labels, namely fluorescein derivatives (Ekeke and Exley, 1978; Hirschfeld *et al.*, 1979). Polyamines such as poly(lysine) and poly(ethyleneimine), which are available in degrees of polymerization up to several hundred, are suitable for a simple reaction with fluorescein isothiocyanate to produce multi-fluorophore labels. The main problem connected with this approach is concentration quenching of fluorescein in multi-substituted polymers or proteins. Different approaches were proposed to overcome this problem (Hassan *et al.*, 1979).

The principle of label amplification by polymeric carriers was realized in full measure by Azad *et al.* (1984). An antibody or an antigen standard is attached to a water-soluble polymer having attached thereto a plurality of marker substances. Markers are preferably fluorescent dyes, although other substances such as non-fluorescent dyes, radioisotopes, electron opaque substances, enzymes, or a second homolog of differing specificity, may be used for substitution. Still alternatively, fluorescent-containing microspheres could be used. A water-soluble polymer is advantageously chosen so as to avoid non-specific binding, thereby permitting greatest sensitivity and immunological response. Consequently, those polymers which are cationically charged and therefore tend to attach non-specifically are preferably avoided. The preferred water-soluble polymers have an anionic or zero charge and are ideally chosen from the group consisting of poly(acrylic acid), poly(methacrylic acid), poly(acrylamide), poly(vinyl alcohol), poly(allyl alcohol), etc. It is noted, however, that many of the marker substances tend to be cationically charged, especially dye-type marker substances. Since this tends to increase unwanted non-specific binding, the water-soluble polymer may be advantageously adjusted to increase its anionic character. The water-soluble polymer, typically having a molecular weight in the range of 10^3 to 10^8 Da, provides the opportunity to attach numerous dye-type substances to an antibody. Of great importance is the fact that the attachment is at a sufficient distance from the antibody thus miniminizing deleterious effects on the antibody's reactivity. Protocols of fluorescein-based immunodetection of T-cells were given as examples of applications of the proposed approach.

Lamture and Wensel (1995) used the same approach to obtain intensely luminescent immunoconjugates. Partial alkylation of poly(lysine) with 4-(iodoacetamido)-2,6-dimethylpyridine dicarboxylate, followed by exhaustive reaction with succinic anhydride, yielded polymers (PLDS, polymer of lysine, dipicolinate, and succinate) containing 50–100 of 4-substituted dipicolinic acid moieties per molecule, with the remaining lysyl side chains succinylated. Competition experiments showed that PLDS binds Tb(III) ions with much higher affinity than EDTA does and strongly enhances the visible luminescence they emit when excited with ultraviolet light. Carbodiimide-mediated coupling to proteins yielded conjugates whose Tb(III) chelates displayed intense green luminescence. These conjugates retained sufficient immunoreactivity to allow their use in luminescence-based immunoassays. Thus, the limit of visual detection for ovalbumin is 10 ng. The ease of preparation of PLDS-protein-Tb(III) conjugates, and their favorable luminescence properties, make them promising reagents for use in time-resolved luminescence immunoassays and other ultrasensitive detection schemes.

10.4.6. *Other Applications*

The effect of charged synthetic polymers (such as poly(N-ethyl-4-vinylpyridinium)) and specific antibodies on the peroxidase activity in the enhanced chemiluminescent reaction was studied by Vlasenko *et al.* (1989). The close approach of an effector molecule to the peroxidase active site was found to inhibit the chemiluminescent reaction. Homogeneous luminescent immunoassays for testosterone, insulin, antibodies to insulin, and antibodies to trinitrophenyl group are proposed on the basis of these regulatory facilities. The assays have detection limits about 10^{-10} M, their overall time being 7–15 min.

Guzov *et al.* (1991) studied the influence of water-soluble polymers to the antigen-binding capacity of soluble and immobilized antibodies. Substantial increase of this parameter by ficoll and dextran was found. The dextran T70 enhanced the sensitivity of competitive radio immunoassays for carcino embrionic antigen and β_2-microglobulin 3–4-fold. Authors indicate that this procedure may also be applied to shorten the incubation period when performing an assay.

Immunoanalysis of tissue and cell preparations is connected with the problem that cellular Fc receptors can interfere with the specificity of binding (receptor–antibody complexes are formed against detected antigen-antibody ones). To obviate binding to Fc receptors, it is often necessary to prepare Fab or F(ab)₂ fragments of the antibody. In general, these relatively time consuming proteolytic procedures result in low yields of relatively unstable antibody fragments. Anderson and Tomasi (1988) proposed to eliminate Fc binding and other non-specific interactions of antibodies by their covalent modification with the water soluble polymer monomethoxypoly(ethylene glycol). The results demonstrate that modification with the polymer of less than 20% of antibody exposed lysine residues eliminates Fc-dependent binding of fluoresceinated antibodies to mouse splenocytes. In addition to flow cytometry, the modified antibodies may also find applications in several other procedures involving antibodies including immunoenzyme techniques and immunohistopathology.

10.5. Conclusions

The examples given above of water-soluble polymers' use in immunoassay demonstrate essential advantages of these reactants. The ability of the polymers to undergo fast structural changes under different weak exposures allows one to develop rapid assay techniques with effective separation of reactants. The described approaches can be applied to different antigens, and so the polymer-based assay techniques are a subject of much current interest for application in medical, ecological, and industrial testing and control.

10.6. Acknowledgments

The described work was made possible in part by Award No. RN2-426 of the U.S. Civilian Research & Development Foundation for the Independent States of the Former Soviet Union (CRDF) and Award No. 02.04/117 of the Russian State Research Program "New Generation of Vaccines and Medical Diagnostic Systems of the Future".

10.7. Abbreviations

AMPGD	3-(2'-spiroadamantane)-4-methoxy-4-(3''-β-D-galactopyranosyloxy)phenyl-1,2-dioxetane
AMPPD	3-(2'-spiroadamantane)-4-methoxy-4-(3'-phosphoryloxy)phenyl-1,2-dioxetane
ELISA	enzyme-linked immunosorbent assay
HBsAg	hepatitis B surface antigen
IgG	immunoglobulin G
PALA	partition affinity ligand assay
PLDS	polymer of lysine, dipicolinate, and succinate
PMA	poly(methacrylic acid)
PEC	polyelectrolyte complexes
PEVP	poly(N-ethyl-4-vinylpyridinium bromide)
SOPHIA	water-soluble conductive polymer homogeneous immunoassay.

10.8. References

Abuchowski, A., McCoy, J.R., Palczuk, N.C., van Es, T., and Davis, F.F. (1977) Effect of covalent attachment of polyethylene glycol on immunogenicity and circulating life of bovine liver catalase, *J. Biol. Chem.*, **252**, 3582–3586.

Anderson, W.L. and Tomasi, T.B. (1988) Polymer modification of antibody to eliminate immune complex and Fc binding, *J. Immunol. Meth.*, **109**, 37–42.

Auditore-Hargreaves, K., Houghton, R.L., Monji, N., Priest, J.H., Hoffman, A.S., and Novinski, R.C. (1987) Phase-separation immunoassays, *Clin. Chem.*, **33**, 1509–1516.

Azad, A.R.M., Kirchanski, S.J., and Brown, M.C. (1984) Immunological reagents employing polymeric backbone possessing reactive functional groups. *US Patent* 4, 434, 150.

Blintsov, A.N., Dzantiev, B.B., Bobkova, A.F., Izumrudov, V.A., Zezin, A.B., and Atabekov, I.G. (1995) A new method for enzyme immunoassay of phytoviruses based on interpolyelectrolyte reactions, *Doklady Biochemistry (Moscow)*, **345**, 175–178.

Chen, J.P. and Hoffman, A.S. (1990) Polymer-protein conjugates. II. Affinity precipitation separation of human immunogammaglobulin by a poly(N-isopropylacrylamide)-protein A conjugate, *Biomaterials*, **11**, 631–634.

Dainiak, M.B., Izumrudov, V.A., Muronetz, V.I., Galaev, I.Yu., and Mattiasson, B. (1998a) Conjugates of monoclonal antibodies with polyelectrolyte complexes – an attempt to make an artificial chaperone, *Biochim. Biophys. Acta*, **1381**, 279–285.

Dainiak, M.B., Izumrudov, V.A., Muronetz, V.I., Galaev, I.Yu., and Mattiasson, B. (1998b) Reactivation of glyceraldehyde-3-phosphate dehydrogenase using conjugates of monoclonal antibodies with polyelectrolyte complexes. An attempt to make an artificial chaperone, *J. Mol. Recognit.*, **11**, 25–27.

Drummer, O.H. (1998) Methods for the measurement of benzodiazepines in biological samples, *J. Chromatogr. B. Biomed. Sci. Appl.*, **713**, 201–225.

Dzantiev, B.B., Blintsov, A.N., Tsivileva, L.S., Berezin, I.V., Egorov, A.M., Izumrudov, V.A., Zezin, A.V., and Kabanov, V.A. (1988) Antigens interaction with the antibodies immobilized on synthetic water-soluble polyelectrolyte, *Doklady Biochemistry (Moscow)*, **302**, 279–282.

Dzantiev, B.B., Neustroeva, N.A., Ananiev, N.V., and Izumrudov, V.A. (1989) The interaction of HBsAg with antibodies immobilized on a water-soluble polyanion, *Vopr. Virusol.*, **3**, 28–31 (in Russian).

Dzantiev, B.B., Blintsov, A.N., Izumrudov, V.A., Bobkova, A.F., Atabekov, I.G., Zezin, A.B., and Kabanov, V.A. (1990) Complexes of viruses and synthetic polyelectrolytes and their interactions with antibodies, *Doklady Biochemistry (Moscow)*, **311**, 79–83.

Dzantiev, B.B., Choi, M.J., Park, J., Choi, J., Romanenko, O.G., Zherdev, A.V., Eremin, S.A., and Izumrudov, V.A. (1994a) A new visual enzyme immunoassay of methamphetamine using linear water-soluble polyelectrolytes, *Immunol. Lett.*, **41**, 205–211.

Dzantiev, B.B., Zherdev, A.V., Romanenko, O.G., Izumrudov, V.A., and Zezin, A.B. (1994b) Membrane enzyme immunoassay of testosterone using linear water-soluble polyelectrolytes. In Görög, S. (ed.), *Advances in Steroid Analysis 93*, Akademiai Kiado, Budapest, pp. 119–125.

Dzantiev, B.B., Blintsov, A.N., Bobkova, A.F., Izumrudov, V.A., and Zezin, A.B. (1995) New enzyme immunoassays based on interpolyelectrolyte reactions, *Doklady Biochemistry (Moscow)*, **342**, 77–80.

Dzantiev, B.B., Blintsov, A.N., Gaponova, N.K., Osipov, A.P., Atabekov, I.G., Izumrudov, V.A., Zezin, A.B., Bobkova, A.F., Natsvlishvili, N.M., Mikhajlov, M.I., Kartvelishvili, M.E., Snegireva, N.S., Vovchuk, A.I., Knyazeva, V.P., and Trofimets, L.N. (1996) Technique for quantitative determination of viruses, *Russian Patent 2,065,497*.

Dzantiev, B.B., Zherdev, A.V., Romanenko, O.G., and Trubaceva, J.N. (1997) Development of various enzyme immunotechniques for pesticides detection, *Amer. Chem. Soc. Symp. Ser.*, **657**, 87–96.

Ekeke, G.I. and Exley, D. (1978) The assay of steroids by fluoroimmunoassay. In Pal, S.B. (ed.), *Enzyme Labelled Immunoassay of Hormones and Drugs*, Berlin, de Gruyter, pp. 195–205.

Englebienne, P. and Weiland, M. (1994) Indicator reagents for detecting or measuring an analyte, kit containing the same and method of detection or measuring, *European Patent 0 623 822 A1*.

Englebienne, P. and Weiland, M. (1996a) Water-soluble conductive polymer homogeneous immunoassay (SOPHIA). A novel immunoassay capable to automation, *J. Immunol. Meth.*, **191**, 159–170.

Englebienne, P. and Weiland, M. (1996b) Synthesis of water-soluble carboxylic and acetic acid-substituted poly(thiophenes) and the application of their photochemical properties in homogeneous competitive immunoassays, *Chem. Commun.*, 1651–1652.

Englebienne, P. (1999) Synthetic materials capable of reporting biomolecular recognition events by chromic transition, *J. Mater. Chem.*, **9**, 1043–1054.

Ferjancic, A., Puigserver, A., and Gaertner, H. (1990) Subtilisin-catalysed peptide synthesis and transesterification in organic solvents, *Appl. Microbiol. Biotechnol.*, **32**, 651–657.

Forsgren, A. and Sjöquist, J. (1969) Protein A from *Staphylococcus aureus*. VII. Physicochemical and immunological characterization, *Acta Pathol. Microbiol. Scand.*, **75**, 466–480.

Galaev, I.Yu. and Mattiasson, B. (1993) Thermoreactive water-soluble polymers, nonionic surfactants, and hydrogels as reagents in biotechnology, *Enzyme Microb. Technol.*, **15**, 354–366.

Galaev, I.Yu., Gupta, M.N., and Mattiasson, B. (1996) Use smart polymers for bioseparation, *ChemTech*, N **12**, 19–25.

Gee, S.J., Hammock, B.D., and Van Emon, J.M. (1994) *A User's Guide to Environmental Immunochemical Analysis*; EPA/540/R-94/509; U.S. EPA, Office of Research & Development, Environmental Monitoring Systems Laboratory: Las Vegas.

Gosling, J.P. (1990) A decade of development in immunoassay methodology, *Clin. Chem.*, **36**, 1408–1427.

Guzov, V.M., Usanov, S.A., and Chashchin, V.L. (1991) Immunoassay in the presence of water-soluble polymers, *J. Immunol. Meth.*, **145**, 167–174.

Hassan, M., Landon, J., and Smith, D.S. (1979) Multi-fluorescein-substituted polymers as potential labels in fluoroimmunoassay, *FEBS Lett.*, **103**, 339–341.

Hirschfeld, T. and Eaton, D. (1979) Antigen detecting reagents. *US Patent* 4, 169, 137.

Hock, B. (1996) Advances in immunochemical detection of microorganisms, *Ann. Biol. Clin. (Paris)*, **54**, 243–252.

Huang, P.Y. and Lee, C.S. (1992) Mechanistic studies of electrostatic potentials on antigen–antibody complexes for bioanalyses, *Anal. Chem.*, **64**, 977–980.

Inada, Y., Nishimura, H., Takahashi, K., Yoshimoto, T., Saha, A.R., and Saito, Y. (1984) Ester synthesis catalyzed by polyethylene glycol-modified lipase in benzene, *Biochem. Biophys. Res. Commun.*, **122**, 845–850.

Izumrudov, V.A., Zezin, A.B., and Kabanov, V.A. (1991) Equilibria in interpolyelectrolyte reactions and the phenomenon of molecular "recognition" in solutions of interpolyelectrolyte complexes, *Russian Chemical Reviews*, **60**, 792–806.

Izumrudov, V.A. (1996) Competitive reactions in solutions of protein-polyelectrolyte complexes, *Ber. Bunsenges. Phys. Chem.*, **100**, 1017–1023.

Kabanov, V.A. and Zezin, A.B. (1984) Soluble interpolymeric complexes as a new class of synthetic polyelectrolytes, *Pure Applied Chemistry*, **56**, 343–354.

Kabanov, V.A. (1994) Basic properties of soluble interpolyelectrolyte complexes applied to bioengineering and cell transformations. In Dubin, P.L. (ed.), *Macromolecular Complexes in Chemistry and Biology*, Chapt. 10, Springer-Verlag, Berlin-Heidelberg, pp. 151–174.

Kharenko, O.A., Kharenko, A.V., Kalyugnaya, R.I., Izumrudov, V.A., Kasaikin, V.A., Zezin, A.B., and Kabanov, V.A. (1979) Nonstoichiometric polyelectrolyte complexes as a new water-soluble macromolecular compounds, *Vysokomolekulyarnie Soedineniya*, **21A**, 2719–2725 (in Russian.)

Kiselev, M.V., Gladilin, A.K., Melik-Nubarov, N.S., Sveshnikov, P.G., Miethe, P., and Levashov, A.V. (1999) Determination of cyclosporin A in 20% ethanol by a magnetic beads-based immunofluorescence assay, *Anal. Biochem.*, **269**, 393–398.

Lamture, J.B. and Wensel, T.G. (1995) Intensely luminescent immunoreactive conjugates of proteins and dipicolinate-based polymeric Tb (III) chelates, *Bioconjug. Chem.*, **6**, 88–92.

Langone, J.J. (1982) Protein A of *Staphylococcus aureus* and related immunoglobulin receptors produced by streptococci and pneumonococci, *Adv. Immunol.*, **32**, 157–252.

Ling, T.G., Ramstorp, M., and Mattiasson, B. (1982) Immunological quantitation of bacterial cells using a partition affinity ligand assay: a model study on the quantitation of streptococci B, *Anal. Biochem.*, **122**, 26–32.

Ling, T.G.I. and Mattiasson, B. (1983) A general study of the binding and separation in partition affinity ligand assay. Immunoassay of β_2-microglobulin, *J. Immunol. Meth.*, **59**, 327–337.

Ling, T.G.I. and Mattiasson, B. (1984) Partition affinity ligand assay (PALA): a concept for immunoassays and for preparative applications, *Eur. Congr. Biotechnol.*, **1**, 685–690.

Liu, F., Liu, F.H., Zhuo, R.X., Peng, Y., Deng, Y.Z., and Zeng, Y. (1995) Development of a polymer-enzyme immunoassay method and its application, *Biotechnol. Appl. Biochem.*, **21**, 257–264.

Margolin, A.L., Izumrudov, V.A., and Svedas, V.K. (1981) Preparation and properties of penicillin amidase immobilized in polyelectrolyte complexes, *Biochim. Biophys. Acta*, **660**, 359–365.

Margolin, A.L., Izumrudov, V.A., Svedas, V.K., and Zezin, A.B. (1982) Soluble–insoluble immobilized enzyme, *Biotechnol. Bioeng.*, **24**, 237–240.

Margolin, A.L., Sherstyuk, S.F., Izumrudov, V.A., Zezin, A.B., and Kabanov, V.A. (1985) Enzymes in polyelectrolyte complexes. The effect of phase transition on thermal stability, *Eur. J. Biochem.*, **146**, 625–632.

Marshall, D.L. (1985) Soluble insoluble polymers in enzymeimmunoassay, *US Patent 4, 530, 900.*

Mattiasson, B. and Ling, T.G. (1980) Partition affinity ligand assay (PALA). A new approach to binding assays, *J. Immunol. Meth.*, **38**, 217–223.

Mattiasson, B., Ling, T.G., and Ramstorp, M. (1981) Application of partition affinity ligand assay (PALA) in quick test for quantitation of *Staphylococcus aureus* bacterial cells, *J. Immunol. Meth.*, **41**, 105–114.

Mattiasson, B., Ramstorp, M., and Ling, T.G. (1982) Partition affinity ligand assay (PALA): applications in the analysis of haptens, macromolecules and cells, *Adv. Appl. Microbiol.*, **28**, 117–147.

Mattiasson, B. and Ling, T.G.I. (1983) Carrying out assaying methods involving biospecific reactions, *US Patent 4, 312, 944.*

Mitchell, J.M., Griffiths, M.W., McEwen, S.A., McNab, W.B., and Yee, A.J. (1998) Antimicrobial drug residues in milk and meat: causes, concerns, prevalence, regulations, tests, and test performance, *J. Food Prot.*, **61**, 742–756.

Monji, N., and Hoffman, A.S. (1987) A novel immunoassay system and bioseparation process based on thermal phase separating polymers, *Appl. Biochem. Biotechnol.*, **14**, 107–120.

Monji, N., Hoffman, A.S., Priest, J.H., and Houghton, R.L. (1988) Thermally induced phase separation immunoassay, *US Patent 4, 780, 409.*

Monji, N. and Cole, C.A. (1993a) Rapid membrane affinity concentration assays, *US Patent 5, 206, 136.*

Monji, N. and Cole, C.A. (1993b) Membrane affinity concentration immunoassay, *US Patent 5, 206, 178.*

Monji, N., Cole, C.A., and Hoffman, A.S. (1994) Activated, N-substituted acrylamide polymers for antibody coupling: Application to a novel membrane-based immunoassay, *J. Biomater. Sci. Polym. Ed.*, **5**, 407–420.

Nowinski, R.C. and Hoffman, A.S. (1987) Polymerization-induced separation immunoassays, *US Patent 4, 711, 840.*

Nowinski, R.C. and Hoffman, A.S. (1989) Polymerization-induced separation immunoassays, *US Patent 4, 843, 010.*

Oster, G. and Davis, B.J. (1990) Redox polymerization diagnostic test composition and method for immunoassay and nucleic acid assay, *European Patent 0 383 124 A2.*

Oster, G. and Davis, B.J. (1991a) Photopolymerization diagnostic test composition and method for immunoassay and nucleic acid assay, *US Patent 5, 019, 496.*

Oster, G. and Davis, B.J. (1991b) Redox polymerization diagnostic test composition and method for immunoassay and nucleic acid assay, *US Patent 5, 035, 997.*

Papisov, I.M. and Litmanovich, A.D. (1989) Molecular recognition in interpolymer interactions and matrix polyreactions, *Adv. Polym. Sci.*, 90, 139–179.

Parija, S.C. (1998) A review of some simple immunoassays in the serodiagnosis of cystic hydatid disease, *Acta Trop.*, 70, 17–24.

Patil, A.O., Heeger, A.J., and Wudl, F. (1988) Optical properties of conducting polymers, *Chem. Rev.*, 88, 183.

Pergushov, D.V., Izumrudov, V.A., Zezin, A.B., and Kabanov, V.A. (1995) Stability of interpolyelectrolyte complexes in aqueous saline solutions, *Polymer Sci.*, 37A, 1081–1087.

Sasaki, S., Tokitsu, Y., Ikebukuro, K., Yokoyama, K., Masuda, Y., and Karube, I. (1997) Biosensing of a herbicide using a chemically modified antibody in organic solvent, *Anal. Lett.*, 30, 429–433.

Shiba, K. (1998) The rapid method in microbiology using chemical method and immunochemical method, *Rinsho Biseibutshu Jinsoku Shindan Kenkyukai Shi*, 9, 73–81.

Sugiura, M. and Mizuoka, K. (1995) Kinds and characteristics of solid-phase enzyme immunoassay, *Nippon Rinsho*, 53, 2188–2191.

Takei, Y.G., Matsukata, M., Aoki, T., Sanui, K., Ogata, N., Kikuchi, A., Sakurai, Y., and Okano, T. (1994) Temperature-responsive bioconjugates. 3. Antibody-poly(N-isopropylacrylamide) conjugates for temperature-modulated precipitations and affinity bioseparations, *Bioconjug. Chem.*, 5, 577–582.

Thomas, E.K., Schwartz, D.E., Priest, J.H., Nowinski, R.C., and Hoffman, A.S. (1988) Polymerization-induced separation assay using recognition pairs, *US Patent* 4, 749, 647.

Tijssen, P. (1985) *Practice and Theory of Enzyme Immunoassay*, Elsevier Sci. Publ., New York.

Tsuchida, E., Osada, Y., and Sanada, K. (1972) Interaction of poly(styrene sulfonate) with polycations carrying charges in the chain backbone, *J. Polym. Sci.*, 10, 3397–3403.

Tsuchida, E. and Abe, K. (1986) Polyelectrolyte complexes. In Wilson, A.D. (ed.), *Developments in ionic polymers*, Elsevier Sci. Publ., New York, pp. 191–266.

Vlasenko, S.B., Arefyev, A.A., Klimov, A.D., Kim, B.B., Gorovits, E.L., Osipov, A.P., Gavrilova, E.M., and Yegorov, A.M. (1989) An investigation on the catalytic mechanism of enhanced chemiluminescence: immunochemical applications of this reaction, *J. Biolumin. Chemilumin.*, 4, 164–176.

Watson, I.D. (1998) Laboratory support for the poisoned patient, *Ther. Drug Monit.*, 20, 490–497.

Wild, D. (1994) *The Immunoassay Handbook*, Stockton Press, New York.

Yazynina, E.V., Zherdev, A.V., Dzantiev, B.B., Izumrudov, V.A., Gee, S.J., and Hammock, B.D. (1999a) Microplate immunoassay technique with use of polyelectrolyte carriers: kinetic studying and application for detection of herbicide atrazine, *Anal. Chim. Acta*, 399, 151–160.

Yazynina, E.V., Zherdev, A.V., Dzantiev, B.B., Izumrudov, V.A., Gee, S.J., and Hammock, B.D. (1999b) New immunoassay techniques for detection of the herbicide simazine based on use of oppositely charged water-soluble polyelectrolytes, *Anal. Chem.*, 71, 3538–3543.

11 Enzymes immobilized in smart hydrogels

Jyh-Ping Chen

11.1. Introduction

The advantages that immobilized enzymes will offer, include ease of recovery for reuse; continuous operation; different shapes for specific processes; and in some cases improved properties. A number of publications on the preparation and characterization of immobilized enzymes, together with their applications in a variety of fields have already been published (Kennedy and Cabral, 1987; Taylor, 1990; Bickerstaff, 1997). The technology of immobilizing enzymes is believed to be relatively mature now. In addition, the nature of immobilized enzymes has become somewhat understandable to us. The key now is to come up with new uses and novel systems of immobilized enzymes that can fulfill specific needs. A recent interest in this research field is seeking the possibility of constructing immobilized enzymes where the enzyme process can be controlled by externally applied stimuli such as light, electric fields, pH, temperature, and mechanical forces. What is crucial in system construction here is not to rely on a possible alternation in the properties of the enzyme, which have frequently been expected as a result of immobilization, but to develop a new capabilities of the enzyme system as a result of a rational design process.

Since the first use of crosslinked acrylamide gel for enzyme immobilization by physical entrapment, polymer hydrogels have frequently been employed for preparation of immobilized biocatalysts. The best known example of such cases may be calcium alginate or chemically modified calcium alginate gels for entrapment of enzymes and cells. In conventional gel-entrapped biocatalysts, the gel material itself played only the role as a support for maintaining or holding the biocatalysts, no functional properties of the gels were employed. Recently, dramatic development in the field of polymer gels has resulted in the successful synthesis of many smart (sometimes called stimuli-responsive, intelligent, or environmentally sensitive) hydrogels that undergo continuous or discontinuous volume change in response to a change in an external stimulus (Dusek, 1993; Osada and Rose-Murphy, 1993; Hoffman, 1995; Galaev and Mattiasson, 1999). This marvelous property of polymer hydrogels may be one of the most interesting and important subjects in the development of immobilized enzymes systems. Novel applications and utilization of the volume-phase transition of smart hydrogels in the construction of immobilized enzyme systems will be discussed here with the emphasis of hydrogels prepared from a thermo-sensitive monomer, N-isopropylacrylamide (NIPAAm).

11.2. Phase Transitions of Smart Hydrogels

Tanaka (1978) observed the discontinuous volume change of a covalently crosslinked acrylamide (AAm) gel in acetone/water mixture when varying the temperature or composition of the mixture. This phenomenon is now called the volume-phase transition of gels and is observed in many gels made of synthetic and natural polymers. These hydrogels undergo reversible discontinuous or continuous volume change, as large as several hundred times, in response to small variations in the external physical or chemical conditions surrounding a gel. There are many different stimuli that have been applied, including temperature, solvent composition, pH, ion concentration, small electric field, and light.

There is a lot of reports describing the investigation of the mechanism of phase transition of hydrogels using the Flory-Huggins theory (Tanaka, 1978; Hirotsu *et al.*, 1987; Suzuki *et al.*, 1996; Hino and Prausnitz, 1998). It can be shown that a gel undergoes either a continuous volume change or a first-order discontinuous phase transition depending on the proportion of ionizable groups in the polymer network and on the stiffness of the polymer chains constituting the network. The counterions of the ionized groups and the stiffness of the polymer chains increase the osmotic pressure acting to expand the polymer network, resulting in a discontinuous volume change.

Other attempts have also been made to account theoretically for the phase behavior of thermo-sensitive polymers in terms of the effect of solvents, additives, chemical structure of constituents, and intermolecular structure on the thermo-sensitive swelling (Inomata and Saito, 1993). In addition, Rathbone *et al.* (1990) proposed a lattice model to explain the swelling curves of gels consisting of crosslinked copolymers, which were measured as a function of the degree of crosslinking and the concentrations of salt. A more generalized explanation of the phase transition of gels was provided by Ilmain *et al.* (1991). They hypothesized a balance between the repulsion and attraction of the crosslinked polymer chains in the networks, which arise from a combination of four intermolecular forces: ionic, hydrophobic, van der Waals and hydrogen bonding. When a repulsive force overcomes an attractive force, gel volume should increase discontinuously in some cases or continuously in others. Conversely, a decrease in gel volume may occur when the attractive force becomes dominant. The variables that trigger the transition influence these intermolecular forces and thereby the balanced state of the attractive and repulsive forces.

11.3. Enzymes and Cells Immobilized in Temperature-sensitive Hydrogel

11.3.1. Hydrogels Prepared by Inverse Suspension Polymerization

Hoffman's group performed a series projects on the use of smart hydrogels for immobilization of enzymes and cells by entrapment using thermally reversible crosslinked copolymer hydrogel of NIPAAm and AAm (Park and Hoffman, 1988, 1990a and b, 1991, 1993). The enzymes were asparaginase and β-galactosidase (βgal) while the cell was *Arthrobacter simplex* for bioconversion of hydrocortisone to prednisolone. The hydrogel was prepared in the form of beads with diameters in the range of 200–400 μm in an inverse suspension polymerization step using paraffin oil as the continuous phase and Pluronic L-81 as a surfactant. A typical immobilization procedure is described below briefly. Enzymes or cells were dissolved or suspended in buffer solution containing varying molar ratios of two monomers (NIPAAm and AAm), crosslinker (N,N'-methylene-bis-acrylamide, MBAAm), and one redox initiator, ammonium persulfate (APS). This aqueous phase was suspended in oil phase containing the surfactant and then polymerized under a nitrogen atmosphere by injecting the other initiator (or accelerator) (N,N,N',N'-tetramethylenediamine, TEMED) into the continuous oil phase where aqueous droplets of the solution with both the monomers and the biocatalyst had been formed by agitation (500 rpm). After 30 min of polymerization in an ice-water bath, the beads were separated, washed, and then freeze-dried.

The thermally reversible hydrogel exhibits a sharp volume-phase transition near its phase transition temperature, which is called the lower critical solution temperature (LCST). This is shown in Figure 11.1 when swelling ratio of the hydrogel beads was plotted as a function of temperature. As the temperature is raised above the LCST, the gel collapses and shrinks with concomitant squeezing out of a large fraction of the water from inside the gel. When the temperature is lowered below the LCST, the gel reswells reversibly, imbibing the aqueous solution surrounding it (Figure 11.2). The collapse and re-expansion of the gel matrix volume could be reversibly effected over a narrow range of temperature near the LCST of the gel matrix. Since collapse of the hydrogel results in the reduction of gel pore volume, mass transfer of a solute through this thermo-sensitive matrix is believed to be a strong function of temperature. The activities of immobilized enzyme in the hydrogel depend on the degree of gel swelling, suggesting that the diffusion rates of the substrate in and the product out of the gel matrix control the overall kinetic properties. The shrinkage and collapse of the pore structure above the LCST will reduce the mass transfer rates, resulting in an overall decrease in immobilized enzyme activity. It is notable, however, that the enzyme did not lose its intrinsic activity within the collapsed and significantly

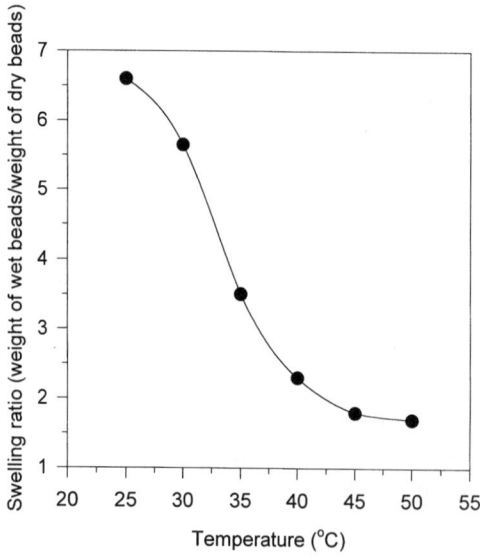

Figure 11.1 Effect of temperature on the equilibrium swelling ratio of NIPAAm/AAm (molar ratio 9 : 1) temperature-sensitive hydrogel beads.

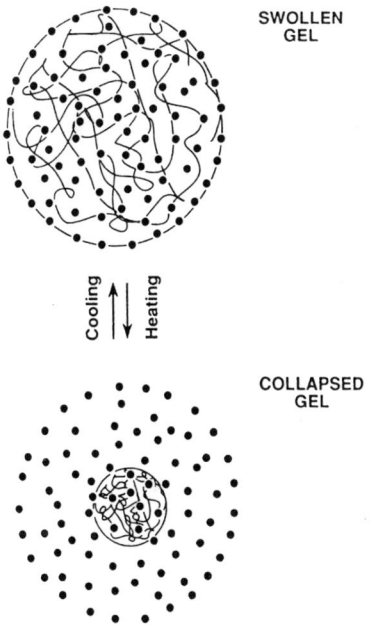

SWOLLEN
GEL

Cooling ↑↓ Heating

COLLAPSED
GEL

Figure 11.2 A schematic diagram showing water molecules (represented by •) in a swollen and collapsed temperature-sensitive hydrogel beads.

dehydrated gel matrix and its apparent activity can be fully restored after re-hydration of the hydrogel. Immobilized enzymes in such temperature-sensitive hydrogels therefore exhibited 'on/off' activity control in response to temperature, as have been shown for asparaginase (Dong and Hoffman, 1986, 1987) and βgal (Park and Hoffman, 1990a).

The pore sizes of these temperature-sensitive hydrogel beads are affected significantly by both the temperature and the gel composition, but not much by crosslinker concentration (Park and Hoffman, 1994). The incorporation of AAm, the hydrophilic monomer, into PNIPAAm gel raises the LCST and broadens the phase transition. The molecular weight cutoff of hydrogel beads as measured by gel filtration method using several probe molecules with known molecular weight was in the order of 10^4 for swollen gel. This put an upper limit on the size of substrates that can diffuse into the gel and a lower limit on the molecular weight of enzyme molecules that can be retained in the gel.

Furthermore, a thermal cycling operation around the LCST of hydrogels (around 32–33 °C), compared to isothermal operations, significantly increased the activity and overall conversion in both an immobilized enzyme packed-bed reactor (Figure 11.3) (Park and Hoffman, 1988, 1993) and in an immobilized cell batch reactor (Figure 11.4) (Park and Hoffman, 1990b, 1991). It was postulated that the mass transfer rates within the hydrogel bead matrix are greatly enhanced by the cyclic 'pumping' of substrate in and product out during swelling and deswelling of the hydrogel polymer chain network as temperature is cycled. This is like a forced, continuous 'breathing' of the immobilized biocatalysts within the hydrogel beads. The increased conversion during

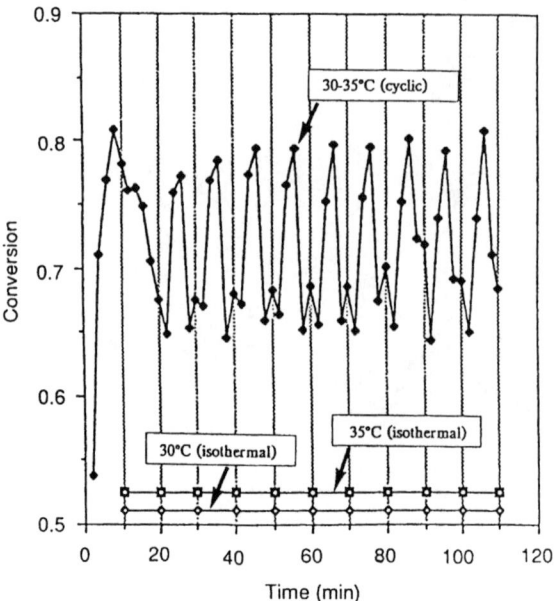

Figure 11.3 The influence of isothermal and temperature cycling operations on the hydrolysis of *o*-nitrophenyl-β-D-galactopyranoside (ONPG) by β-galactosidase entrapped in NIPAAm/AAm (molar ratio 9:1) hydrogel beads in a packed-bed reactor. ○, 30 °C isothermal; □, 35 °C isothermal; ●, 30–35 °C temperature cycling at $1\,°C\,min^{-1}$. Redrawn from Park and Hoffman, 1988.

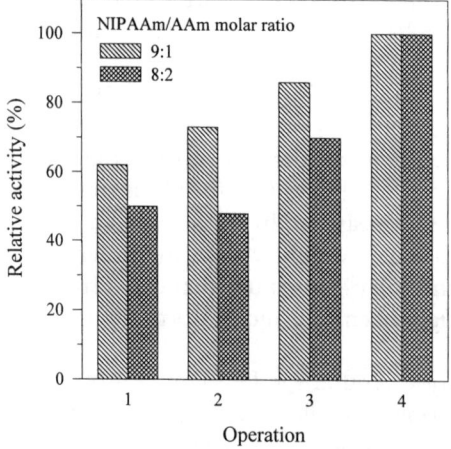

Figure 11.4 Comparison of isothermal and temperature cycling operations on bioconversion of hydrocortisone to prednisolone by *Arthrobacter simplex* cells immobilized in NIPAAm/AAm hydrogel beads in a batch reactor. Operations: 1, 25 °C isothermal; 2, 30 °C isothermal; 3, 35 °C isothermal; 4, 30–35 °C temperature cycling at $1\,°C\,min^{-1}$. Data from Park and Hoffman, 1990b.

Table 11.1 Comparisons of bioreactor performance for β-galactosidase immobilized in temperature-sensitive hydrogel beads[a]

Reactor type[b] and operation condition	Conversion (%)	Activity ($\mu mol\, g^{-1}\, min^{-1}$)
CSTR, isothermal, 30 °C	66.3	484
CSTR, isothermal, 35 °C	72.5	523
CSTR, 30–35 °C cycle, 1 °C min^{-1}	65.5	478
PFR, isothermal, 30 °C	51.1	373
PFR, isothermal, 35 °C	52.6	442
PFR, 30–35 °C cycle, 1 °C min^{-1}	71.6	523
PFR, 30–35 °C cycle, 0.5 °C min^{-1}	71.9	525
PFR, 30–35 °C cycle, 0.1 °C min^{-1}	59.2	432
PFR, 35–40 °C cycle, 1 °C min^{-1}	70.9	518
PFR, 35–40 °C cycle, 0.5 °C min^{-1}	69.1	505
PFR, 30–40 °C cycle, 1 °C min^{-1}	71.4	521
PFR, 30–40 °C cycle, 0.5 °C min^{-1}	57.5	420

a Data from Park and Hoffman, 1993.
b CSTR, continuous stirred-tank reactor; PFR, plug flow reactor.

thermal cycling operation may also be due in part to reduced product inhibition. This is demonstrated in the case of bioconversion of steroid with immobilized cells (Park and Hoffman, 1990b). The conversion could reach 100% within a relatively short time under a thermal cycling operation, which could not be accomplished due in part to product inhibition, even with a long reaction time under isothermal operation. A detailed analysis of a series at thermal cycling operations with immobilized βgal has been conducted. It was concluded that the thermal cycling operation for the continuous stirred tank reactor does not enhance the activity, but that for the plug flow reactor increases the activity significantly over isothermal operation (see Table 11.1). Thermal cycling operation should be performed just below the LCST, with fast heating/cooling rates in order to maximize the overall conversion of the immobilized enzyme (Park and Hoffman, 1993).

It should be noted that the surfactant (Pluronic L-81) used in polymerizing the hydrogel beads has a rather low hydrophile–lipophile balance ratio of 2. To the best of our knowledge, the manufacturer (BASF Wyandott Corp.) may have discontinued production of this product. It is therefore important to choose a suitable surfactant for forming oil/water emulsion during the polymerization step. Wang *et al.* (1996) prepared PNIPAAm hydrogel beads with a different surfactant, octylphenylether. The authors explored the pressure-sensitivity of PNIPAAm hydrogels. The gel beads were prepared similar to Park and Hoffman (1990b), however, pressure cycling was used to enhance the catalytic activity of entrapped βgal. The enzyme reaction was carried out in a magnetically stirred batch reactor at a constant temperature. The pressure inside the reactor was regulated with an oil cylinder system. During pressure cycling operation, a pulsatile pressure variation strategy was adopted, which resulted in a tooth-shaped swelling/deswelling curve of the supporting matrix. The conversion of ONPG substrate by the immobilized enzyme increased 52% with pressure cycling between 1 and 120 atm every 10 min when compared with isobaric operation, where conversion was pressure-independent. Enhancement of mass transfer rate within the gel bead and possible reduction of product inhibition during the pressure cycling operation, similar to temperature cycling, was believed to be responsible. The authors also concluded that pressure cycling amplitude rather than the pressure cycling range is important, with higher conversion observed under higher pressure cycling amplitude. This paper in interesting from the viewpoint that it provides an alternative stimulus other than temperature based on the same principle, which will broaden the application of PNIPAAm hydrogel. However, from a practical point of view, instantaneous pressure change is more difficult to achieve than temperature change.

Murakata *et al.* (1993) prepared thermo-sensitive hydrogels in organic solvents. Monomer solution containing NIPAAm, MBAAm, and enzyme (lipase) was mixed with initiator APS solution and poured into an organic phase in a polymerization vessel. The organic phase consisted of 73% (v/v) toluene and 27% (v/v) chloroform and contained 1% (v/v) of a surfactant, sorbitan sesquioleate. The polymerization was carried out in an ice-water bath for 1 h with stirring under nitrogen atmosphere, followed by washing with buffers. The hydrogel beads had an average diameter of $60 \pm 5\,\mu\text{m}$. The immobilized enzyme was used for esterification reaction in organic medium. The activity showed discontinuity around the phase transition temperature of the gel, increasing significantly with the phase transition at which the gel altered from swollen to shrunken state. Since the reactants (lauric acid and lauryl alcohol) are hydrophobic in nature, by squeezing out of water at phase transition, pores in the hydrogel will become more hydrophobic to promote diffusion of the substrates and to favor the reaction. Such decrease in diffusion resistance of substrates and the increase in hydrophobicity in the pores possibly enhance the reaction rate.

We have compared the above two inverse suspension polymerization methods, either with oil or organic solvents as the continuous phase, for immobilization of glucoamylase. The immobilized enzyme shows 3.1 times higher specific activity and 2.1 times higher protein immobilization efficiency when prepared in paraffin oil than in organic solvents. These values regarding the influence of the polymerization medium, however, may be sensitive to the type of enzyme chosen, and need to be verified from experiments when choosing the best hydrophobic medium forming the emulsion.

11.3.2. Hydrogels Prepared by Other Methods

The complete on/off control of enzyme activity using a thermo-sensitive hydrogel has also been performed by Kokufuta *et al.* (1992), who prepared a gel with immobilized exo-1, 4-α-D-glucosidase by crosslinking an aqueous poly(vinyl methyl ether) solution using γ-ray irradiation. The resultant gel exhibited a thermo-sensitive characteristic: it shrank above 38 °C and swelled below this transition temperature. This behavior was reversible. The immobilized preparation obtained displayed an excellent capacity for on/off control of the enzymatic hydrolysis of maltose. When the immobilized enzyme was used, glucose formation from maltose halted at 42 °C but recommenced immediately when the temperature was lowered to 32 °C. Such initiation-termination control could be repeatedly reversible throughout a single run of the measurements and reproduced without a serious loss of activity for at least 20 runs with a freshly prepared substrate solution.

Bayhan and Tuncel (1998) have used a modified procedure for preparing large uniformly spherical, and thermally reversible hydrogel beads. The hydrogel beads were prepared in an aqueous dispersion medium by using Ca-alginate gel as the polymerization mold. The continuous medium was prepared by dissolving potassium persulfate (KPS) and calcium chloride in water. It was purged with nitrogen before adding the dispersed phase, which includes monomer (NIPAAm), crosslinker (MBAAm), initiator

(TEMED), enzyme (α-chymotrypsin), and stabilizer (sodium alginate). The dispersed phase was dropped through a syringe into the continuous phase which was kept at 20 °C and stirred magnetically at 250 rpm. The polymerization was conducted for 4 h. The uniform and thermally reversible enzyme-gel beads were 3 mm in size. The highest swelling ratio (weight of wet bead/weight of dry bead) is much higher for hydrogel beads prepared in this way, increasing from 6.7 (as shown in Figure 11.1) to 38. However, strong internal mass transfer resistance was observed where K_m value increased from 0.063 to 2.543 mM with BAEE as the substrate. Additional mass transfer resistance may come from alginate. Previously, Park and Hoffman (1992) have developed a similar procedure for preparing large (2 mm diameter) hydrogel beads by using alginate as a mold, however, alginate was removed from the polymerized gel beads with calcium-chelating agents (0.1 M EDTA in phosphate buffer) at the end. This step can be expected to reduce mass transfer resistance at the expense of a weaker gel structure.

A random NIPAAm and 2-hydroxyethylmethacrylate (HEMA) copolymer gels having thermo-responsive characteristic was prepared by a redox copolymerization method in cylindrical geometry (Çiçek and Tuncel, 1998a). PEG 4000 was included in the copolymerization recipe as a diluent to increase the thermo-sensitivity of the copolymer gel. The copolymerization of NIPAAm with HEMA yielded a thermo-responsive gel matrix with higher elasticity and mechanical strength than PNIPAAm gels. Furthermore, some undesired properties of conventional PNIPAAm gels, such as the coexistence of shrunken and swollen phases and the formation of water pockets and cracks on the matrix surfaces, can be avoided. α-chymotrypsin was immobilized by physical entrapment in the poly(NIPAAm-*co*-HEMA) copolymer hydrogels (Çiçek and Tuncel, 1998b). By investigating the kinetic behavior of enzyme-gel cylinders in a batch reactor, the authors concluded that the overall reaction rate was controlled by substrate diffusion through the gel matrix. The maximum enzyme activity was achieved at 30 °C for the immobilized enzyme in contrast to 40 °C for the free enzyme due to shrinkage of the hydrogel above its LCST (around 30–35 °C). The performance of the thermo-sensitive enzyme-gel system was also investigated in a continuous stirred tank reactor (Tuncel, 1998). A mathematical model was developed to explain the kinetic behavior of the thermally reversible enzyme gel cylinders. Effective diffusion coefficient of the substrate (benzoyl-L-tyrosin ethyl ester) within the gel matrix can be calculated at different reaction temperatures by this model, which makes estimation of Thiele modulus and effectiveness factors possible (Bailey and Ollis, 1986). A sudden increase in Thiele modulus was obtained around the LCST of the hydrogel matrix. The effectiveness factor determined at different reaction temperatures indicated that the overall hydrolysis rate was controlled by mass transfer resistance within the gel matrix. The shrinkage of gel induced by increasing temperature was significantly effective on reducing substrate diffusion coefficient above 30 °C.

Liu and Zhuo (1993) prepared temperature-sensitive hydrogels by crosslinking copolymer of NIPAAm and N-acryloxysuccinimide (NAS). Thus, α-chymotrypsin can be immobilized to the backbone of the polymer chain by covalent binding rather than by physical entrapment in the gel matrix as in other studies, through the active ester groups of NAS. The gel particle was formed by crosslinking NIPAAm-NAS copolymer with ethylenediamine in aqueous solution and followed by grinding in a mortar to give irregular particles having a diameter of 20–80 μm. The enzyme can be immobilized to the crosslinked gel or immobilized to the copolymer first and followed by crosslinking. The former method gave immobilized enzyme with higher catalytic activity and better properties. This method will be useful for immobilization of small enzyme molecules within temperature-sensitive hydrogels, where enzyme leakage will be a major concern.

Takahashi *et al.* (1997) prepared NIPAAm/AAm gels entrapping γ-Fe_2O_3 particles (8–24%) and invertase for hydrolysis of sucrose. NIPAAm, AAm, MBAAm and γ-Fe_2O_3 were mixed with buffer. After bubbling with nitrogen, enzyme solution, APS and TEMED solutions were added to the mixture. The mixture was transferred into silicone tubes and both ends were closed. The tubes were placed at the center of the gap between two magnet poles. The polymerization was carried out in an ice-water bath under exposure to a magnetic field to avoid sedimentation of the γ-Fe_2O_3 powder in the reaction tubes. The shrinking or swelling of gel rods was registered at around 30–45 °C measuring their lengths at various temperatures. Thus, volume of the gel can be regulated by a magnetic field due to magnetic heating of the thermo-sensitive hydrogel. Since the gel was heated from inside to outside, a dehydrated, dense 'skin' could not be formed on the outer surface of the gel as reported by Park and Hoffman (1993). By exposing column-packed gel rods (2 cm long and 1 mm in diameter) to a magnetic field, the column temperature rose and became constant within 10 min due to heat generated from magnetic hysteresis loss. The overall concentration of reducing sugars in the outlet solution increased initially and then decreased due to thermal shrinkage of the gel support. The increase in reducing sugars concentration resulted from both promotion of the enzyme reaction with increase in temperature and squeezing out of the products from the swollen region with shrinkage of the gel immediately after application of the magnetic field. After removal of the magnetic field, the temperature decreased to its baseline value, and the overall concentration of reducing sugars decreased immediately and then gradually increased again due to swelling of the gel. This system demonstrates an alternative way of controlling immobilized enzyme reactions in thermo-sensitive hydrogels, i.e. via magnetic heating. Compared with circulating water in the jacket of column, magnetic heating will not give uniform temperature distribution in the column, temperature at the bottom of the column is expected to be lower than that in the upper region due to accumulation of heat of hysteresis loss. Temperature control will be also more difficult and sluggish. Nonetheless, magnetic heating with swelling and shrinkage of the gel matrix controlled by removal and application of a magnetic field will be more energy-saving than heating the whole column.

11.4. Enzyme Immobilized in Hydrogel with Biochemical–mechanical Functions

The purpose of all studies shown in previous sections was to use immobilized enzymes in chemical conversions. On the other hand, it is possible to convert the 'energy' of immobilized enzyme reactions into mechanical work. A system with such a biochemical–mechanical function can be prepared by entrapping enzymes into polymer gels that undergo volume-phase transitions in response to enzymatic changes (Kokufuta and Tanaka, 1991). Biochemical–mechanical systems differ from chemo-mechanical systems by that biochemical changes such as enzymatic reactions are used in place of the usual chemical changes for creating mechanical energies. Since biochemical reactions are generally more specified, the creation of mechanical energies in a biochemical–mechanical system is expected to be responsive to a specific kind of molecule.

Among the variables triggering phase transition in smart hydrogel, solution pH seems to be the best candidate to be controlled enzymatically. Since that phase transition occurs as a result of a change in the balance between the ionic (repulsive) and hydrophobic (attractive) forces, it can be regulated conveniently through enzyme-induced change in pH. One example of such systems is a copolymer hydrogel of acrylic acid (AA) and NIPAAm into which urease has been entrapped (Kokufuta *et al.*, 1994). The gel was obtained by gelling an aqueous solution containing NIPAAm, AA, MBAAm, and urease. APS and TEMED were used as the initiator and accelerator, respectively. The gelation was carried out at $0\,°C$ for 1 h in a test tube into which glass capillaries with an inner diameter of 0.1 mm had previously been inserted and the gel samples were cut into cylinders of *ca.* 2 mm length after washing with pH 4.0 buffer. Swelling curves between 25 and $40\,°C$ were measured in the same buffer solutions both with and without urea as the substrate, by measuring the diameter of gel fiber. In the absence of urea, the gel underwent a discontinuous phase transition at $32.4\,°C$ (LCST). In contrast, the presence of urea (1 M) brought about a continuous transition, and the transition temperature was shifted to a higher range. Taking advantage of these results, reversible regulation of gel volume was possible by alternatively immersing the gel into buffer solutions with and without urea. At $33.4\,°C$ which is higher than LCST, the gel shrank in the buffer in the absence of urea. However, it began to swell as soon as the buffer was replaced with 1 M urea solution and the swelling process completed within 15 min. The swollen gel collapsed again when the ambient solution was replaced with the urea-free buffer. The time needed for a complete collapse of the gel was approximately 90 min. The swelling and collapsing process is reversible with intermittent change of solution into which the gel fiber was immersed.

The result obtained can be interpreted as shown in Figure 11.5. The immobilized urease catalyzes the hydrolysis of urea into ammonia and carbon dioxide. The increase in ammonium ion (NH_4^+) concentration leads to a pH increase in the gel phase, which de-protonated the carboxyl pendant groups of the gel. The electric force arising from the negatively charged carboxylate group (COO^-) overcame the hydrophobic

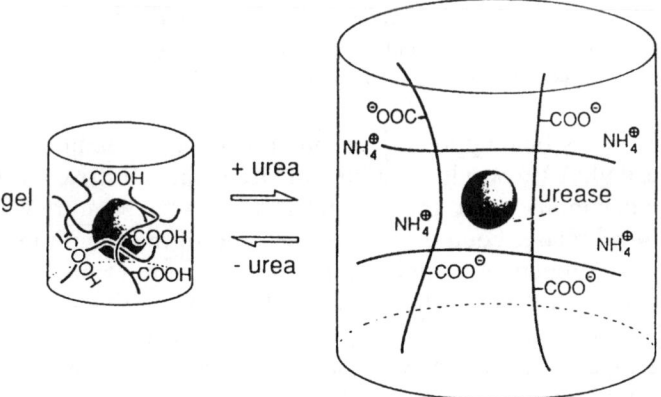

Figure 11.5 The mechanism of controlling reversible swelling and collapse of NIPAAm/AA hydrogel fibers by biochemical reaction. Ammonium ions generated from the hydrolysis reaction of entrapped urease cause volume transition of the hydrogel in response to the presence and absence of the substrate (urea).

interaction between the network chains and brought about an increase in the transition temperature. As a result, the gel was swollen when a certain amount of urea was present, but collapsed in its absence if the temperature was kept within a suitable range above the LCST.

11.5. Enzyme Immobilized onto Composite Temperature-sensitive Hydrogel Membranes

11.5.1. *Preparation of Composite Hydrogel Membrane*

Composite hydrogel membranes made of crosslinked poly(N-isopropylacrylamide-*co*-N-acryloxysuccinimide-*co*-2-hydroxyethylmethacrylate) [P(NIPAAm-NAS-HEMA)], with starch as a macropore forming agent, on non-woven polyester can be prepared (Sun *et al.*, 1999a). The composite membrane is temperature-sensitive with a LCST around 35 °C. It responds to temperature change by swelling below the LCST and shrinking above the LCST, corresponding to opening and closing of the membrane pores. A conventional hydrogel of PNIPAAm or its copolymer hydrogels is weak and fragile. The composite hydrogel membrane retains the same temperature-sensitivity characteristic as conventional hydrogels but is more robust, it is as strong mechanically as the nonwoven original.

The composite hydrogel membrane was prepared by casting NIPAAm onto polyester nonwoven support. A square piece of nonwoven support was placed between two polyester films after spreading monomer solutions on both sides of the support. The monomer solution was prepared by mixing NIPAAm, MBAAm, NAS, HEMA, AIBN, and a suitable amount of soluble starch in DMF. The membrane was placed in a 60 °C vacuum oven to initiate the polymerization and continued for another 24 h at 60 °C under nitrogen. The membrane was removed form the low-adhesion films by soaking in water. Composite membrane prepared in this way was 152 µm thick and with a surface density of 0.012 g/cm^2.

The hydrogel phase of the composite membranes was formed in the void space within the nonwoven skeleton. As shown from Scanning Electron Microscope (SEM) pictures of the non-woven sheet and composite membrane in Figure 11.6, the fibers of the nonwoven sheet were distributed within the gel phase in the interior of the composite membrane, which serves as a support. There is also hydrogel on the exterior of the membrane.

HEMA can provide better binding for the gel phase when the membrane is dry; otherwise dehydrated PNIPAAm would be peeled off from the nonwoven skeleton easily. The composite membrane can respond to the change in temperature with the network of PNIPAAm gel phase. Equilibrium swelling ratio of the composite membrane was determined as a function of temperature, as reported in Figure 11.7. The swelling ratio changed from 2.45 at 5 °C to 1.19 at 75 °C, with half change occurring at about 32 °C. This temperature also gives a region where the composite membrane is most sensitive to temperature change. Compared to other PNIPAAm hydrogels reported in the literature (Park and Hoffman, 1990a and b), the magnitude of swelling of the composite membrane is relatively small, possibly due to the limitation of nonwoven skeleton and the presence of other co-monomers in the gel structure. The insert in Figure 11.7 shows the kinetics of swelling and deswelling process. Both processes responded quickly to temperature change, with the time for half change being within

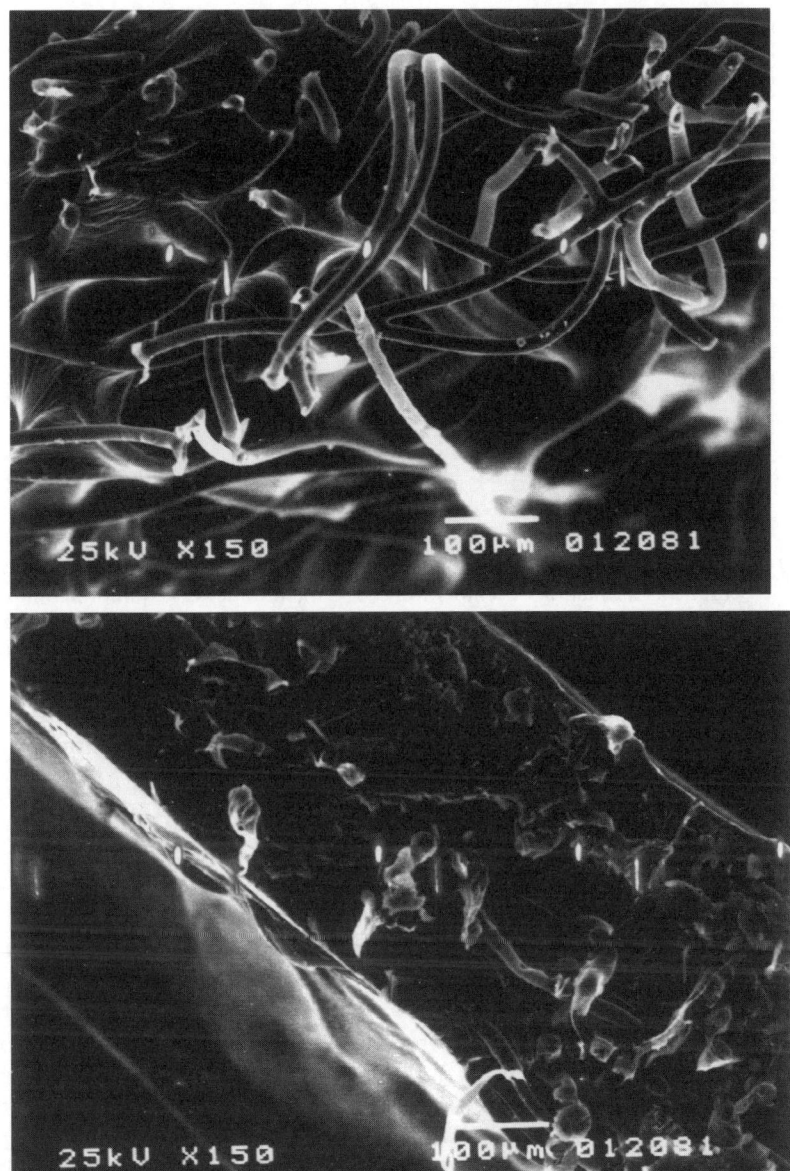

Figure 11.6 Scanning electron microscope pictures of the cross sections of non-woven polyester sheet (top) and composite temperature-sensitive hydrogel membrane (bottom).

1 min and the time for 90% change being between 4 and 5 min. It can be noted that response time of the composite membrane was comparable to that of a macroporous PNIPAAm hydrogel using hydroxy-propylcellulose as a porogen for preparing the hydrogel (Wu *et al.*, 1992). The response time is also at least one order of magnitude smaller than those of conventional PNIPAAm hydrogels. Such a fast temperature

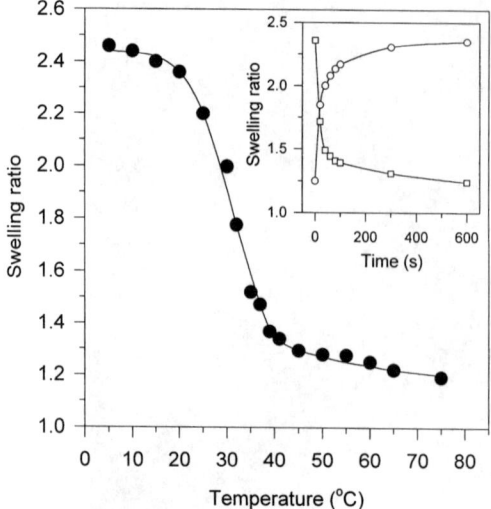

Figure 11.7 Equilibrium swelling ratio of temperature-sensitive composite hydrogel membrane as a function of temperature. The membrane was equilibrated in water
at the temperature indicated for 24 h. The insert shows the swelling (○)
and de-swelling (□) kinetics of the membrane as the temperature changed
instantaneously from 60 to 20 °C and from 20 to 60 °C, respectively.

response will efficiently enhance the transport of solute through the membrane during
a temperature cycling operation.

11.5.1.1. Immobilization of α-amylase

Since NIPAAm does not have any functional groups to react with enzyme, entrapment of the enzyme within NIPAAm-based hydrogels was employed in all previous
works investigating immobilized β-galactosidase in temperature-sensitive hydrogels
(Park and Hoffman, 1988). This has placed a limit on the size of the substrate (lactose)
that can diffuse into the gel matrix, as pores of the hydrogel must be small enough
to prevent enzyme leakage. To hydrolyze the high molecular weight soluble starch
substrate by using the temperature-sensitive membrane, NAS and soluble starch were
added when preparing the composite hydrogel membrane. NAS has highly reactive
ester groups reacting with amino groups of the enzyme, making enzyme immobilization by stable covalent binding possible. The concept of adding soluble starch during
the polymerization step is to make macroporous hydrogels. The starch molecules may
act to physically separate the aggregated PNIPAAm chains from each other, resulting
in large pores in the hydrogel after removing the starch from the gel by dissolving it in
water during the last membrane preparation step. Any residual starch in the hydrogel
can also be conveniently hydrolyzed by α-amylase during the enzyme immobilization
step to generate pores large enough for penetration of starch.

The best condition for enzyme immobilization was by reacting in pH 8.0 buffer
at 4 °C for 24 h; the activity retention of the immobilized enzyme was 32% and
the composite membrane contained 5.04 mg/g of protein. The activity retention of

immobilized enzyme increased with the amount of starch added and leveled off after 40% (w/w). The amount of enzyme immobilized was independent of the amount of starch added. The composite membrane prepared with NIPAAm/NAS = 9 (mol/mol), NIPAAm/starch = 1.5 (w/w), and NIPAAm/HEMA = 9 (w/w) was the best formulation.

The optimum reaction temperature of the immobilized enzyme was 60 °C, which was 5 °C higher than that of the free enzyme. The immobilized enzyme was more stable at high temperature (>60 °C) than the free enzyme. The K_m values of immobilized enzyme are 0.92, 4.52, and 12.03% (w/w) at 20, 30, and 40 °C, respectively, compared with 0.67, 2.54, and 5.70% (w/w) for free enzyme. The large increase in K_m indicated that the membrane-immobilized enzyme had somewhat lower affinity for the substrate than the native enzyme at temperature close to or above LCST, where the hydrogel started to shrink and hindered mass transfer of the substrate.

From the same consideration as for thermo-sensitive hydrogel beads, it would be best to take advantage of the temperature-sensitivity of the composite membrane and operate a membrane reactor with temperature cycling, as shown in Figure 11.8, for enhancing the hydrolysis rate of starch (Chen *et al.*, 1998). Temperature in the reactor should be changing periodically between two values, one below the LCST and the other above it. For this purpose, temperature cycling between 20 and 40 °C and between 20 and 50 °C with a cycle half-time of 30 min was employed. As shown in Figure 11.9, the conversion increased with a higher high-end temperature. Also, the temperature cycle should be started with the high-end temperature (50 °C) first and then switched to the low-end temperature (20 °C). The effect of the length of the cycle half-time is also shown in Figure 11.9. The conversion decreased with increasing cycle half-time. By using the same membrane in the membrane reactor for 8 successive starch hydrolysis

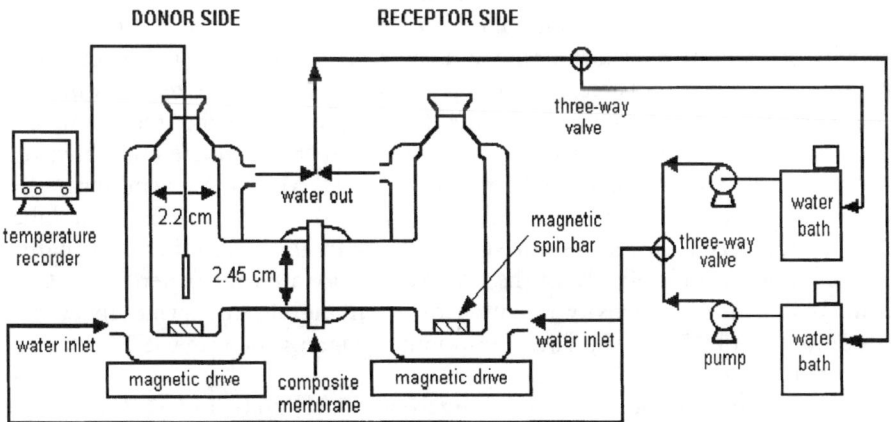

Figure 11.8 Hydrolysis of starch in a membrane reactor with α-amylase-immobilized composite hydrogel membranes under temperature cycling conditions. Instantaneous temperature change of the reactor to either of the two water baths was accomplished by switching the three-way valve back and forth at a constant time interval. The temperature was maintained in both baths for the same period of time, which was defined as the cycle half-time. Redrawn from Chen *et al.* (1998).

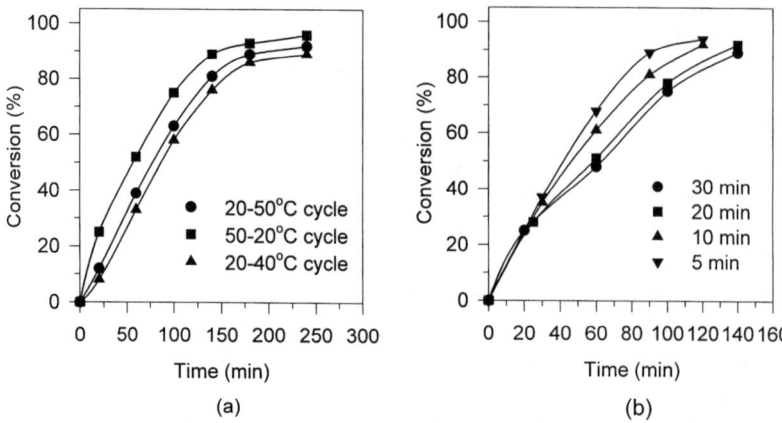

Figure 11.9 Effects of cycling temperature and frequency on the hydrolysis of 5% (w/w) soluble starch in the membrane reactor with temperature cycling. (a) Temperature was cycled with a step change between the indicated first and the second temperature with a cycle half-time of 30 min. (b) The temperature was cycled with a step change between 50 and 20 °C with the indicated cycle half-time.

reactions, the conversion changed little during the first 5 runs and started to decrease gradually after that.

11.5.1.2. Immobilization of Urease

Recently, we also immobilized urease on composite temperature-sensitive hydrogel membrane for hydrolysis of urea (Chen and Chiu, 2000). As a comparison of previous studies of membrane-immobilized urease, the amount of protein and enzyme activity per unit area of membrane and the resultant relative specific enzyme activity are summarized in Table 11.2 (Chen and Chiu, 2000). The temperature-sensitive composite hydrogel membrane gives good loading of urease that retains high specific activity after immobilization when compared to other membrane-immobilized urease systems. For urease immobilized to P(NIPAAm-HEMA-NAS) membrane, the optimum reaction temperature increased from 60 to 70 °C after enzyme immobilization. This immobilized urease also showed enhanced stability against thermal denaturation. Free enzyme retains only 5% of the original activity after incubated at 70 °C for 5 h, compared with 67% activity retention of the immobilized enzyme under the same condition (in pH 7 buffer). The pH optimum remains the same at 7.5 after enzyme immobilization.

The K_m values of the immobilized urease are 7.81, 20.0, and 52.3 mM at 20, 30, and 40 °C, respectively, compared with 2.84, 7.92 and 17.4 mM for free enzyme. The activation energies calculated from the Arrhenius equation using the temperature dependence of k_{cat} values are 6.7 and 3.4 kcal/mol for free and immobilized enzyme, respectively. Large increases of K_m values have been observed frequently for urease immobilized to various matrices. For membrane-immobilized urease K_m increase is also commonly observed, where K_m values can increase up to 8 folds after enzyme immobilization, as shown in Table 11.2. Structural change of enzyme after

Table 11.2 Comparison of immobilization efficiency and kinetic parameter for immobilization of urease to different membranes[a]

Membrane type[b]	Loading of protein $(mg\,cm^{-2})$	Activities of enzyme $(U\,cm^{-2})$	Relative specific activity $(\%)^c$	$K_{m_{imm}}/K^d_{m_{free}}$
P(NIPAAm-HEMA-NAS)	0.102	5.71	55	2.5
PS	NA[e]	1.40	NA	4.4
Copolymer of HEMA and VP	NA	5.62, 4.67, 3.28	NA	NA
PS	0.323	0.071	0.14	8.82
PA	1.015	5.55	3.38	8.40
EVA copolymer	0.63×10^{-4}	6.8×10^{-4}	1.68	0.40
AN	0.016	0.73	40.7	NA
AN modified with AMPSA	0.021	0.99	41.9	NA
AN modified with DMAEM	0.018	1.31	65.1	NA
Gelatin	NA	$3.0 \sim 6.0$	NA	NA
Chitosan	0.049	1.56	94	5.28

a Data from Chen and Chiu, 2000.
b Abbreviations: PS, polysulfone; VP, N-vinyl pyrrolidone; PA, polyamide; EVA, ethylene vinyl alcohol; AN, acrylonitrile; AMPSA, diacrylamido-2-methylpropanesulfonic acid; DMAEM, 2-dimethylaminoethyl methacrylate.
c Specific activity of immobilized enzyme/specific activity of free enzyme.
d $K_{m_{imm}}$: kinetic constant of immobilized enzyme; $K_{m_{free}}$: kinetic constant of free enzyme.
e NA = not available.

immobilization, which will influence the affinity between substrate and enzyme, will most probably account for such behavior.

Hydrolysis of urea was carried out isothermally and with temperature cycling in the same reactor shown in Figure 11.8. Temperature cycling could substantially enhance urea hydrolysis as shown in Figure 11.10 for temperature cycling between 60 to 20 °C and 50 to 20 °C and cycle half-time of 10 min, with the larger temperature range giving better performance. The ammonia concentration in the receptor side after 3 h with 60 to 20 °C temperature cycling is 3.8, 3.0, and 2.1 times that with isothermal operations at 20, 40, and 60 °C, respectively. Increasing mass transfer rates of substrate and products by temperature swing will be the most probable cause for the observed activity increase. Alleviation of product inhibition will be important in view of the large increase of reaction rate by temperature cycling. Ammonium ion is a product from urease reaction and it was reported to be a noncompetitive inhibitor of the reaction (Moynihan *et al.*, 1989). The frequency of temperature cycling also influences urea hydrolysis. Decreasing cycle half-time from 20 to 10 min resulted in 1.7 folds increase in ammonia concentration after 3 h. However, this trend was not observed when cycle half-time was further decreased to 5 min. Therefore, a high frequency of temperature change will not necessarily give better performance of the membrane reactor. This can be related to the time lag required for solutions in the reactors responding to temperature change. The same composite membrane has been used repeatedly in the membrane reactor by washing the membrane with buffer between runs to test for its reusability. There was no significant change of ammonia concentration profiles between the first and the seventh runs, indicating the stability of the immobilized enzyme.

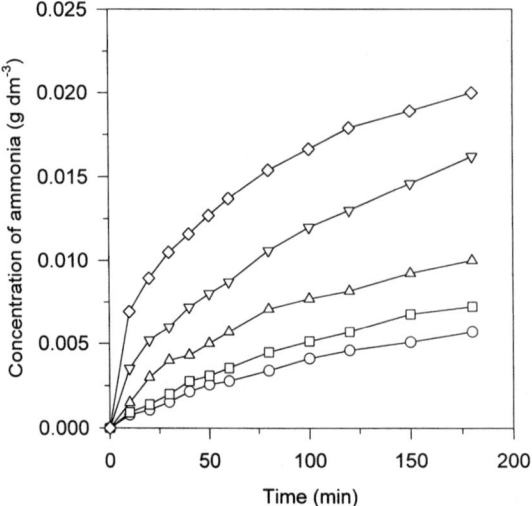

Figure 11.10 Hydrolysis of 1 mg cm^{-3} urea in the membrane reactor containing an anion-exchange membrane and a urease-immobilized composite hydrogel membrane under isothermal and temperature cycling (cycle half-time = 10 min) conditions. Isothermal: O, 20 °C; □, 40 °C; △, 60 °C. Temperature cycling: ▽, 50–20 °C; ◇, 60–20 °C.

11.6. Enzyme Immobilized onto Temperature-sensitive Hydrogel Microsphere

11.6.1. *Regulation of Enzyme Activity*

Mono-dispersed thermo-sensitive hydrogel microspheres were prepared by precipitation polymerization in water and used for immobilization of trypsin (Shiroya *et al.*, 1995). NIPAAm, AAm, and MBAAm were dissolved in water and polymerized at 70 °C under nitrogen with KPS as the initiator. The hydrogel microspheres have a large specific surface area on which to immobilize a large amount of enzyme and is expected to have a fast response to external stimuli. The microsphere has a diameter of 0.5 μm in the dried state. The hydrodynamic size was temperature-dependent and the diameter shrank from about 750 to 550 nm when crossing the LCST (around 32 °C). Carboxyl groups were introduced to the microsphere by hydrolysis with 1 N NaOH for 2 weeks, where 6.3% of AAm was hydrolyzed. The hydrodynamic size of hydrogel microsphere increased after the hydrolysis (to 900 and 700 nm below and above the LCST, respectively) due to presence of charged groups on the surface. Immobilization of trypsin was carried out via peptide bond formation using l-ethyl-3-(3-dimethylaminopropyl)-carbodiimide and *N*-hydroxysuccinimide as activating reagents. Enzyme molecules were thought to be on the surface of the microsphere as hydrolysis reaction occurring more frequently near the surfaces than in the inner part of the microsphere. Immobilization of trypsin shifts the LCSTs to higher temperatures with increasing amounts of immobilized trypsin. Enzymatic activity of the immobilized enzyme decreased above the transition temperature. This was attributed to suppression of substrate diffusion and confinement of enzyme with the surface layer of the microsphere after shrinkage of the microsphere. In an attempt to overcome these effects, enzyme can be immobilized via a hydrophilic spacer. By the presence of a spacer between enzyme molecule and hydrogel surface, the immobilized enzyme show activity independent of temperature, although these conjugates still show phase transition around the LCST. This result suggests that even microspheres were in a shrunken state above the LCST, immobilized enzyme molecules were not confined close to the surface of the microsphere. Therefore, contact with the substrates would not be hindered in this case and the activity of trypsin immobilized via a spacer was temperature-independent, a distinct feature different from the preparation without a spacer.

11.6.2. Recovery of Enzyme by Thermo-flocculation

A different kind of hydrogel microsphere, nonporous latex particles have been synthesized and used for enzyme immobilization (Kondo *et al.* 1994; Kondo and Teshima, 1995). Applications of non-porous polymer latex particles to enzyme immobilization have been investigated actively, because they have large and clean surfaces with various functional groups (Kamei *et al.*, 1987; Hayashi and Ikada, 1990). However, smaller particles are needed to achieve a sufficiently large surface area for enzyme immobilization. Nonetheless, particle separation becomes more difficult in this case. Thermal-sensitive latex particles composed of styrene and N-alkylacrylamides such as NIPAAm were synthesized by emulsifier-free emulsion polymerization (Hoshino *et al.*, 1987a and b). These submicrometer-size thermo-sensitive latex particles were expected to be effective for overcoming the above problem of particle separation, because they can be flocculated and separated easily by raising temperature (thermoflocculation) and/or ionic strength. Figure 11.11 shows the schematic illustration of an enzyme reaction system based on the thermo-sensitive latex particles. After completion of the enzymatic reaction, the immobilized enzyme is separated by thermoflocculation, and the product is recovered. Then, the separated immobilized enzyme is resuspended in cooled buffer solution and reused in the next cycle after readjusting temperature.

The thermo-sensitive latex particles composed of poly(styrene/N-isopropylacrylamide/glycidyl methacrylate) [P(St/NIPAAm/GMA)] and poly(styrene/N-isopropylacrylamide/methacrylic acid) [P(St/NIPAAm/MAA)] were prepared by emulsifier-free emulsion polymerization (Sun *et al.*, 1999b). The latex particles have epoxy and carboxyl groups, respectively, for direct covalent immobilization of enzyme after activation by carbodiimide. Polymerization was carried out for 6 h at 70 °C under a nitrogen atmosphere in the system containing the monomers and water. APS was used as the initiator. The scanning electron microscope pictures show that the latex particles were monodispersed and spherical in shape (Figure 11.12). The hydrodynamic diameters of P(St/NIPAAm/GMA) and P(St/NIPAAm/MAA) latex particles were 0.35 and 0.40 μm, respectively, as estimated by light scattering measurements.

Effect of temperature on the colloidal stability of the suspensions of the latex particles was studied by suspending the particles in pH 7 buffer solutions with various NaCl concentrations and incubated for 2 h at the desired temperature. The minimum NaCl concentration for flocculation of the latex particles was defined as the critical flocculation concentration (CFC). The temperature dependence of CFCs of the latex particles were measured and reported in Figure 11.13. The CFC decreased significantly with increasing temperature and reached a constant value. These latex particles become unstable and are susceptible to flocculation at high temperature. This reduction in

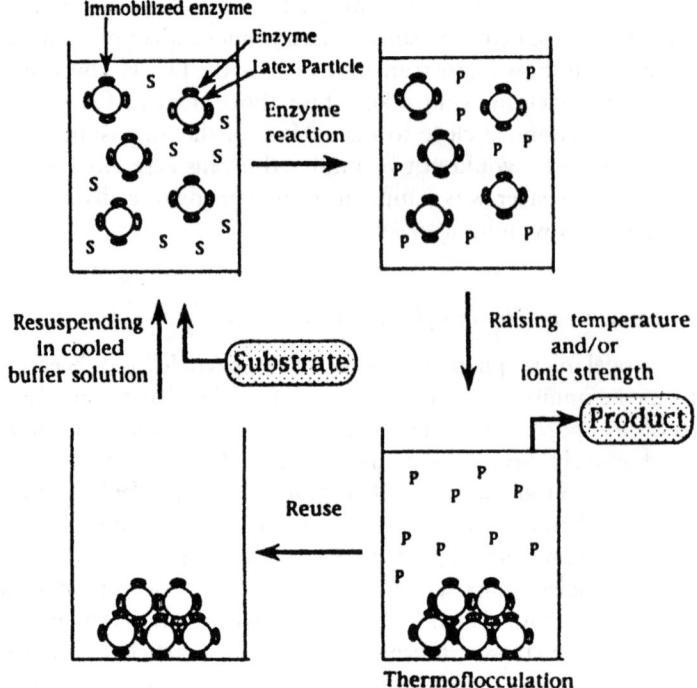

Figure 11.11 A schematic illustration of the reaction system using enzyme-immobilized thermo-sensitive latex particles for repeated batch reactions. S, substrate; P, product.

colloidal stability of the latex particles at high temperature is attributed to the dehydration of the NIPAAm layers on their surfaces. The difference in CFC values between the latex particles indicates that they are affected by surface properties. At a high ionic strength, all the latex particles started flocculating when the temperature was raised, and, conversely, the flocculated latex particles were dispersed completely by lowering temperature. These results indicate that the reversible transitions of dispersion and flocculation of these latices can be controlled by temperature and/or ionic strength.

Papain was immobilized on the latex particles of poly(methyl methacrylate/N-isopropylacrylamide/methacrylic acid) [P(MMA/NIPAAm/MAA)] and poly(styrene/N-isopropylacrylamide/methacrylic acid) P(St/NIPAAm/ MAA) (Kondo *et al.*, 1994). The hydrodynamic diameters of the P(MMA/NIPAAm/ MAA) and P(St/NIPAAm/MAA) latex particles were 0.53 and 0.46 µm, respectively. The temperature dependence of the latex size was not observed. These latex particles have negative charges, which arose from the initiator fragments and MAA. The surface of the P(MMA/NIPAAm/MAA) latex particles is expected to be more hydrophobic than that of the P(St/NIPAAm/MAA) latex particles because of higher hydrophobicity of MAA than that of St.

The relative activities (relative to that free papain) of papain immobilized on P(MMA/NIPAAm/MAA) latex particles can reach 90%, which is higher than those of papain immobilized on P(St/NIPAAm/MAA) latex particles (45%). The result

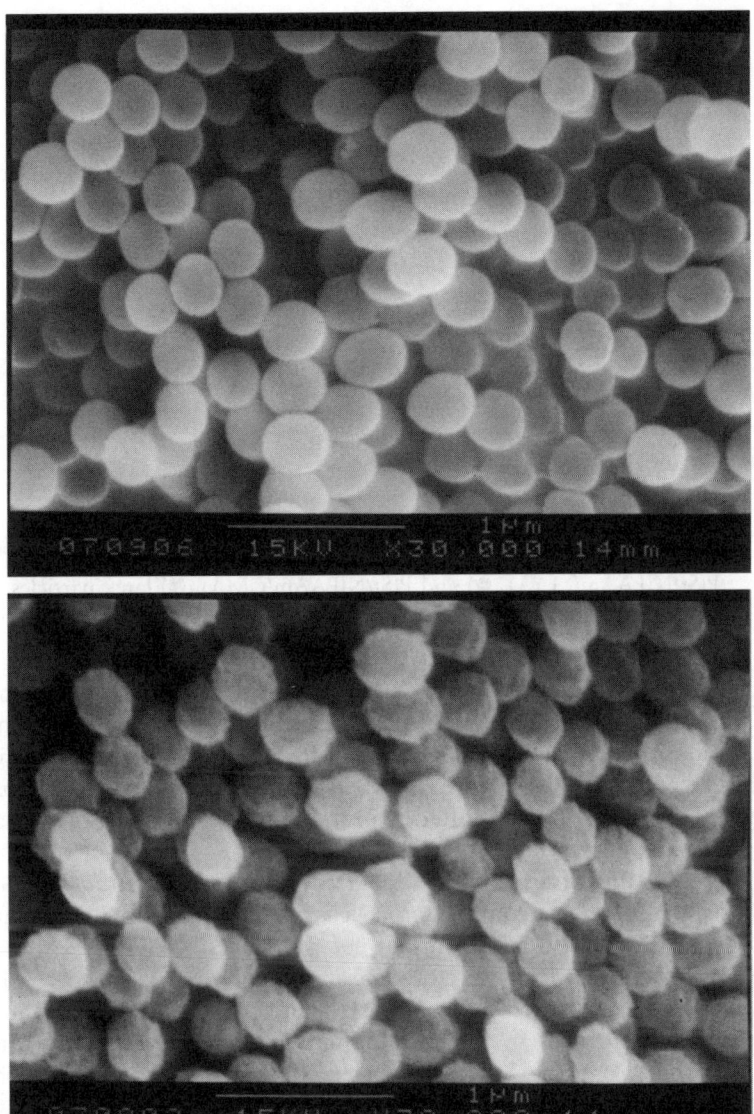

Figure 11.12 Scanning electron microscope pictures of P(St/NIPAAm/MAA) (top) and P(St/NIPAAm/GMA) (bottom) latex particles. The bar represents 1 μm.

suggests larger structural changes in papain molecules upon immobilization on the P(St/NIPAAm/MAA) latex particles and is consistent with the previous observation that the extent of structural changes in the protein molecules increases with increasing hydrophobicity of the adsorbent surfaces (Norde, 1986). The relative activity of papain immobilized on the P(MMA/NIPAAm/MAA) latex particles toward casein showed the maximum at around the immobilization amount of 1–1.5 mg/m^2, and was much higher than those reported previously. Since casein is a high molecular weight substrate, the high relative specific activity will result from the fact that most

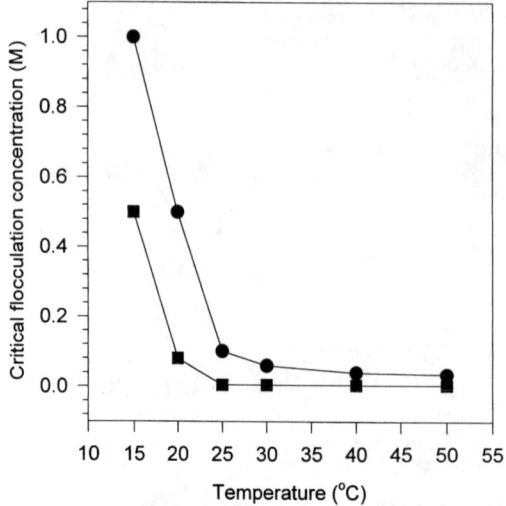

Figure 11.13 Effect of temperature on the critical flocculation concentrations (CFC) of P(St/NIPAAm/MAA) (●) and P(St/NIPAAm/GMA) (■) latex particles. The CFC is the minimum NaCl concentration for flocculation of the latex particles.

enzyme molecules were immobilized on the surfaces of the latex particles and mass transfer resistance of substrate can be reduced. The effect of thermal cycle, namely, thermoflocculation and subsequent dispersion lead to little loss of enzyme activity. Papain immobilized on the P(MMA/NIPAAm/MAA) latex particles retained its activity after 20 cycles, indicating that repeated hydration and dehydration of the latex surfaces does not reduce enzyme activity.

Affinity tag AG consisting of IgG binding domains of protein from *Staphylococcus aureus* and those of protein G from *Streptococcus* were used to facilitate immobilization of β-galactosidase (βgal) from *E. coli*. P(MMA/NIPAAm/MAA) and P(St/NIPAAm/MAA) latex particles were used as supports to prepare affinity adsorbents. Human γ-globulin was used as an affinity ligand and was covalently immobilized onto both latex particles by the carbodiimide method. The fusion protein, AGβgal, was immobilized at pH 7.3 by the specific binding of affinity tag to these affinity adsorbents. The efficiency of ligand utilization reached maximum in the affinity adsorbents prepared at pH 6.0–7.0, and that in the P(MMA/NIPAAm/MAA) latex particles was higher. Immobilized enzyme retained approximately 75% of its activity in solution and the binding is stable enough to allow repeated use (Kondo and Teshima, 1995).

11.6.3. Recovery of Enzyme by Thermo-flocculation Followed by Magnetic Separation

Thermo-sensitive magnetic hydrogel microsphere can be prepared based on NIPAAm, MBAAm, and MAA containing ultrafine magnetite particles (Kondo and Fukuda, 1997). A relative large amount of crosslinking reagent MBAAm was used to increase the mechanical strength of the hydrogel microspheres. The thermo-flocculated microspheres can be separated quickly from solutions in a relatively low magnetic field. A suspension of ultrafine magnetite particles coated with oleic acid and sodium dodecylbenzenesulfonate was formed by the conventional co-precipitation method (Shinkai *et al.*, 1991). The average hydrodynamic diameter of the ultrafine magnetite was approximately 40 nm.

Magnetite (0.2 g), NIPAAm and MAA (total 1.95 g), and MBAAm (0.17 g) were added to 100 ml water with stirring (180 rpm). APS (0.1 g) was used as an initiator, and polymerization was carried out for 6 h under nitrogen atmosphere. The resulting magnetic hydrogel microspheres were purified by dialysis. The magnetite content was around 8–10% and the average particle diameter of the microsphere measured by dynamic light scattering was from 148 to 252 nm, decreasing with increasing MAA content. The thermo-flocculated hydrogel microsphere were separated very quickly, although the dispersed microspheres were not magnetically separated. Therefore, thermal-flocculation is effective to recover the magnetic hydrogel microspheres. Enzymes were immobilized onto these thermo-sensitive magnetic hydrogel microspheres by two different methods. Trypsin was covalently immobilized onto the magnetic hydrogel microspheres in high yield by the carbodiimide method. In the case of microspheres with higher MAA content, immobilized trypsin demonstrated a high relative specific activity, close to 80% with casein as the substrate. On the other hand, the fusion enzyme AGβgal was immobilized onto the magnetic hydrogel microspheres with covalently immobilized human γ-globulin by affinity interaction. The immobilized AGβgal retained high activity even in the case of microspheres with a low MAA content.

We have synthesized non-magnetic and magnetic poly(styrene/N-isopropylacrylamide/N-acryloxysuccinimide) [P(St/NIPAAm/NAS)] temperature-sensitive latex particles for immobilization of α-chymotrypsin (Chen and Su, 1998). Emulsion polymerization was carried out by reacting NIPAAm, styrene, and NAS in water under nitrogen atmosphere at 70 °C with APS as the initiator. Ultra-fine magnetite particles were first prepared by co-precipitation method. Magnetic latex particles was synthesized by two-step emulsifier-free emulsion polymerization. St and NAS were first polymerized in water in the presence of magnetite for 6 h and then NIPAAm, additional NAS and styrene were added and polymerized for another 24 h. Latex particles were purified by centrifugation and dialysis.

The diameter distribution of non-magnetic and magnetic particles was measured and the average diameters were 0.34 and 0.25 μm, respectively, with sharp distribution curve. The CFCs decreased very sharply around the lower critical solution temperature of NIPAAm (33 °C), and they are higher for magnetic particles than for non-magnetic particles, which may be caused by differences in polymerization conditions. Transmission electron microscope pictures are shown in Figure 11.14, where magnetite can be seen to be fully enclosed within the latex particles, as shown by the dark regions. Latex particles formed stable emulsion in a buffer, but flocculated and settled by gravity after adding salt and raising temperature. As can be seen in Figure 11.15, under the same conditions (30 °C, 2 M NaCl), clear supernatant solutions can be obtained in less than 30 min by sedimentation with gravity for non-magnetic latex particles and the magnetic particles can be separated almost 6 times faster than the non-magnetic particles in a low magnetic field.

Non-magnetic particles can give specific activity even higher than that of free enzyme while magnetic particles retained 30% of specific activity. The optimum

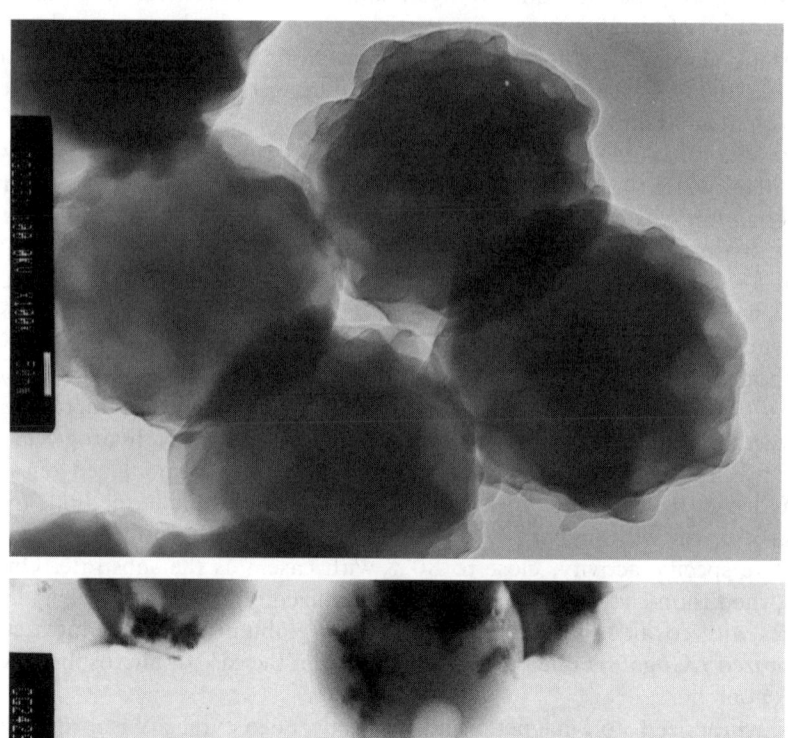

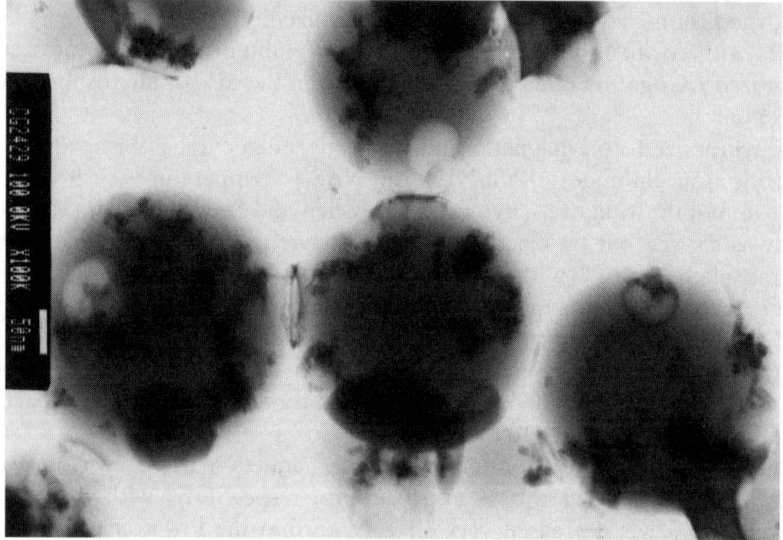

Figure 11.14 Transmission electron microscope pictures of nonmagnetic (top) and magnetic (bottom) P(St/NIPAAm/NAS) latex particles. The bar represents 50 nm.

reaction temperatures for highest enzyme activity are 40, 45, and 50 °C for enzyme immobilized to non-magnetic particles, free enzyme and enzyme immobilized to magnetic particles, respectively. The optimum reaction pH for highest enzyme activity occurs at 8.5 for free enzyme and all immobilized enzymes. Kinetic parameters determined from Michaelis-Menten equations are shown in Table 11.3. Similar K_m values were found for free and immobilized enzymes regardless of the size of the substrates. These results indicate that the immobilized enzyme is not limited by diffusion even for substrate as large as casein

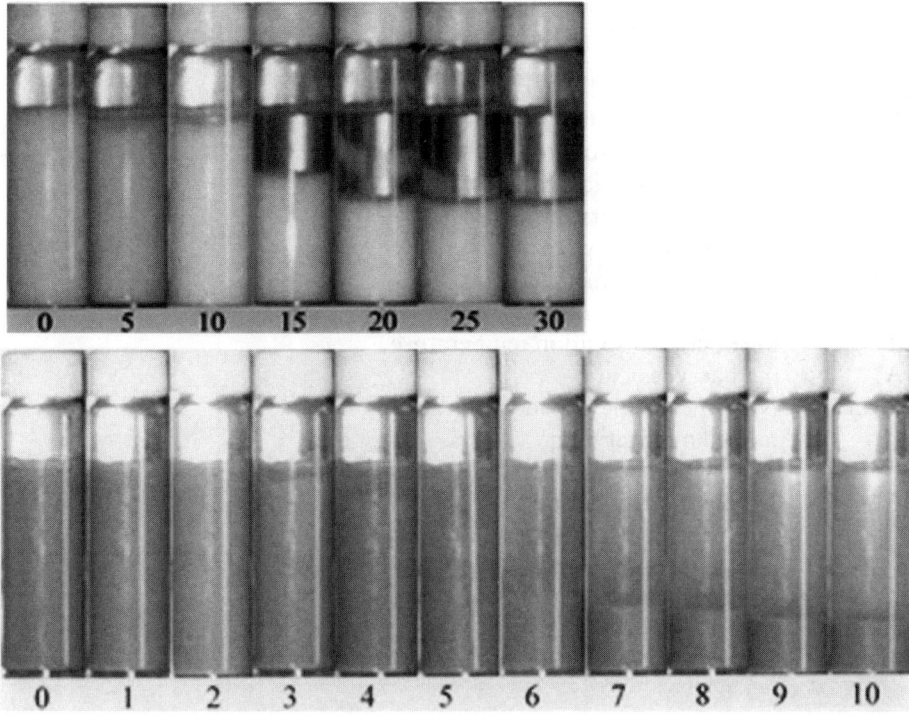

Figure 11.15 Progress of thermo-flocculation of non-magnetic (top) and magnetic (bottom) P(St/NIPAAm/NAS) latex particles. The latex particles were prepared in 2 M NaCl solution and incubated at 30 °C. A magnet (4500 G) was placed at the bottom of each vial. The numbers indicate the time elapsed in minutes since preparing the latex solutions.

Table 11.3 Kinetic parameters of free α-chymotrypsin and α-chymotrypsin immobilized on non-magnetic and magnetic latex particles

Enzyme system	Substrates					
	ATEE		Hemoglobin		Casein	
	$k_{cat}(\text{min}^{-1})$	K_m (mM)	$k_{cat}(\text{min}^{-1})$	K_m (%, w/v)	$k_{cat}(\text{min}^{-1})$	K_m(%, w/v)
Free	1.01×10^4	2.44	5.23×10^2	0.18	3.13×10^3	2.00
Immobilized non-magnetic	1.05×10^4	2.67	5.96×10^2	0.21	3.33×10^3	1.70
Immobilized magnetic	0.33×10^4	2.52	1.61×10^2	0.19	0.954×10^3	1.89

(micelle in the size of 50–300 nm), which may be due to the non-porous nature of the particles. Reusability of the immobilized enzymes was good, over 70% of enzyme activity were found after 8 flocculation-dispersion cycles of the immobilized enzyme.

11.7. Abbreviations

βgal	β-galactosidase
AA	acrylic acid
AAm	acrylamide
AIBN	2,2'-azobis(isobutyronitrile)
APS	ammonium persulfate
CFC	critical flocculation concentration
DMF	*N,N*-dimethylformamide
GMA	glycidyl methacrylate
HEMA	2-hydroxyethylmethacrylate
KPS	potassium persulfate
LCST	lower critical solution temperature
MAA	methacrylic acid
MBAAm	*N,N'*-methylene-bis-acrylamide
MMA	methyl methacrylate
NAS	*N*-acryloxysuccinimide
NIPAAm	*N*-isopropylacrylamide
ONPG	*o*-nitrophenyl-β-D-galactopyranoside
PNIPAAm	poly-*N*-(isopropylacrylamide)
PEG	polyethylene glycol
St	styrene
TEMED	*N,N,N',N'*-tetramethylenediamine

11.8. References

Bailey, J.E. and Ollis, D.F. (1986) *Biochemical Engineering Fundamentals*, 2nd ed., McGraw-Hill, New York, pp. 210–212.

Bayhan, M. and Tuncel, A. (1998) Uniform poly(isopropylacrylamide) gel beads for immobilization of α-chymotrypsin, *J. Appl. Polym. Sci.*, **67**, 1127–1139.

Bickerstaff, G.F. (1997) *Immobilization of Enzymes and Cells*, Humana Press, Totowa.

Chen, J.P. and Su, D.R. (1998) Preparations and characterizations of non-magnetic and magnetic temperature-sensitive latex particles for enzyme immobilization, *Proc. 4th Young Asian Biochemical Engineers' Community Symposium*, Hualien, November 1999.

Chen, J.P., Sun, Y.M., and Chu, D.S. (1998) Immobilization of α-amylase to a composite temperature-sensitive membrane for starch hydrolysis, *Biotech. Prog.*, **14**, 473–478.

Chen, J.P. and Chiu, S.H. (2000) A poly(N-isopropylacrylamide-co-N-acryloxysuccinimide-co-2-hydroxyethylmethacrylate) composite hydrogel membrane for urease immobilization to enhance urea hydrolysis rate by temperature swing, *Enzyme Microb. Technol.*, **26**, 359–367.

Çiçek, H. and Tuncel, A. (1998a) Preparation and characterization of the thermoresponsive isopropylacrylamide-hydroxyethylmethacrylate copolymer gels, *J. Polym. Sci., Part A: Polym. Chem.*, **36**, 527–541.

Çiçek, H. and Tuncel, A. (1998b) Immobilization of α-chymotrypsin in thermally reversible isopropylacrylamide-hydroxyethylmethacrylate copolymer gels, *J. Polym. Sci., Part A: Polym. Chem.*, **36**, 543–552.

Dong, L.C. and Hoffman, A.S. (1986) Thermally reversible hydrogels: III. Immobilization of enzymes for feedback reaction control, *J. Contr. Rel.*, **4**, 223–227.

Dong, L.C. and Hoffman, A.S. (1987) Thermally reversible hydrogels: IV. Swelling and deswelling characteristics and activities of copoly(NIPAAm-AAm) gels containing immobilized enzymes. In Russo, P. (ed.) *ACS Symposium Series No. 350, Reversible Polymeric Gels and Related Systems*, ACS, Washington, D.C., pp. 236–244.

Dusek, K. (1993) *Advances in Polymer Sciences, Responsive Gels: Volume Trasition II*. Springer Verlag, Berlin.

Galaev, I.Yu. and Mattiasson, B. (1999) "Smart" polymers and what they could do in biotechnology and medicine, *Trends Biotechnol.*, **17**, 335–340.

Hayashi, T. and Idada, Y. (1990) Protease immobilization onto polyacrolein microspheres, *Biotechnol. Bioeng.*, **35**, 518–524.

Hino, T. and Prausnitz, J.M. (1998) Molecular thermodynamics for volume-change transitions in temperature-sensitive polymer gels, *Polymer*, **39**, 3279–3283.

Hirotsu, S., Hirokawa, Y., and Tanaka, T.J. (1987) Volume phase transitions of ionized N-isopropylacrylamide, *J. Chem. Phys.*, **87**, 1392–1395.

Hoffman, A.S. (1995) "Intelligent" polymers in medicine and biotechnology, *Artif. Organs*, **19**, 458–467.

Hoshino, F., Fujimoto, T., Kawaguchi, H., and Ohtsuka, Y. (1987a) N-substituted acrylamide-styrene copolymer latices. II. Polymerization behavior and thermosensitive stability of latices, *Polymer J.* **19**, 241–247.

Hoshino, F., Kawaguchi, H., and Ohtsuka, Y. (1987b) N-substituted acrylamide-styrene copolymer latices. III. Morphology of latex particles, *Polymer J.*, **19**, 1157–1164.

Ilmain, F., Tanaka, T., and Kokufuta, E. (1991) Volume transition in a gel driven by hydrogen bonding, *Nature*, **349**, 400–410.

Inomata, H. and Saito, S. (1993) Studies on volume phase transition of N-substituted acrylamide hydrogels, *Fluid Phase Equilibr.*, **82**, 291–302.

Kamie, S., Okubo, M., and Matsumoto, T. (1987) Adsorption of trypsin onto poly(2-hydroxethyl methacrylate)/polystrene composite microspheres and its enzymatic activity, *J. App. Polym. Sci.*, **34**, 1439–1446.

Kennedy, J.F. and Cabral, J.M.S. (1987) Enzyme immobilization. In Kennedy, J.F. (ed.), *Biotechnology*, vol. 7a *Enzyme Technology*, VCH, Weinheim, pp. 347–404.

Kokufuta, E. and Tanaka, T. (1991) Biochemically controlled thermal phase transition of gels, *Macromolecules*, **24**, 1605–1607.

Kokufuta, E., Ogane, O., Ichijo, I, Watanabe, S., and Hirasa, O. (1992) Poly(vinyl methyl ether) gel for the construction of a thermosensitive immobilised enzyme system exhibiting controllable reaction initiation and termination, *J. Chem. Soc., Chem. Commun.*, **30**, 416–418.

Kokufuta, E., Zhang, Y.Q., and Tanaka T. (1994) Biochemo-mechanical function of urease-loaded gels, *J. Biomater. Sci., Polymer Ed.*, **6**, 35–40.

Kondo, A., Imura, K., Nakama, K., and Higashitani, K. (1994) Preparation of immobilized papain using thermosensitive latex particles, *J. Ferment. Bioeng.*, **78**, 241–245.

Kondo, A. and Teshima, T. (1995) Preparation of immobilized enzyme with high activity using affinity tag based on proteins A and G, *Biotechnol. Bioeng.*, **46**, 421–428.

Kondo, A. and Fukuda, H. (1997) Preparation of thermo-sensitive magnetic hydrogel microspheres and its application to enzyme immobilization, *J. Ferment. Bioeng.*, **84**, 337–341.

Liu, F. and Zhuo, R.X. (1993) A convenient method for the preparation of temperature-sensitive hydrogels and their use for enzyme immobilization, *Biotechnol. Appl. Biochem.*, **18**, 57–65.

Moynihan, H.J., Lee, C.K., Clark, W., and Wang, N.-H.L. (1989) Urea hydrolysis by immobilized urease in a fixed-bed reactor: analysis and kinetic parameter estimation, *Biotechnol. Bioeng.*, **34**, 951–960.

Murakata, T., Liu, X.B., and Sato, S. (1993) Esterification activity of immobilized lipase entrapped in thermal-phase transition gel, *J. Chem. Eng. Japan*, **26**, 681–685.

Norde, W. (1986) Adsorption of protein from solution at the solid-liquid interface, *Adv. Colloid Interface Sci.*, 25, 267–340.

Osada, Y. and Rose-Murphy, S.B. (1993) Intelligent gels, *Sci. Amer.*, 268, 82–87.

Park, T.G. and Hoffman, A.S. (1988) Effect of temperature cycling on the activity and productivity of immobilized β-galactosidase in a thermally reversible hydrogel bead reactor, *Appl. Biochem. Biotechnol.*, 19, 1–9.

Park, T.G. and Hoffman, A.S. (1990a) Immobilization and characterization of β-galactosidase in thermally reversible hydrogel beads, *J. Biomed. Mater. Res.*, 24, 21–38.

Park, T.G. and Hoffman, A.S. (1990b) Immobilization of *Arthrobacter simplex* in a thermally reversible hydrogel: effect of temperature cycling on steroid conversion, *Biotechnol. Bioeng.*, 35, 152–159.

Park, T.G. and Hoffman, A.S. (1991) Immobilization of *Arthrobacter simplex* in thermally reversible hydrogels: effect of gel hydrophobicity on steroid conversion, *Biotechnol. Prog.*, 7, 383–390.

Park, T.G. and Hoffman, A.S. (1992) Preparation of large, uniform size temperature-sensitive hydrogel beads, *J. Polym. Sci., Part A: Polym. Chem.*, 30, 505–507.

Park, T.G. and Hoffman, A.S. (1993) Thermal cycling effects on the bioreactor performance of immobilized β-galactosidase in temperature-sensitive hydrogel beads, *Enzyme Microb. Technol.*, 15, 476–482.

Park, T.G. and Hoffman, A.S. (1994) Estimation of temperature-dependent pore size in poly(N-isopropylacrylamide) hydrogel beads, *Biotechnol. Prog.*, 10, 82–86.

Rathbone, S.J., Haynes, C.A., Blanch, H.B., and Prausnitz, J.M. (1990) Thermodynamic properties of dilute aqueous solutions from low-angle laser-light-scattering measurements, *Macromolecules*, 23, 3944–3947

Shinkai, M., Honda, H., and Kobayashi, T. (1991) Preparation of fine magnetic particle and application for enzyme immobilization, *Biocatalysis*, 5, 61–69.

Shiroya, T., Tamura, N., Yasui, M., Fujimoto, K., and Kawaguchi. (1995) Enzyme immobilization on thermal sensitive hydrogel microspheres, *Colloids Surfaces B: Biointerfaces*, 4, 267–274.

Sun, Y.M., Chen, J.P., and Chu, D.H. (1999a) Preparation and characterization of α-amylase-immobilized thermal-responsive composite hydrogel membrane, *J. Biomed. Mater. Res.*, 45, 125–132.

Sun, Y.M., Yu, C.W., Liang, H.C., and Chen, J.P. (1999b) Temperature-sensitive latex particles for immobilization of α-amylase, *J. Disper. Sci Technol.*, 20, 907–920.

Suzuki, A., Yamazaki, M., and Kobini, Y. (1996) Direct observation of polymer gel surfaces by atomic force microscope, *J. Chem. Phys.*, 104, 1751–1757.

Takahashi, F., Sakai, Y., and Mizutani, Y. (1997) Immobilized enzyme reaction controlled by magnetic heating: γ-Fe$_2$O$_3$ loaded thermosensitive polymer gels consisting of N-isopropylacrylamide and acrylamide, *J. Ferment. Bioeng.*, 83, 152–156.

Tanaka, T. (1978) Collapse of gels and the critical endpoint, *Phys. Rev. Lett.*, 40, 820–823.

Taylor, R.F. (1990) *Protein Immobilization*, Marcel Dekker, New York.

Tuncel, A. (1998) An engineering analysis for the continuous reactor behavior of α-chymotrypsin-immobilized thermosensitive gel cylinders, *J. Biotechnol.*, 63, 41–54.

Wang, Y., Zong, X., and Wang, S. (1996) Pressure cycling to enhance an immobilized enzyme reaction by enzyme entrapment in a pressure-sensitive gel, *J. Chem. Technol. Biotechnol.*, 67, 243–247.

Wu, X.H., Hoffman, A.S., and Yager, P. (1992) Synthesis and characterization of thermally reversible macroporous poly(N-isopropylacrylamide) hydrogels, *J. Polym. Sci., Part A: Polym. Chem.*, 30, 2121–2126.

12 Stimuli responsive polymers in bioprocessing

Kazuhiro Hoshino and Masayuki Taniguchi

12.1. Introduction

Stimuli responsive polymers are functional polymers that change their status due to various environmental changes, such as changes in pH (Margolin *et al.*, 1985; Fujimura *et al.*, 1987; Chen and Chang, 1994), temperature (Nguyen and Luong, 1989; Chen *et al.*, 1990; Chen and Hoffman, 1990), and ionic strength (Charles *et al.*, 1974; Leemputten and Horisberger, 1976). Stimuli responsive polymers have recently received significant attention because polymers have potential applications as tools for assembling novel bioprocesses, such as enzyme reactions, chromatography, and drug delivery systems (DDS). Among bioprocesses that involve the use of these polymers, their utilization as a carrier for the immobilization of enzyme is expected to be applied immediately in several manufacturing industries. In general, immobilization of an enzyme on water-insoluble carriers, such as polysaccharide derivatives, ion-exchange resins, or ceramics, is useful for increasing the total productivity and stability of these biocatalysts. Therefore, these immobilized enzymes have been utilized to manufacture foods or pharmaceuticals (Bailey and Ollis, 1986; Sato and Tosa, 1993). However, the use of conventional immobilized enzymes has presented many difficult problems. Tosa (1987) pointed out the following problems when the application requires such immobilized enzymes in actual bioprocess:

(1) degradation of enzymatic activity in continuous reactions;
(2) removal of byproducts from reactions;
(3) design of a reactor for exothermic reactions;
(4) design of a reactor for gas–liquid–solid reactions;
(5) optimization of a reactor for conjugated reaction systems;
(6) design of a reactor for high-concentration substrates;
(7) access by substrate and/or product is limited by diffusion;
(8) decrease in apparent activity of high molecular weight substrate and solid substrate.

Several researchers proposed concrete solutions to problems 1–6, and implementation of the proposed solutions in the actual systems in which the respective problems occurred resulted in successful application of the immobilized enzyme. However, problems 7 and 8 have not been solved for the application of conventional immobilized enzyme in a reaction system containing water-insoluble substrate and/or products, i.e., in the heterogeneous reaction system. Table 12.1 lists the problems related to the utilization of the conventional immobilized enzyme in the heterogeneous reaction

Table 12.1 Problems in heterogeneous reaction using enzymes immobilized on insoluble polymers

Reaction pattern	Condition of reactants		Problems	
	Substrate	*Product*	*Reactivity*[a]	*Separation*[b]
A	S[c]	S	O	O
B	I[d]	S	X	O
C	S	I	X	X
D	S	I	X	X

a Problem on reactivity of immobilized enzymes against a substrate. O: Good, X: Bad.
b Problem on separation of between immobilized enzyme and product after reaction was finished. O: Possible, X: Impossible.
c Water-soluble material.
d Water-insoluble material.

system. As shown in Table 12.1, the enzymatic reaction using the conventional immobilized enzyme can be classified as one of four types. When the immobilized enzyme was used in the homogeneous reaction (Type A), high reactivity was achieved because the resistance of substrate diffusion within the carrier is small. Moreover, separation of the immobilized enzyme and products is easy because the product is in a water-soluble state. On the other hand, in the Type B–D heterogeneous reaction, the problems of enzymatic reactivity, the separation between immobilized enzyme, and unreacted solid residue and/or product occur. These problems are very serious. The substrate cannot diffuse within the carrier due to the water-insoluble nature of substrate and/or products (Matsuno, 1985). Furthermore, in heterogeneous reaction systems that produce a water-insoluble product, the apparent activity of the enzyme is gradually reduced because the pores in the carrier are clogged by the water-insoluble product that is formed (Epton *et al.*, 1977). Repeated operation of packed or fluidized bed reactors is not possible due to clogging caused by water-insoluble substrate and/or a water-insoluble product (Chen and Hoffman, 1993). Another problem is the separation and recovery of the immobilized enzyme. It is very difficult to separate an enzyme that is immobilized on water-insoluble carrier from a reaction mixture that contains unreacted water-insoluble substrate and/or a water-insoluble product (Rao *et al.*, 1983).

In order to develop novel immobilized enzymes for use in the heterogeneous reaction system (Type B–D in Table 12.1) and to develop applications to novel bioprocesses, we have investigated the application of stimuli responsive polymers (primarily, pH-responsive polymers and thermo-responsive polymers) as immobilization carriers since 1986. In this chapter, we introduce our proposed bioprocesses using several pH-responsive polymers and thermo-responsive polymers.

12.2. Preparation of Reversibly Soluble–insoluble Enzymes

Recently, stimuli responsive polymers which are able to change their solubility depending on the surrounding conditions, such as pH, temperature, or ionic strength, were developed as carriers for immobilization of enzymes or a tool for purification of bioproducts. Although there are many other applications of stimuli responsive polymer, the utilization as a carrier for immobilization of enzyme is given mainly an outline as a representative example (Figure 12.1). Soluble immobilized enzyme, i.e., enzyme immobilized on such a stimuli responsive polymer, has higher reactivity than enzyme

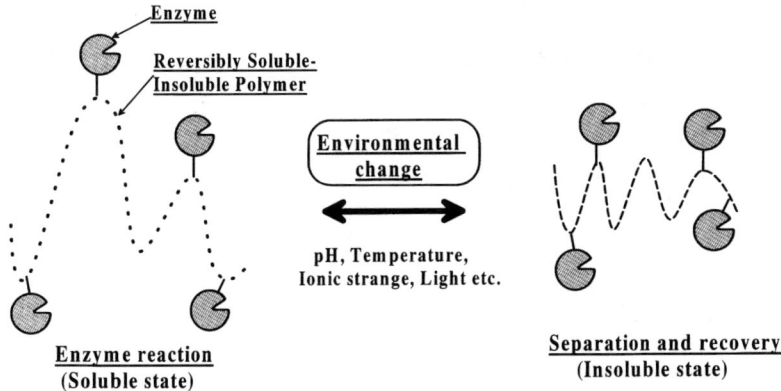

Figure 12.1 Reversibly soluble–insoluble enzymes.

immobilized on water-insoluble carrier because the substrate can contact smoothly the active site due to the extremely slight steric hindrance between substrate and carrier. Moreover, if such an immobilized enzyme is insolubilized by the surrounding stimulation after the reaction, it is possible to separate the immobilized enzyme and water-insoluble product; however, separation is not possible when using conventional immobilized enzymes. The application of a stimuli responsive polymer is considered to be a promising method for overcoming the problems associated with reactivity and separation in the heterogeneous reaction. The enzyme immobilized on such a functional polymer referred to herein as reversibly soluble–insoluble (S–IS) enzyme. As an example applied to such an immobilized enzyme, Fujimura *et al.* reported in 1987 that for repeated enzymatic synthesis of insoluble dipeptide in a reaction medium containing a water-miscible organic solvent, reversibly S–IS chymotrypsin was prepared by immobilization on an entric coating polymer (ECP) as a pH-responsive polymer. Taking their report into consideration, we attempted to verify the usability of reversibly S-IS enzyme in the heterogeneous reaction by using the hydrolysis reaction of raw starch as a model reaction containing water-insoluble substrate. In addition, we attempted to achieve a novel bioprocess using reversibly S–IS enzyme. In order to find a carrier which can be used as a reversibly S–IS polymer, many ECPs were searched by referring to a report by Fujimura *et al.* (1987).

12.2.1. *Selection of Reversibly Soluble–insoluble Polymer*

Figure 12.2 shows the relative turbidity at each pH of nine ECPs used as carrier. Each maximum turbidity of the suspension containing ECP was expressed as 1.0. Since the optimal pH of a raw-starch-digesting amylase, (Dabiase K-27, Daikin Kogyo Co.) is found to be around 5.0, CAP (cellulose acetate phthalate, Wako Pure Chemical Industries Co.), HP-55 (hydroxyropyl methylcellulose phthalate, Shin-Etsu Chemical Co.), Eudragit L (methacrylic acid-methylmethacrylate copolymer, Rohm Pharma Co.), and AS-L (hydroxyropyl methylcellulose acetate succinate, Shin-Etsu Chemical Co.) that was insoluble (i.e. became opaque) near pH 5.0 and soluble (i.e. clear) at pH 5.0 were

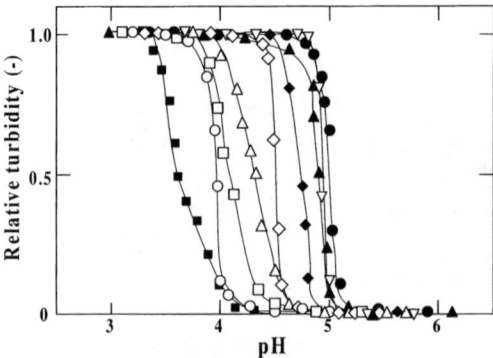

Figure 12.2 Solubility response of nine entric coating polymers to changes in pH of medium. Symbols: ○, Eudragit L; ●, Eudragit S-100; □, HP-55; ■, HP-50; △, CAP; ▲, MPM-06; ◇, AS-L; ◆, AS-M; ▽, AS-H.

selected from among the polymers that were tested. For these four polymers, the enzymatic activities per amount of protein bound to the polymers by immobilization were compared under the same conditions. The specific activity was higher when Eudragit L and AS-L were used as a carriers, the values of which correspond to greater than 85% of native Dabiase. Therefore, although two ECPs were selected as the best carriers, the results using Eudragit L are omitted in this chapter. For details, please refer to Hoshino *et al.* (1989a) and Taniguchi *et al.* (1989a).

12.2.2. *Preparation and Properties of Reversibly Soluble–insoluble Enzyme*

12.2.2.1. *Preparation of pH-responsive Amylase*

A raw-starch-digesting amylase, Dabiase, was immobilized covalently on AS-L (hydroxypropyl methylcellulose acetate succinate) as a carrier that is reversibly soluble–insoluble depending on pH as determined by the water-soluble carbodiimide method, as described in previous papers (Hoshino *et al.*, 1989b). The abbreviation for the prepared immobilized enzyme is D-AS.

12.2.2.2. *Response of Solubility*

Figure 12.3 shows the solubility of D-AS as a function of the pH of the buffer solution. The relative activity of D-AS in the supernatant obtained by centrifuging, i.e., the relative activity of D-AS existing in a soluble state at each pH, is also shown in Figure 12.3. Above pH 5.0, the mixture containing D-AS was completely soluble and became clear. The total activity was retained in the clear solution. Between pH 5.0 and 4.0, the mixture gradually became opaque as the pH of the buffer solution was reduced. Below pH 4.0, the suspension showed maximum turbidity and little activity was detected in the supernatant. The sharp response to pH change of the activity in the supernatant is the opposite of that of the turbidity. In another experiment involving repetitive cycling of pH, the relationships between the solubility and activity of D-AS shown in Figure 12.3 were followed. Consequently, it is confirmed that the solubility

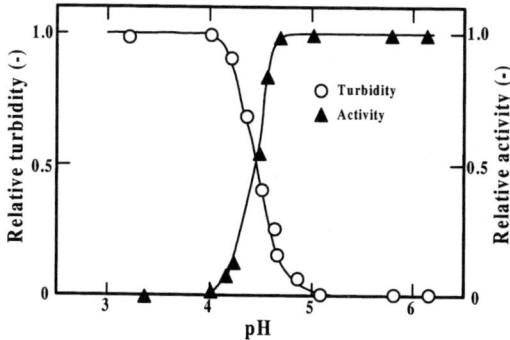

Figure 12.3 Solubility response of pH-responsive Dabiase to changes in pH of medium. Symbols: ○, turbidity of D-AS; ▲, relative activity in supernatant.

of D-AS can be adjusted reversibly by a slight change in pH without decrease of enzymatic activity, suggesting that the pH-responsive polymer is applicable as a carrier for immobilization of enzyme.

12.2.2.3. Substrate Specificity

Next, in order to clarify the effectiveness of the use of solublized D-AS (pH 5.0), the specific activities per amount of immobilized enzyme for raw starch and three kinds of soluble substrates were compared to our previous results (Hoshino *et al.*, 1989a) for the native Dabiase (ND) and Dabiase immobilized on the insoluble carrier CM-Toyopearl (Tosoh Co., D-CMT) as an enzyme immobilized on a water-insoluble carrier. As shown in Table 12.2, there was little significant difference in the specific activity for maltose between ND and the two types of immobilized Dabiase. The apparent activities of D-AS for not only dextrin and soluble starch, but also for raw starch, were greater than 70% of those of ND. However, when dextrin and soluble starch were used as substrates, the relative activities of D-CMT decreased considerably with increases in the molecular weight or the substrate compared with those of ND and D-AS. Moreover, when using raw starch, the apparent activity of D-CMT was 32% of that of ND, as reported previously, whereas the activity of D-AS was 74%. Therefore, D-AS in soluble form was confirmed to be more suitable for hydrolysis of raw starch than D-CMT in the heterogeneous reaction containing high-molecular weight substrate and solid substrate. This suggests that the unresolved problems described above can be overcome using the enzyme immobilized on pH-responsive polymer as a carrier.

12.2.2.4. Sedimentation Properties of D-AS

We discovered by chance that insolublized D-AS is self-sedimenting, i.e., has autoprecipitating properties that are dependent on pH. The degree of sedimentation of D-AS was measured with a graduated cylinder (Hoshino *et al.*, 1989b). Figure 12.4 shows the effects of pH on the sedimentation. Below pH 4.5, D-AS in suspension began to sink after agitation was stopped. The lower the pH of the suspension, the higher the

Table 12.2 Substrate specificity of Dabiase preparations

Dabiase precipitation	Specific activity (U/mg-protein)			
	Maltose	Dextrin	Soluble starch	Raw starch
Native Dabiase (ND)	2.43 (100)[a]	6.51 (100)	6.23 (100)	1.68 (100)
D-AS[b]	2.17 (89)	5.62 (86)	4.94 (79)	1.24 (74)
D-CMT[c]	1.86 (77)	3.62 (56)	2.87 (46)	0.53 (32)

The reaction was carried out at pH 5.0 and 30 °C. Each substrate concentration was 0.5%.
The results for ND and D-CMT are quoted from the paper of Hoshino *et al.* (1989a).
a　Relative activity when the activity of native Dabiase was expressed as 100%.
b　Dabiase immobilized on AS-L.
c　Dabiase immobilized on CM-Toyopearl.

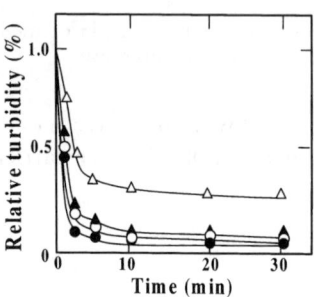

Figure 12.4 Effect of pH of medium to sedimentation properties of insolubilized D-AS. Symbols: △, pH 4.5; ▲, pH 4.25; ○, pH 4.0; ●, pH 3.5.

rate of decrease of the turbidity in the upper phase. The time required for settling a large portion of D-AS was about 10 min, below pH 4.0. Taking the application of D-AS to ethanol production into consideration, the degree of decrease in the turbidity was also examined under the conditions of coexistence of D-AS and flocculating yeast cells in culture medium at pH 4.0 and 30 °C. As a result, the culture medium formed above the layer of sediment after settling for 10 min at pH 4.0 was quite clear and almost free from D-AS and the flocculating yeast cells. The soluble-autoprecipitation (S-AP) characteristic is thought to be useful for the achievement of energy-saving bioprocesses such as brewing.

12.3. Application of pH-responsive Polymer in Bioprocesses with Heterogeneous Reaction System

12.3.1. *Repeated Hydrolysis of Raw Starch*

Hydrolysis of solid substrate (raw starch) using D-AS, and ethanol production from raw starch by simultaneous saccharification and fermentation (SSF) using a combination of D-AS and flocculating yeast cells have been studied. First, the case in which D-AS was applied to hydrolysis of raw starch is introduced. A flow chart of the procedure for repeated hydrolysis of raw starch using D-AS and ethanol production is shown in Figure 12.5. In method A, raw starch was hydrolyzed in batches in a cylindrical reactor with a conical bottom (300 ml). The re-use of D-AS was performed as follows: every 48 h during the reaction at 30 °C, the pH of reaction mixture was adjusted from

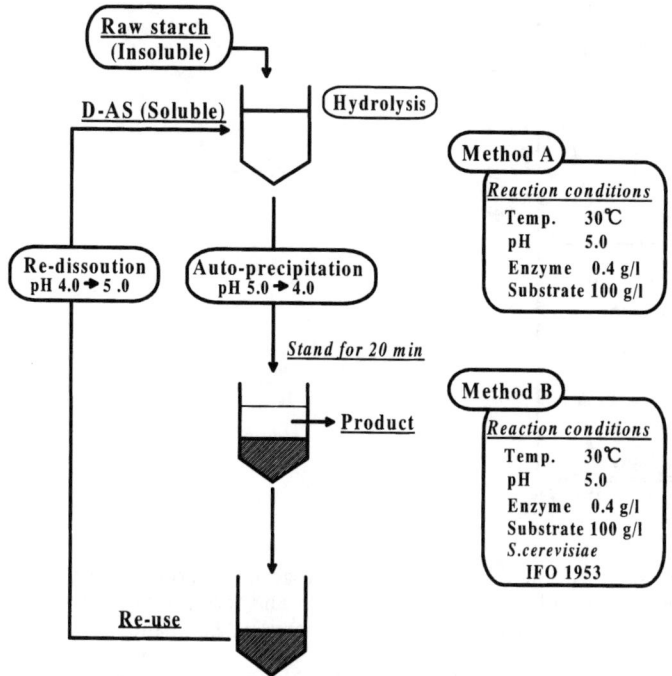

Figure 12.5 Flow diagram of repeated use by using auto-precipitation property of D-AS.

5.0 to 4.0 using a pH controller equipped reactor. The reaction mixture was left undisturbed for 20 min in order to allow the D-AS to settle in an insoluble state at the conical bottom of the reactor. The clear product solution formed above the sedimenting layer of D-AS was removed using a glass tube. After D-AS in an insoluble state was redissolved by changing the pH to 5.0 via a pH-controller, the next hydrolysis reaction was started by supplying a fresh substrate suspension. However, in method B, repeated production from raw starch via SSF was done by adding D-AS and *S. cerevisiae* IFO 1953. The flocculating yeast cells were inoculated at a level of $3-4 \times 10^8$ cells/ml. The other procedures for re-use of the yeast, as well as D-AS, were similar to those of method A, except fermentation medium was used instead of acetate buffer.

Figure 12.6(A) shows the results of the repeated batch hydrolysis of 100 g/l raw starch by method A. The conversion ratio after the third reaction cycle reduced with increasing number of reaction cycles. The amount of glucose produced in the fifth batch reaction was about 66% that produced in the first batch. The total quantity of glucose obtained from 150 g (30 g × 5 cycles) of raw starch through 5-batch reactions was 101 g. Based on the amount of immobilized enzyme, D-AS recovered after the fifth batch cycle of the saccharification was very high, 96% of the initial amount. Therefore, the gradual decrease in the amount of glucose produced in five consecutive batches of hydrolysis appears to be a result of inactivation of D-AS during repeated operations. This result indicates that D-AS could be repeatedly used in the heterogeneous reaction system containing solid substrate. Next, in order to investigate the possibility of bioprocess by combination of pH-responsive enzyme and microorganisms, we attempted direct SSF using raw starch.

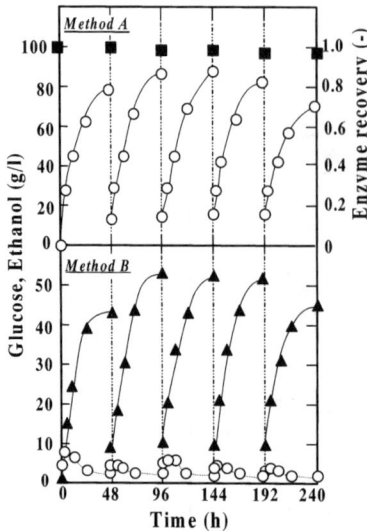

Figure 12.6 Repeated utilization of D-AS. A: Repeated saccharification of 10% raw starch. B: Repeated simultaneous saccharification and fermentation from 10% raw starch. Symbols: ○, glucose; ■, enzyme recovery; ▲, ethanol.

Figure 12.6(B) shows the repeated ethanol production from 100 g/l raw starch by a combination of D-AS and flocculating cells. Little or no glucose was detected, except during the early part of each batch culture. Therefore, the ethanol production from raw starch was found to be limited by the saccharification step of raw starch. The quantity of glucose produced in the fifth batch culture was about 80% of that of the first batch culture. The total amount of glucose obtained from 150 g of raw starch in five repeated batches of ethanol production was 61 g. The value was higher than the quantity (52 g) that was estimated based on the assumption that the glucose formed in the repeated hydrolysis shown in Figure 12.6(A) was converted completely in ethanol, showing that the inhibition of the enzymatic action by ethanol was less than that of glucose. The number of living yeast cells removed along with ethanol after each batch culture averaged 1.8×10^6 cells/ml. Constituting less than 0.5% of that of the total living cells in the reactor. The results show that D-AS could be reused successfully in repeated ethanol production, as well as in repeated hydrolysis, by using sedimentation for separating D-AS from the product solution. Thus, the bioprocess using a reversible S-AP enzyme in the heterogeneous reaction system containing not only solid substrate but also microorganism is considered to be an alternative to the reaction method using an ultrafiltration membrane for separating product and enzyme. Moreover, where the product can be separated from the biocatalyst bioprocesses by sedimentation of a reversibly S–IS enzyme is expected to be widely applied to other heterogeneous reaction systems.

12.3.2. Continuous Simultaneous Saccharification and Fermentation of Raw Starch

Next, we investigated the continuous SSF of raw starch by utilizing the S-AP character of D-AS (Hoshino *et al.*, 1990). The schematic diagram of the reactor used for

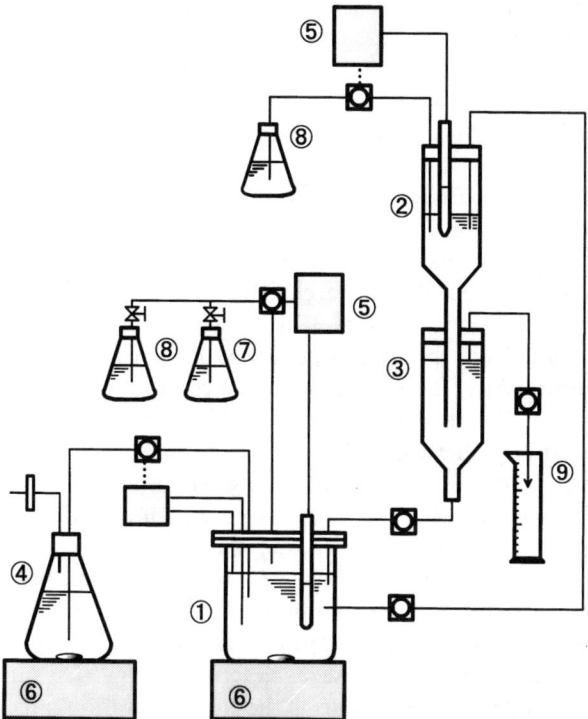

Figure 12.7 Schematic diagram of a reactor for continuous use of D-AS. 1, main reactor; 2, mixing vessel; 3, separation vessel; 4, stock tank for fresh medium; 5, pH controller; 6, stirrer; 7, NaOH solution (2 N); 8, HCl solution (1 N); 9, stock tank for product.

continuous SSF is shown in Figure 12.7. This reactor is composed of a main reactor, 1, with a working volume of 600 ml, a mixing vessel, 2, with a working volume of 150 ml, and a separation vessel, 3, with a working volume of 250 ml. The pH in the main reactor was controlled at 5.0 in order to keep the D-AS soluble. The pH in the mixing vessel was adjusted to 4.0 for the purpose of insolubilizing the D-AS and separating it from the reaction mixture. Nitrogen gas supplied through an air filter was sparged in the mixing vessel for agitation of the mixture. The slurry containing the insolubilized D-AS was allowed to flow down from the bottom of the mixing vessel to the separation vessel. The insoluble D-AS separated from the product solution within the separation vessel was returned to the main reactor. The product solution was withdrawn from the top of the separation vessel. A fresh suspension containing raw starch was supplied to keep the working volume in the reactor constant, using a feed pump connected to a level controller. The circulation rate of the mixture in the reactor system was maintained at 0.5 l/h. Continuous ethanol production from raw starch was performed by adding flocculating yeast cells and D-AS at a concentration of 0.4 g-protein/l. The flocculating cells were inoculated at a level of 4×10^8 cells/ml. The fermentation temperature was 30 °C. For the primary step of the start-up of continuous production, batchwise SSF was carried out for 24 h using 100 g/l of raw starch, and then continuous SSF was performed by supplying fresh medium containing 50 g/l of raw starch. The dilution rate in the continuous saccharification operation was maintained at 0.1 h^{-1}. The pH of the reaction mixture in the main reactor was maintained constant at 5.0 with a peristaltic pump connected to a pH controller. Moreover, in order to separate D-AS from the reaction mixture containing ethanol, the pH of the mixing vessel was maintained constant at 4.0 by adding acid solution using a pH controller.

Figure 12.8 shows the results of continuous SSF form raw starch by the method described above. In the batch operation for 24 h, little or no glucose was detected except for the initial period. Thus, the ethanol production from raw starch was found to be limited by the saccharification step of raw starch. The quantity of ethanol produced at the end of the batch operation (24 h) was 37.1 g/l, which corresponds to an ethanol productivity of 1.55 g/l/h. As in the batch operation, little or no glucose was detected in the continuous culture. After 24 h, the ethanol concentration was decreasing rapidly and reached a minimum value of 18.7 g/l at 36 h. Then, the value increased gradually and was maintained at 20 g/l between 72 and 192 h. In this stationary phase, the ethanol concentration of 20 g/l corresponds to an average ethanol productivity of 2.0 g/l/h, which is 2.4-fold higher than that of the repeated ethanol production, as shown in Figure 12.6. Moreover, it is noteworthy that the ethanol concentration (20 g/l) in the stationary phase was higher than that (17–19 g/l) estimated based on the assumption that the glucose produced in the continuous saccharification (data not shown) was converted completely to ethanol, indicating that the inhibition of enzymatic action by ethanol was less than that of glucose, as described below. The number of living yeast cells in the reactor did not increase during the entire operation, probably due to a balancing of cell growth and cell death in addition to the flowing-out of the cells, and remained constant at 4.0×10^8 cells/ml. The sediment of the flocculating yeast cells with the insoluble D-AS was separated stably from the medium containing ethanol in the separation vessel. The living cells removed along with ethanol in the product solution from the separation vessel averaged 4.8×10^6 cells/ml. Raw starch and D-AS were barely observed in the removed product solution. On the other hand,

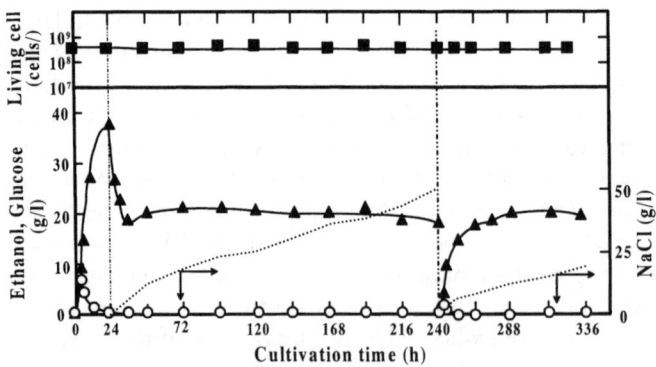

Figure 12.8 Continuous simultaneous saccharification and fermentation from raw starch using D-AS. The removal of accumulated NaCl in the reactor was carried out at 240 h. The dashed line indicates the concentration of NaCl accumulated in the reactor. Symbols: ○, glucose; ▲, ethanol; ■, number of living cells.

the amount of accumulated NaCl increased with culture time. Since the NaCl concentration became 52 g/l at 240 h, which reduced the D-AS activity to about 87% of the maximum value (data not shown), the NaCl was removed. The time required for this operation was about 60 min. The quantity of ethanol formed after this removal operation recovered to the same level (20 g/l) as that in the stationary phase between 72 and 192 h.

12.3.3. Comparison of Ethanol Productivity

We compared the bioprocess shown in Figure 12.6(B) and Figure 12.8 using a combination of D-AS and flocculating yeast to various systems for ethanol production from starch, as shown in Table 12.3. In batch cultures, the ethanol productivity (1.55 g/l/h) of this system (No. 1, Hoshino *et al.*, 1989b) is superior to those (0.54–1.04 g/l/h) of mixed (No. 3, Abouzied and Reddy, 1986; No. 4, Dostalek and Häggström, 1983) or coimmobilized culture (No. 6, Tanaka *et al.*, 1986), in which one microorganism produces enzymes for liquefying and saccharifying starch and another strain converts sugar to ethanol. Moreover, among the systems (No. 1, Nos. 2 and 5, Nam *et al.*, 1988) using a raw-starch-digesting enzyme and an ethanol-producing microorganism, the highest ethanol productivity was obtained in our system (No. 1) because D-AS in a soluble state acts effectively toward insoluble starch, as described above. We have found no report on continuous ethanol production from raw starch materials, and there are few studies on the system of ethanol fermentation from soluble starch (No. 8, Inlow *et al.*, 1988; No. 10, Kurosawa *et al.*, 1989). It was found that the productivity (2.0 g/l/h) of this system (Figure 12.8, No. 7) from raw starch was higher than those (0.30–0.98 g/l/h) of other fermentation systems (Nos. 8 and 10) from soluble starch. When starch liquefied by α-amylase was supplied in a coimmobilized system (No. 9, Rhee *et al.*, 1986) and a system using a membrane (No. 11, Lee *et al.*, 1983), ethanol productivity was enhanced remarkably. However, since these systems (Nos. 9 and 11) contain an additional process for the liquefaction of soluble starch, as well as the saccharification and fermentation process of subsequent liquefied starch, their ethanol productivity cannot be compared directly with that of other systems, including our system. Nevertheless, using our system, which contains D-AS and flocculating yeast cells, ethanol productivity from raw starch was enhanced significantly.

From the above results, the novel immobilized enzyme prepared using pH-responsive polymer as a carrier, which is not only reversible S–IS, dependent on pH, but also auto-precipitating (S-AP) in an insoluble state, was found to be useful as a tool for the achievement of new bioprocesses. The bioprocess involving separation of the product by sedimentation is expected to be widely applied to other heterogeneous reaction systems. Therefore, we investigated the applications of this bioprocess using pH-responsive enzymes, such as hydrolysis of cellulose materials, lactic acid production from raw starch, and ethanol production from rice straw. The details of these investigations can be obtained by referring to our papers (Hoshino *et al.*, 1991, 1994b, 1996a, 1997a; Taniguchi *et al.*, 1989b, 1992).

12.4. Application of Thermo-responsive Polymer to Hydrolysis

In the previous section, we presented applications of pH-responsive polymer as an immobilized carrier. The results of these studies indicate that the same bioprocess can

Table 12.3 Comparison of ethanol productivity among different fermentation systems

System No.	Combination Enzyme	Microorganism Hydrolysis	Microorganism Fermentation	Substrate[a]	Starch concentration (g/l)	Ethanol concentration (g/l)	Ethanol productivity (g/l/h)
Batch							
1.	D-AS	–	S. cerevisiae	IS	100	37	1.55
2.	Glucoamylase	–	S. cerevisiae	IS	200	85	0.88
3.	–	A. niger	S. cerevisiae	IS	100	≈50	≈1.04
4.	–	S. fibuliger	Z. mobilis	SS	30	9.7	0.54
5.	Glucoamylase	–	S. cerevisiae	IS	200	88	0.91
6.	–	A. awamori	Z. mobilis	SS	100	22	0.61
Continuous							
7.	D-AS	–	S. cerevisiae	IS	50	20	2.0
8.	–	–	S. cerevisiae[b]	SS	60	20	0.98
9.	Glucoamylase	–	Z. mobilis	LS	100	21	17.7
10.	–	A. awamori	S. cerevisiae	SS	50	≈9	≈0.39
11.	Glucoamylase	–	Z. mobilis	LS	130	60	60

System nos. 1 and 7: use of a reversibly soluble-autoprecipitating enzyme (D-AS) and flocculating yeasts.
System no. 2: use of a native enzyme and free yeasts.
System nos. 3 and 4: mixed culture of microorganisms producing starch-digesting enzymes and ethanol-producing microorganisms.
System nos. 5 and 9: coimmobilized system of an enzyme and an ethanol-producing microorganisms.
System nos. 6 and 10: coimmobilized system of microorganisms producing starch-digesting enzymes and ethanol-producing microorganisms.
System no. 8: use of S. cerevisiae with glucoamylase.
System no. 11: membrane reactor containing glucoamylase and Z. mobilis.
a IS, insoluble starch; SS, soluble starch; LS, starch liquefied by α-amylase.
b S. cerevisiae harboring the glucoamylase gene from A. awamori.

be achieved using other stimuli responsive polymers that depend on changes in the surroundings, such as temperature or ionic strength. Among stimuli responsive polymers, thermo-responsive (TR) polymers have been vigorously examined. In particular, poly(*N*-isopropylacrylamide) (pNIPAm) is a thermoreversible water-soluble polymer, and aqueous solutions of this polymer exhibit a lower critical solution temperature (LCST) about 32 °C (Heskins and Guillet, 1968; Schild and Tirrell, 1990). The bioprocess using the TR polymers consisting of *N*-isopropylacrylamide has been investigated by many researchers. For example, Nguyen and Luong (1989) reported the preparation of trypsin immobilized on a copolymer composed of *N*-isopropylacrylamide (NIPAm) and *N*-hydroxysuccinimide. Chen *et al.* (1990a and b) have prepared a thermo-responsive adsorbent with Protein A and reported that the purification of human IgG. In addition, Otake *et al.* (1990) investigated the application of thermosensitive gel composed of NIPAm and butylmethacrylate for drug delivery. However, since most research has been based on the use of NIPAm as a monomer, the temperature range for use of thermo-responsive polymer is invariably limited. Therefore, we describe here a strategy for solving the problem.

12.4.1. *Application of Thermo-responsive Polymer to Hydrolysis*

LCST of pNIPAm is not always optimum for the activity of all enzymes. For example, the optimum temperature of α-amylase is usually greater than 50 °C. When such enzymes are immobilized on the conventional TR polymer, a suitable reaction cannot be accomplished because the TR enzyme is in an insoluble state at the optimum temperature. Hence, one method for solving this problem is to develop a high-LCST thermo-responsive polymer as a carrier for immobilization of enzymes possessing higher optimum temperatures than a LCST of pNIPAm. Recently, thermo-responsive polymers synthesized by polymerization of an *N*-alkyl acrylamide such as *N*-ethylmethacrylamide or *N*-ethylacrylamide have been shown to have higher LCSTs (about 50 and 70 °C, respectively) than conventional thermo-responsive polymers containing NIPAm (Ito, 1989). Since these *N*-alkyl acrylamide monomers are not commercially available and are very difficult to synthesize, we synthesized a high-LCST thermo-responsive polymer by copolymerization of NIPAm with several commercially available monomers for immobilization of enzyme possessing a high optimum temperature (Hoshino *et al.*, 1997b).

12.4.2. *Preparation Thermo-responsive Enzyme with High-LCST*

We reported previously two TR enzymes which were reversibly S-IS, depending on the temperature of the solution (Hoshino *et al.*, 1992, 1994a). However, although the TR enzymes exhibited a high level of activity toward a high-MW substrate (soluble starch) and/or a solid substrate (uncooked starch) at 30 °C they were completely insoluble above 40 °C and their activity levels at 40 °C were lower than those of the soluble forms of these enzymes at 30 °C. This seems to have been caused by a decrease in the contact frequency between the enzymes and substrates due to the insolubilization of the immobilized enzymes. In order to immobilize enzymes which exhibit maximum activity at temperature above 32 °C we selected thermolysin, which has a high optimum temperature of about 60 °C, as an enzyme, and attempted to prepare a new TR polymer possessing a LCST higher than 32 °C. Methacrylamide (MAm) was chosen as a monomer for copolymerization with NIPAm based on the previous finding that the LCST of the polymer, poly(NIPAm/MAm), is higher than that of a NIPAm homopolymer (Hoshino *et al.*, 1994a). Glycidyl methacrylate (GMA) containing an epoxy group was selected as another monomer. The epoxy group is essential for covalent conjugation of free amino groups existing on the surface of enzymes and copolymers. The copolymers composed of three monomers, GMA, NIPAm, and MAm (GMNs), were prepared by solvent polymerizations.

Figure 12.9 shows the solubilities of GNMs prepared as a function of the temperature of the Tris-HCl buffer (pH 8.0). GNM-0 (TR polymer prepared by polymerization of the mole percentage ratio of GMA : NIPAm : MAm = 15 : 85 : 0), the polymer without MAm, was completely soluble below 32 °C, and the mixture was clear. Between 32 and 42 °C, the suspension containing GNM-0 gradually became opaque as the temperature increased. Above 42 °C, the suspension showed a minimum transmittance and the GNM-0 was completely insoluble. Several investigators have reported that the insolubilization of thermo-responsive polymers containing NIPAm upon an increase in temperature is due to a phase transition from the random-coil structure of the polymers to compact globular structures upon dehydration (Heskins and Guillet 1968; Yamamoto *et al.*, 1990; Kubota *et al.*, 1990). The higher the mole percentage ratio of MAm in the polymer, the higher the LCST of the polymer at which the transmittance exhibited a sharp change. The temperature at which the transmittance was a half-maximum increased by about 3.9 °C whenever the mole percentage ratio of the initially added MAm was increased by 2.5% . Introduction of the hydrophilic MAm into the polymer chain made the LCST of the polymer higher than that of the homopolymer of NIPAm described previously (Hoshino *et al.*, 1992). This result is consistent with the observations of Steinke and Vorlop (1987). When the polymer suspensions were centrifuged at the temperature at which these polymers were completely insoluble, the insolubilized polymers were recovered with very high efficiency as a hard paste without GNM-100. (GMA : NIPAm : MAm = 15 : 75 : 10), since the optimum temperature of thermolysin activity is about 60 °C, it is desirable that the

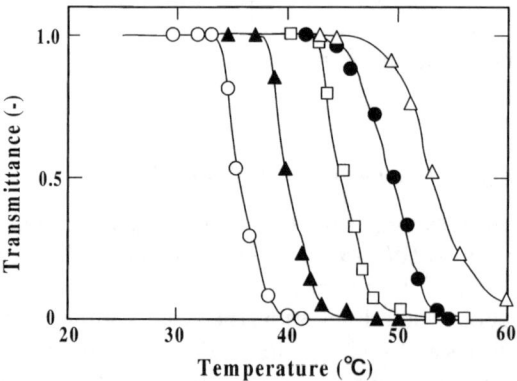

Figure 12.9 Solubility response of thermo-responsive polymers to changes in temperature. The polymer concentration was 4 g/l. Symbols: ○, GNM-0 (polymer polymerized at mole percentage ratio at GMA : NIPAm : MAm = (15 : 85 : 0); ▲, GNM-25 (15 : 82.5 : 2.5); ■, GNM-50 (15 : 80 : 10); ●, GNM-75 (15 : 77.5 : 7.5); △, GNM-100 (15 : 75 : 10).

polymer on which the thermolysin is immobilized contains as much MAm as possible. Unfortunately, as the mole percentage of the initially added MAm increased, the solubility response of the polymer to changes in temperature became weak and the polymer could not be recovered completely. Taking the heat stability of thermolysin and recovery of the polymer into consideration, it is desirable to select a TR polymer which has as sharp a thermo-response range as possible. We selected GNM-75, which was produced when the mole percentage ratio of initially added monomers was 15 : 77.5 : 7.5 (GMA : NIPAm : MAm). The immobilization of thermolysin on GNM-75 was carried out at pH 10.5 and 4 °C for at least 72 h using 50 mg-protein per gram-polymer and the prepared TR thermolysin has the highest specific activity.

12.4.3. *Response of Solubility to Changes in Temperature*

Figure 12.10 shows the solubility of TR thermolysin as a function of temperature and concentration of ammonium sulfate in the mixture. The TR thermolysin in the basic buffer without ammonium sulfate was completely soluble below 44 °C. That is, the mixture was clear and the LCST of the immobilized thermolysin was 44 °C. Above 54 °C, the transmittance of the suspension was zero and the immobilized thermolysin was completely insoluble. The solubility response of the TR thermolysin to changes in temperature was reversible, and the response profile agreed with that of GNM-75, as shown in Figure 12.9. These results show the following: the LCST of the immobilized thermolysin was about 12 °C higher than those of the immobilized enzymes described previously (Hoshino *et al.*, 1992, 1994a, 1996b), the solubility of the immobilized thermolysin can be adjusted reversibly due to changes in temperature, the amount of immobilized enzyme recovered by centrifugation at 55 °C was 98% or more, and the insoluble immobilized thermolysin was in the same state as the insoluble TR polymer. On the other hand, when ammonium sulfate was added to the mixture, the higher the ammonium sulfate concentration in the mixture, the lower the temperature at which the mixture became opaque. The temperature at which the transmittance was

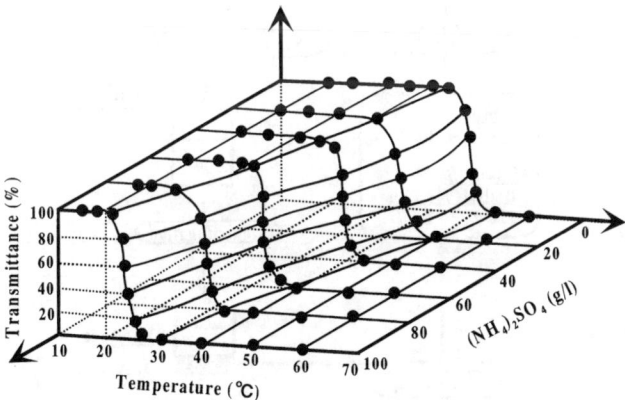

Figure 12.10 Solubility response of the thermo-responsive thermolysin to changes in temperature and effect of ammonium sulfate concentration on the solubility response. The TR thermolysin was prepared by mixing thermolysin and GNM-75 for 72 h at 4 °C and pH 10.5. The experiment was carried out in the basic buffer (0.1 M Tris-HCl buffer, pH 8.0, containing 5 mM CaCl$_2$).

half-maximum decreased by about 4.8 °C for each 20 g/l increase in the ammonium sulfate concentration. The sharp response of solubility of the immobilized thermolysin is similar to that of GNM-75, and is considered to be favorable for its recovery from a reaction mixture during repeated use of the immobilized enzyme. Moreover, the immobilized enzyme in the mixture with 60 g/l ammonium sulfate could be recovered as a precipitate by centrifugation at 40 °C. The fact that the solubility of TR thermolysin can be adjusted by changing the salt concentration, as well as the temperature, is very important information that is useful for achieving novel bioprocesses using TR polymer.

12.4.4. Repeated Use of Thermo-responsive Thermolysin in Hydrolysis of α-Casein

A flow diagram of the procedure for repeated use of TR thermolysin is shown in Figure 12.11. α-Casein was hydrolyzed as follows. In method A, hydrolysis was started in the basic buffer at 40 °C. At 6 h intervals during the hydrolysis, the reaction mixture was centrifuged at 4 °C to remove insoluble material. Then, the reaction mixture was heated to 55 °C for the insolubilization of the immobilized enzyme and centrifuged at 55 °C for recovery of the insolubilized enzyme from the solution containing products. After separation of the supernatant, the precipitate of the insolubilized immobilized enzyme was re-dissolved by addition of the basic buffer. By mixing the resolubilized immobilized enzyme with fresh substrate, the next hydrolysis reaction was initiated at 40 °C. In method B, α-casein was hydrolyzed with the immobilized enzyme at 40 °C, and after 6 h the insoluble materials were removed by the procedure used in method A. Recovery of the immobilized enzyme at the end of one hydrolysis reaction was performed by adding 120 g/l ammonium sulfate solution of the same volume as the

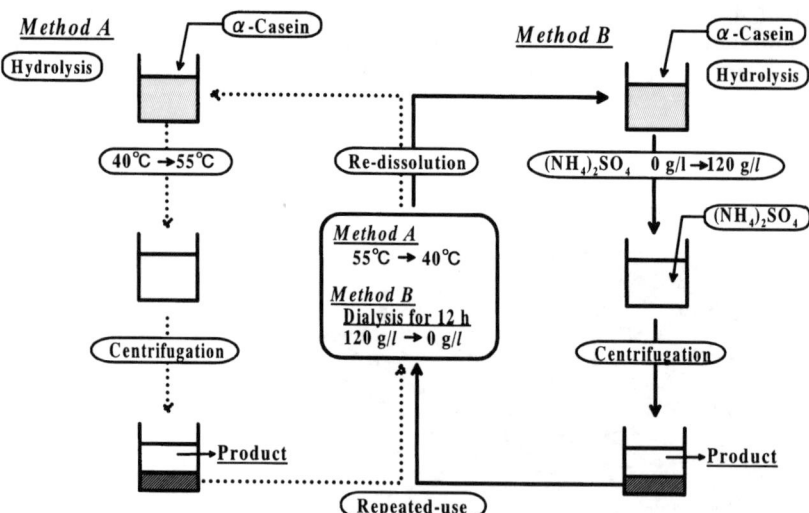

Figure 12.11 Flow diagram of repeated hydrolysis of α-casein using thermo-responsive thermolysin. Method A: repeated-use of TR thermolysin by up-and-down of temperature. Method B: repeated of TR thermolysin by change of $(NH_4)_2SO_4$ concentration. The desalting was done by dialyzing against the basic buffer (0.05 M Tris-HCl buffer, pH 8.0, with containing 5 mM $CaCl_2$) for 12 h.

reaction mixture at 40 °C, rather than elevating the temperature of the reaction mixture to 55 °C. The suspension was then centrifuged at 40 °C for collection of the insolubilized enzyme. After separation of the supernatant, the precipitate was re-dissolved by addition of the cold basic buffer and the solution was dialyzed against the basic buffer for 12 h. By mixing of the solution of the immobilized enzyme with fresh substrate, the next hydrolysis reaction was initiated at 40 °C. Figure 12.12 shows the results of the repeated batch hydrolysis of 10 g/l α-casein by methods A and B. In the hydrolysis performed according to method A, the amount of acid-soluble peptides produced by the end of each batch reaction decreased gradually from the sixth cycle of the reaction. Although the amount of acid-soluble peptide obtained in the sixth batch of hydrolysis was 79% of that obtained in the first batch, the total quantity of acid-soluble peptide produced from 60 g of α-casein through five batches of hydrolysis was 19.2 g. Furthermore, the enzyme recovery after the sixth batch cycle of hydrolysis decreased only to about 88% of the initial amount. On the other hand, in the case of method B, the amount of acid-soluble peptide obtained at the end of each batch reaction was greater than those obtained by method A. The amount of acid-soluble peptide obtained in the sixth batch cycle of hydrolysis was 90% of that obtained in the first batch and higher than that obtained in the case of the hydrolysis by method A. The total quantity of acid-soluble peptide produced from α-casein through the five batches of hydrolysis was 20.9 g. The enzyme recovery decreased with increasing reaction cycle number, similar to the case for method A, and after the sixth batch cycle of hydrolysis was about 88% of the initial amount. Accordingly, the difference in the total amount of acid-soluble peptide produced by these methods is thought to be mainly due not to the decrease for enzyme recovered by centrifugation, but to the inactivation of thermolysin during the repeated operations (raising and lowering of temperature or the ammonium

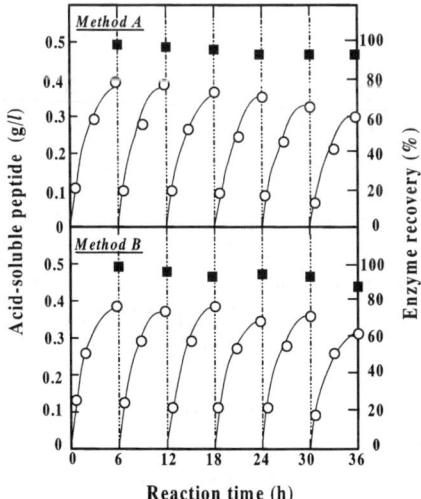

Figure 12.12 Hydrolysis of α-casein with repeated use of thermo-responsive thermolysin. Method A: repeated-use of TR thermolysin by up-and-down of temperature. Method B: repeated of TR thermolysin by change of (NH₄)₂SO₄ concentration. Symbols: ○, acid-soluble peptide; ■, enzyme recovery.

sulfate concentration). However, in method B the removal of the ammonium sulfate by dialysis required a relatively long time (12 h).

Here, we have examined several problems associated with TR enzyme and the methods for its utilization. We proposed a novel thermo-responsive polymer for the immobilization of enzyme which has a high optimum temperature. The solubility of new enzyme immobilized on the thermo-responsive polymer can be changed reversibly by depending on not only the temperature, but also the addition of salt (ammonium sulfate). Therefore, the TR enzyme is expected to be suitable to be used consecutively by the thermo-change and/or salt concentration-change. In the repeated-use of TR enzyme by method A, the TR enzyme exhibited only a slight decrease in the level of activity. Therefore, such repeated use is expected to be applied to fast enzymatic reactions and enzymatic reactions *in vitro*. However, if a relatively large reactor is used for scale-up, method A cannot be used for recovery of the TR enzyme because the temperature of the reaction mixture in the reactor could not be raised and lowered quickly. On the other hand, although dialysis is required to remove of the salt from the reaction mixture in method B, use of method B with a large reactor would not involve the scale-up problem associated with use of method A. Thus, the better of the two recovery methods for effective reuse of the TR enzymes should be selected on a case-by-case basis. However, a new TR polymer that has a higher LCST should be prepared for the development of stimuli responsive polymer in the future. The bioprocess containing such a TR polymer is expected to be related to the effective utilization of enzymes, which exhibit maximum activity at high temperatures, including thermolysin.

12.5. Application of Thermo-responsive Polymers with Low LCST for Protein Precipitation

The TR polymers have been used as not only an immobilized carrier as described above, but also as artificial muscles, sensors, actuators, and as tools for separation and purification of biochemical products (Otake et al., 1990). Application to affinity precipitation for purification of the biological molecules has received attention because the bioprocess is far superior in the heterogeneous system, as compared with the conventional affinity chromatography. The solubility of the TR polymer can be adjusted by slightly changing the temperature of the solution as described above. The solublized TR adsorbent adsorbs easily the target molecule without diffusion resistance. The target molecule can be recovered as a precipitate from an extract containing many contaminants by insolubilizing the TR adsorbent. Several TR adsorbents have been proposed for affinity precipitation of the biological molecules of interest (Schneider et al., 1981; Nguyen and Luong, 1989; Chen et al., 1990; Chen and Hoffman, 1990; Pecs et al., 1991; Galaev and Mattiasson, 1993). Chen and Hoffman (1993) reported that the complex of the thermo-responsive adsorbent (a conjugate of pNIPAm with Protein A) and IgG was recovered by raising the temperature of the mixture from 4 to 37 °C. However, such TR adsorbents could not be used for purification of the thermo-labile biomolecule. In other words, relatively high temperatures were required in order to recover the complex of the TR adsorbent and the target molecule in an insoluble state, as compared with temperature (mainly 4 °C), for common protein chromatography. The thermal inactivation is considered to occur when heat-labile compounds are purified by affinity precipitation. Therefore, we searched for a novel TR polymer having a LCST of around 4 °C and used the polymer as a carrier for preparing adsorbent for the purification of thermolabile biomolecules. The results and application of this TR polymer to the affinity precipitation system between Concanavalin A (Con A) – maltose coupled with a new thermo-responsive polymer are introduced here.

12.5.1. *Preparation of Thermo-responsive Adsorbent*

First, we attempted to develop new thermo-responsive polymers with as low LCST as possible. Several *N*-substituted acrylamides were synthesized using Ito's method (1989) as potential monomers for the preparation of thermo-responsive polymers with LCST of about 4 °C. Then, different poly(*N*-substituted acrylamide)s with an amino group were prepared according to the polymerization procedure reported by Chen and Hoffman (1993) and Takei *et al.* (1994). The detailed synthetic methods are described elsewhere (Hoshino *et al.*, 1998). Table 12.4 shows the LCSTs of poly(*N*-substituted acrylamide)s. Several conjugates of poly(*N*-substituted acrylamide) and cysteamine obtained from alkylamines such as *N*-diethylamine, *N*-propylamine, isopropylamine, pyrrolidine, and piperidine were found to have a LCST between 0 and 52 °C. Fortunately, among these thermo-responsive polymers, LCST of poly(*N*-acryloyl piperidine)-S-$(CH_2)_2$-NH_2, (pAP), was about 4 °C, which was lower than that of pNIPAm-S-$(CH_2)_2$-NH_2 by 27 °C. The affinity precipitation using pAP as a carrier of affinity ligand is expected to be promising for purification of thermolabile molecules because it may be possible to recover the complex of adsorbent and the target molecule in insoluble form at low temperatures. pAP was applied in further studies. Next, adsorbent for affinity precipitation of Con A was prepared by combining pAP with maltose using trimethylamine-boran as a reducing reagent. Figure 12.13 shows the chemical structure of pAP coupled to maltose, TR adsorbent. The average MWs of the TR adsorbents was from 9.6×10^2 to 2.4×10^4 Da. Maltose bound almost stoichiometrically to pAP regardless of the MW of pAP.

Table 12.4 Lower critical solution temperature (LCST) of poly(*N*-substituted acrylamide)s with an amino group

Polymers	*LCST (°C)*
Poly(*N,N*-diethylacrylamide)-S-$(CH_2)_2$-NH_2	52
Poly(*N*-n-propylacrylamide)-S-$(CH_2)_2$-NH_2	22
Poly(isopropylacrylamide)-S-$(CH_2)_2$-NH_2	31
Poly(*N*-n-butylacrylamide)-S-$(CH_2)_2$-NH_2	SP[a]
Poly(isobutylacrylamide)-S-$(CH_2)_2$-NH_2	SP
Poly(*tert.*-butylacrylamide)-S-$(CH_2)_2$-NH_2	SP
Poly(acryloyl pyrrolidine)-S-$(CH_2)_2$-NH_2	52
Poly(acryloyl piperidine)-S-$(CH_2)_2$-NH_2	4
Poly(acryloyl hexamethyleneimine)-S-$(CH_2)_2$-NH_2	IP[b]

Each polymer was prepared using 1 g of *N*-substituted acrylamide and 10 mg of cysteamine.
a The polymer is in a soluble state at 0–80 °C.
b The polymer is in an insoluble state at 0–80 °C.

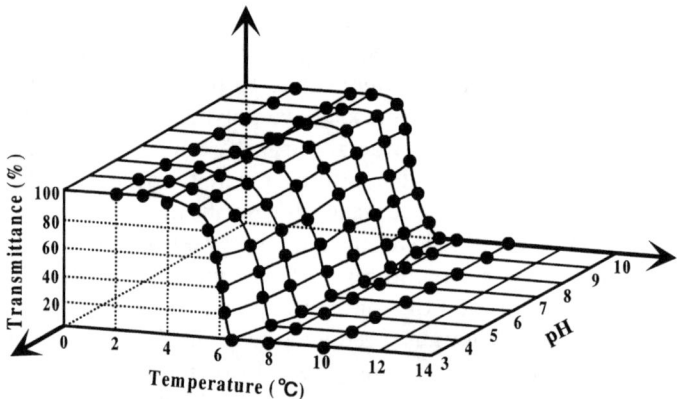

Figure 12.13 Chemical structure of poly(*N*-acryloylpiperidine) with maltose (TR adsorbent).

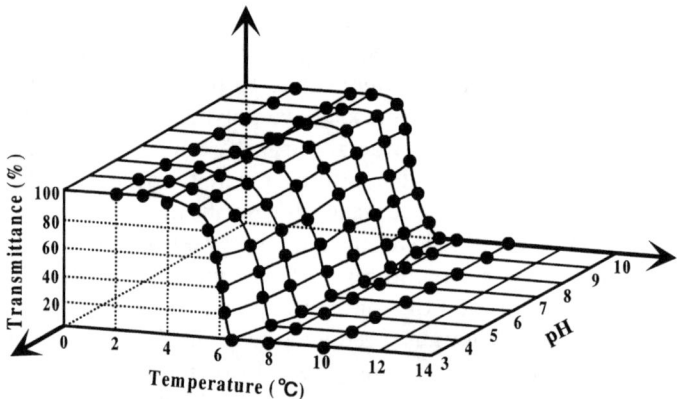

Figure 12.14 Response of TR adsorbent solubility to the changes in temperature and pH of the basal buffer (0.05 M Tris-HCl buffer, pH 7.0, with 150 mM NaCl, 1 mM MnCl$_2$, and 1 mM CaCl$_2$).

12.5.2. Properties of Thermo-responsive Adsorbent

12.5.2.1. Solubility of the Adsorbent

Figure 12.14 shows the solubility of TR adsorbent as a function of pH and temperature. Below 3.5 °C, all of the mixtures containing pAPM had no turbidity at different pH values and became clear. However, the transmittance of the mixture decreased gradually with increasing temperature. Above 8 °C, all of the mixtures containing pAPM were in a completely insoluble state and became opaque. There is a slight difference in the LCST of pAPM depending on the pH of the mixture. The precipitation started at 3.5 °C at pH 3.0 and at 5.2 °C at pH 10. The shift of the TR range seems to be dependent upon the degree of adsorption of hydronium ions toward TR sites of the polymer at each pH because the TR ability of poly(*N*-substituted acrylamide) is attributed to the hydration/dehydration at hydrophobic site of polymer as reported by Kubota *et al.* (1990). Such a sharp response of solubility to temperature and/or pH change is reversible. Furthermore, when the transmittance of the mixture showed a minimum value, pAPM was completely collected as a precipitate by centrifugation.

Consequently, it is confirmed that the TR adsorbent prepared has a LCST of about 4 °C around neutral pH and the solubility of pAPM can be adjusted reversibly by a slight change of temperature between 4 and 8 °C.

12.5.2.2. Properties of Adsorption and Desorption

For affinity purification, the adsorption of Con A to the TR adsorbent and the desorption of Con A from the complex of the adsorbent and Con A is important. Figure 12.15 shows the effects of pH, temperature on the amount of Con A adsorbed to TR adsorbent. As shown in Figure 12.15, the amount of pure Con A adsorbed on soluble TR adsorbent at 4 °C followed by TR precipitation at 10 °C was higher than that on insoluble TR at 10 °C, regardless a pH range of 6–9. At 4 °C, the maximum amount adsorbed Con A was obtained at pH 7, and the maximum value (0.14 µmol Con A per gram of TR adsorbent) was 2.2-fold higher than that at 10 °C. Furthermore, in order to clarify the adsorption capacity of the adsorbent, the apparent association constant against free Con A was compared to those obtained using other ligands. The apparent association constant of TR adsorbent ($1.31 \times 10^3 \, M^{-1}$) is almost as high as that ($1.25 \times 10^4 \, M^{-1}$) between free Con A and glucose (Pai *et al.*, 1992), but is lower than that ($1.44 \times 10^4 \, M^{-1}$) between free Con A and *p*-nitrophenyl-α-mannopyranoside (Lewis *et al.*, 1976). However, this suggests that the soluble TR adsorbent has a high capacity for affinity purification because Con A can contact smoothly the TR adsorbent due to the extremely slight steric hindrance. On the other hand, in order to elute Con A from the complex, the effect of the desorption reagent was investigated. The reagents for dissociation of binding affinity between Con A and sugar moiety are methyl-α-D-glucopyranoside, D-mannose, and methyl-α-D-mannopyranoside (Baues and Gray, 1977). These reagents have a stronger affinity toward Con A than maltose or glucose have and are expected to be able to elute Con A entirely from the complex. Based on the results of these experiments, we selected methyl-α-D-mannopyranoside as the desorption reagent of Con A from the complex.

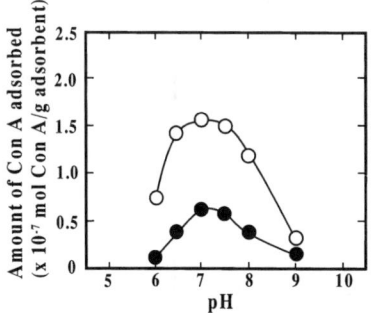

Figure 12.15 Effects of pH and temperature on the amounts of Con A adsorbed to TR adsorbent. Pure Con A (2 mg) was added to the basal buffer containing 10 mg TR adsorbent. The adsorption was carried out at 4 °C (○) and 10 °C (●) for 1 h.

12.5.3. Affinity Precipitation Using Thermo-responsive Adsorbent

A flow diagram of procedure for affinity precipitation of Con A from the crude sample using the adsorbent is shown in Figure 12.16. The mixture of TR adsorbent and jack bean meal extract was stirred at 4 °C for 1 h. After the temperature of the mixture was rapidly elevated to 10 °C, the resulting precipitate was recovered by centrifugation (10 °C) and then washed twice at 10 °C with the basal buffer. The supernatant containing impurities was discarded. The precipitate (the complex of the target molecule and the adsorbent) was redissolved in the basal buffer without salts at 4 °C and the solution was stirred for 10 min. Con A was desorbed from the complex by adding a desorption reagent to the buffer solution. After stirring for 1 h at 4 °C, the temperature of the solution was again raised to 10 °C. The precipitate formed was separated by centrifugation. The supernatant containing Con A was dialyzed against the 0.05 M Tris-HCl buffer (pH 7.0) for 12 h in order to remove the desorption reagent. The insoluble TR adsorbent was dissolved in the basal buffer by lowering the solution temperature to 4 °C and was used for the next cycle of the purification. In repeated use

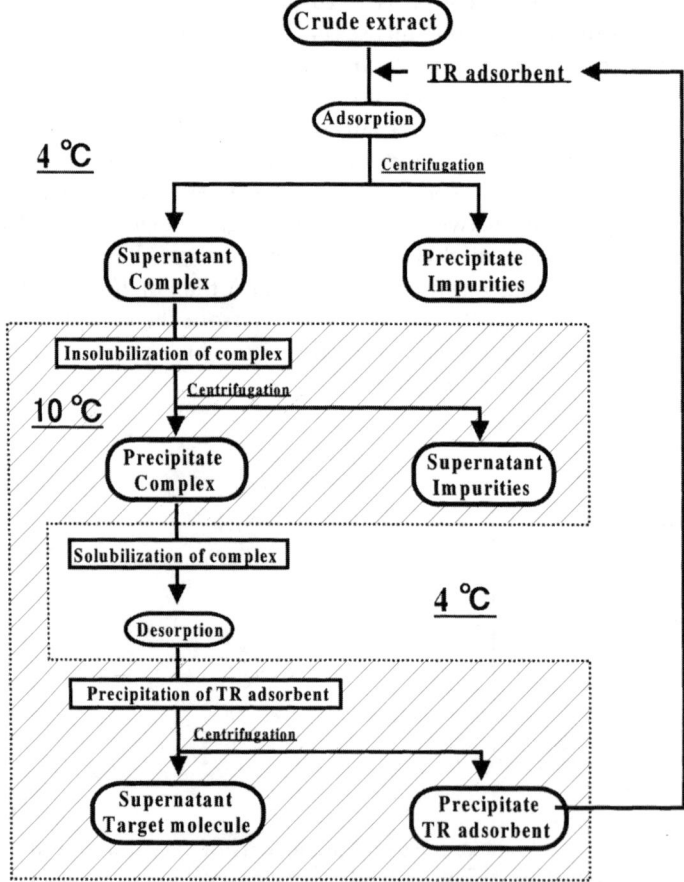

Figure 12.16 Flow diagram of the affinity precipitation procedures using TR adsorbent.

Table 12.5 Purification of Con A from the crude extract with TR adsorbent

Stage	Volume (ml)	Total protein (mg)	Con A (mg)	Con A content (mg/mg-protein)	Recovery (%)
Crude extract[a]	50	89.3	4.39	0.049	100
Adsorption	ND[b]	ND	3.88	ND	88.4
Desorption [c]	10	4.49	3.58	0.797	81.5

The adsorbent prepared from pAP having MW of 1.28×10^4 is used in this experiment.
a Crude Con A extract (88 µg/ml) was added to pAPM (10 mg) solution at 4 °C.
b TR adsorbent combined with Con A was in an insoluble state (precipitate) at 10 °C.
c Desorption with 2.5 ml of the basal buffer (pH 7.0) containing 0.2 M methyl-α-D-mannopyranoside without salts was repeated four times.
ND: not determined.

of the adsorbent, fresh crude extract was applied to the adsorbent solution whenever the purification cycle was repeated.

Table 12.5 shows the results of purification of Con A from the crude extract of jack beans meal using TR adsorbent. In order to recover all of Con A in the extract, 10 mg of the adsorbent was used for affinity precipitation. In the adsorption step, 88% of Con A in the crude extract was adsorbed to TR adsorbent. This incomplete adsorption was probably due to the impurities in the crude extract. In the desorption step, 92% of Con A adsorbed to TR adsorbent was recovered by elution using 0.05 M Tris-HCl buffer (pH 7.0), containing 0.2 M methyl-α-D-mannopyranoside as a desorption reagent. Finally, 81.5% of Con A in the crude extract was collected. The Con A content of the purified solution was enhanced 16.3-fold as compared with that of the crude extract by the single affinity purification procedure. Moreover, a Con A purity of 80% in the purified preparation was achieved. In order to determine the performance under repeated use of TR adsorbent, the purification procedure was repeated four times according to the method shown in Figure 12.16. In each purification cycle, 78–80% of Con A in the crude extract was collected and the Con A content of the purified solution was enhanced about 16-fold as compared with that of the crude extract. The TR adsorbent could be reused satisfactorily in the repeated purification of Con A from the crude extract by the method employing a change in the temperature.

We proposed a novel TR adsorbent, whose solubility changes reversibly between 4 and 10 °C and verified its use for purification of Con A by the affinity precipitation. The TR adsorbent is an excellent tool which is able to not only purify at low temperature, but also to directly target molecules in the heterogeneous system containing water-insoluble contaminants. Although in this investigation maltose was used as a ligand, it is possible that other affinity ligand materials, such as other sugars, antibodies, antigens, and substrates are used. By combining the TR polymer with other affinity ligands, many effective adsorbents applicable to affinity purification of thermolabile biomolecule are expected to be prepared.

12.6. Conclusions

In this chapter, we have introduced primarily bioprocesses that use stimuli responsive polymers that can adjust their solubility in the heterogeneous system. Our results verify that various problems, such as those associated with the diffusion limit within water-insoluble carrier, pointed out by Tosa (1987), can be overcome by bioprocesses that use such polymers. The utilization of stimuli responsive polymers is expected to be advantageous in the bioprocesses containing heterogeneous systems, such as enzymatic reactions containing a solid substrate or product, or recovery of target molecules from suspension. Regretfully, only a few stimuli responsive polymers have been applied to bioprocesses. Therefore, it is appealing to develop polymers that can be quantitatively insolubilized by small decrease in the temperature.

12.7. Abbreviations

AS-L	hydroxyropyl methylcellulose acetate succinate
CAP	cellulose acetate phthalate
Con A	Concanavalin A
D-AS	Dabiase immobilized covalently on AS-L
D-CMT	Dabiase immobilized on the insoluble carrier, CM-Toyopearl
DDS	drug delivery systems
Dabiase	raw-starch-digesting amylase,
ECP	enteric coating polymer
Eudragit L	methacrylic acid-methylmethacrylate copolymer
GMN	copolymers composed of three monomers, GMA, NIPAm, and MAm
HP-55	hydroxyropyl methylcellulose phthalate
LCST	lower critical solution temperature
MAm	methacrylamide
ND	native Dabiase
pAP	poly(N-acryloyl piperidine)-S-$(CH_2)_2$-NH_2
pNIPAm	poly(N-isopropylacrylamide)
S-AP	soluble-autoprecipitation
S–IS	soluble–insoluble
SSF	simultaneous saccharification and fermentation
TR	thermo-responsive

12.8. References

Abouzied, M.M. and Reddy, C.A. (1986) Direct fermentation of potato starch to ethanol by cocultures of *Aspergillus niger* and *Saccharomyces cerevisiae*, *Appl. Environ. Microbiol.*, **52**, 1055–1059.

Bailey, J.E. and Ollis, D.F. (1986) Biochemical engineering fundamentals, 2nd ed., McGraw-Hill Book Co., New York, pp. 157–227.

Baues, R.J. and Gray, G.R. (1977) Lectin purification on affinity columns containing reductively aminated disaccharides, *J. Biol. Chem.*, **10**, 57–60.

Charles, M., Coughlin, R.W., and Hasselbergcr, F.X. (1974) Soluble–insoluble enzyme catalysts, *Biotechnol. Bioeng.*, **16**, 1553–1556.

Chen, J.P., Yang, H.J., and Hoffman, A.S. (1990) Polymer-protein conjugates. I. Effect of protein conjugation on the cloud point of poly(N-isopropylacrylamide), *Biomaterials*, **11**, 625–630.

Chen, J.P. and Hoffman, A.S. (1990) Polymer-protein conjugates. II. Affinity precipitation separation of human immunogammaglobulin by a poly(N-isopropylacrylamide)-protein A conjugate, *Biomaterials*, **11**, 631–634.

Chen, G. and Hoffman, A.S. (1993) Preparation and properties of thermoreversible, phase-separating enzyme-oligo(*N*-isopropylacrylamide) conjugates, *Bioconjugate Chem.*, **4**, 509–514.

Chen, J.P. and Chang, K.C. (1994) Immobilization of chitinase on a reversibly soluble–insoluble polymer for chitin hydrolysis, *J. Chem. Technol. Biotechnol.*, **60**, 133–140.

Dostálek, M. and Häggström, M.H. (1983) Mixed culture of *Saccharomycopsis fibuliger* and *Zymomonas mobilis* on starch-use of oxygen as a regulator, *Eur. J. Appl. Microbiol. Biotechnol.*, **17**, 269–274.

Epton, R., Marr, G., and Morgan, G.J. (1977) Soluble-polymer conjugates. *Polymer*, **18**, 319–323.

Fujimura, M., Mori, T., and Tosa, T. (1987) Preparation and properties of soluble–insoluble immobilized proteases, *Biotechnol. Bioeng.*, **29**, 747–752.

Galaev, I.Yu. and Mattiasson, B. (1993) Affinity thermo precipitation: contribution of efficiency of ligand-protein interaction and access of the ligand, *Biotechnol. Bioeng.*, **41**, 1101–1106.

Heskins, M. and Gulletm, J.E. (1968) Solution properties of poly(*N*-isopropylacrylamide), *J. Macromol. Sci. Chem.*, **A2**, 1441–1455.

Hoshino, K., Taniguchi, M., Nestu, Y., and Fujii, M. (1989a) Repeated hydrolysis of raw starch using amylase immobilized on reversibly soluble–insoluble carrier, *J. Chem. Eng. Jpn.*, **22**, 54–59.

Hoshino, K., Taniguchi, M., Marumoto, H., and Fujii, M. (1989b) Repeated batch conversion of raw starch to ethanol using amylase immobilized on a reversible soluble-autoprecipitating carrier and flocculating yeast cells, *Agric. Biol. Chem.*, **53**, 1961–1967.

Hoshino, K., Taniguchi, M., Marumoto, H., and Fujii, M. (1990) Continuous ethanol production from raw starch using a reversibly soluble-autoprecipitating amylase and flocculating yeast cells, *J. Ferment. Bioeng.*, **69**, 228–233.

Hoshino, K., Taniguchi, M., Marumoto, H., Shimizu, K., and Fujii, M. (1991) Continous lactic acid production from raw starch in a fermentation system using a reversibly soluble-autoprecipitating amylase and immobilized cells of *Lactobacillus casei*, *Agric. Biochem. Chem.*, **55**, 479–485.

Hoshino, K., Taniguchi, M., Katagiri, M., and Fujii, M. (1992) Properties of amylase immobilized on a new reversibly soluble–insoluble polymer and its application of repeated hydrolysis of soluble starch, *J. Chem. Eng. Jpn.*, **25**, 569–574.

Hoshino, K., Taniguchi, M., Katagiri, M., and Fujii, M. (1994a) Hydrolysis of starchy materials by repeated use of an amylase immobilized on a novel thermo-responsive polymer, *J. Ferment. Bioeng.*, **77**, 407–412.

Hoshino, K., Sasakura, T., Sugai, K., Taniguchi, M., and Fujii, M. (1994b) Production of ethanol from reclaimed paper using a combination of a reversibly soluble autoprecipitating cellulase and flocculating cells, *J. Chem. Eng. Jpn.*, **27**, 260–262.

Hoshino, K., Taniguchi, M., Ueoka, H., Ohkuwa, M., Chida, C., Morohashi, S., and Sasakura, T. (1996a) Repeated utilization of β-glucosidase immobilized on a reversibly soluble-insoluble polymer for hydrolysis of phloridzin as a model reaction producing a water-insoluble product, *J. Ferment. Bioeng.*, **82**, 253–258.

Hoshino, K., Akakabe, S., Morohashi, S., and Sasakura, T. (1996b) Immobilization of enzymes on thermo-responsive polymers. In Bickerstaff, G.F. (ed.), Immobilization of Enzymes and Cells, Humana Press, New Jersey, pp. 101–108.

Hoshino, K., Yamasaki, H., Chida, C., Morohashi, S., Taniguchi, M., and Fujii, M. (1997a) Continuous simultaneous saccharification and fermentation of delignified rice straw by a combination of two reversibly soluble-autoprecipitation enzymes and pentose-fermenting yeast cells, *J. Chem. Eng. Jpn.*, **30**, 30–37.

Hoshino, K., Taniguchi, M., Kawaberi, H., Takeda, Y., Morohashi, S., and Sasakura, T. (1997b) Preparation of a novel thermo-responsive polymer and its use as a carrier for immobilization of thermolysin, *J. Ferment. Bioeng.*, **83**, 246–252.

Hoshino, K., Taniguchi, M., Kitao, T., Morohashi, S., and Sasakura, T. (1998) Preparation of a new thermo-responsive adsorbent with maltose as a ligand and its application to affinity precipitation, *Biotechnol. Bioeng.*, **60**, 568–579.

Inlow., D., McRae, J., and Ben-Bassat. A. (1988) Fermentation of corn starch to ethanol with genetically engineered yeast, *Biotechnol. Bioeng.*, **32**, 227–234.

Ito, S. (1989) Phase transition of aqueous solution of poly(N-alkylacrylamide) derivations, *Kobunshi Ronbunshu*, **46**, 437–443.

Kubota, K., Fujishige, S., and Ando, I. (1990) Single-chain transition of poly(N-isopropylacrylamide) in water, *J. Phys. Chem.*, **94**, 5154–5158.

Kurosawa, H., Nomura, N., and Tanaka, H. (1989) Ethanol production from starch by a coimmobilized mixed culture system of *Aspergillus awamori* and *Saccharomyces cerevisiae*, *Biotechnol. Bioeng.*, **33**, 716–723.

Lee, J.H., Pagan. A.J., and Rogers, P.L. (1983) Continuous simultaneous saccharification and fermentation of starch using *Zymomonas mobilis*, *Biotechnol. Bioeng.*, **25**, 659–669.

Leemputtten, E.V. and Horisberger, M. (1976) Soluble-insoluble complexes of trypsin immobilized on acrolein-acrylic acid copolymer, *Biotechnol. Bioeng.*, **18**, 587–590.

Lewis, S.D., Shafer, J.A., and Goldstein, I.J. (1976) Kinetic parameters for the binding p-nitrophenyl α-D-mannopyranoside to concanavalin A, *Arch. Biochem. Biophys.*, **176**, 689–695.

Margolin, A.L., Sherstyuk, S.F., Izumrudov, V.A., Zezin, A.B., and Kabanov, V.A. (1985) Enzymes in polyelectrolyte complexes (the effect of phase transition on thermal stability), *Eur. J. Biochem.*, **146**, 625–632.

Matsuno, R. (1985) Immobilized biocatalysis in chemical enginering. In Chibata, I. (ed), Immobilized biocatalysis, Kodansha, Tokyo, pp. 307–348.

Nguyen, A.L. and Luong, J.H. (1989) Syntheses and applications of water soluble reactive polymers for purification and immobilization of biomolecules, *Biotechnol. Bioeng.*, **34**, 1186–1190.

Nam, K.D., Choi, M.H., Kin, W.S., Kin, H.S., and Ryu, B.H. (1988) Simultaneous saccharification and alcohol fermentation of unheated starch by free, immobilized and coimmobilized systems of glucoamylase and *Saccharomyces cerevisiae*, *J. Ferment. Technol.*, **66**, 427–432.

Otake, K., Inomata, H., Konno, M., and Saito, S. (1990) Thermal analysis of the volume phase transition with N-isopropyl acrylamide gel, *Macromolecules*, **23**, 283–289.

Pai, C.M., Jazobs, H. Bae, Y.H., and Kim, S.W. (1993) Synthesis and characterization of soluble concanavalin A oligomer, *Biotechnol. Bioeng.*, **41**, 957–963.

Rao, M., Seeta, R., and Deshpande, V. (1983) Effect of pretreatment on the hydrolysis of cellulase by *Penicillum funiculosum* cellulase and recovery of enzyme, *Biotechnol. Bioeng.*, **25**, 1863–1873.

Pecs, M., Eggert, M., and Schügerl, K. (1991) Affinity precipitation of extracellular microbial enzymes, *J. Biotechnol.*, **21**, 137–142.

Rhee, S.A., Lee, G.M., Xim, C.H., Abidin, Z., and Ham, M.H. (1986) Simultaneous sago starch hydrolysis and ethanol production by *Zymomonas mobilis* and glucoamylase, *Biotechnol. Bioeng. Symp.*, **17**, 482–493.

Sato, T. and Tosa, T. (1992) Optical resolution of racemic amino acids by aminoacylase. In Tanaka, A., Yosa, T., and Kobayashi, T. (eds.) Industrial application of immobilized biocatalysis, Marcel Dekker Inc., New York, pp. 3–14.

Schild, H.G. and Tirrell, D.A. (1990) Microcalorimetric detection of lower critical solution temperature in aqueous polymer solution, *J. Phys. Chem.*, **94**, 437–443.

Schneider, M., Guillot, M., and Lamy, B. (1981) The affinity precipitation technique. Application to the isolation and purification of trypsin from bovine pancreas, *Ann. NY Acad. Sci.*, **369**, 257–263.

Steinke, K. and Vorlop, K.D. (1987) Water-soluble enzyme-polymer conjugate which can be precipitated by change in temperature, *Proc. 4th Eur. Cong. Biotechnol.*, **2**, 185–188.

Takei, Y.G., Matsukata, M., Aoki, T., Sanui, K., Ogata, N., Kikuchi, A., Sakurai, Y., and Okano, T. (1994) Temperature-responsive bioconjugates 3. Antibody-poly (*N*-isopropylacrylamide) conjugates for temperature-modulated precipitations and affinity bioseparation, *Bioconjugate Chem.*, 5, 577–582.

Tanaka, H., Kurosawa, H., and Murakami, H. (1986) Ethanol production from starch by a coimmobilized mixed culture system of *Aspergillus awamori* and *Zymomonas mobilis*, *Biotechnol. Bioeng.*, 28, 1761–1768.

Taniguchi, M., Kobayashi, M., and Fujii, M. (1989a) Properties of a reversible soluble-insoluble cellulase and its application to repeated hydrolysis of crystalline cellulose, *Biotechnol. Bioeng.*, 34, 1092–1097.

Taniguchi, M., Hoshino, Netsu, Y., and Fujii, M. (1989b) Repeated simultaneous saccharification and fermentation of raw starch by a combination of reversibly soluble–insoluble amylase and yeast cells, *J. Chem. Eng. Jpn.*, 22, 313–314.

Taniguchi, M., Hoshino, K., Watanabe, K., and Fujii, M. (1992) Production of soluble sugar from cellulosic materials by repeated use of a reversibly soluble-autoprecipitating cellulase, *Biotechnol. Bioeng.*, 39, 287–292.

Tosa, T. (1987) Various problems for industrializing of bioreactor. In the Society of Chemical Engineers, Japan, KANTO (ed.), Bioreactor with separation process, Kagaku Kogyosha Inc., Tokyo, pp. 52–64.

Yamamoto, I., Iwasaki, K., and Hirotsu, S. (1990) Light scattering study of condensation of poly(*N*-isopropylacrylamide) chain, *J. Phys. Soc. Jpn.*, 58, 210–215.

Index